Advanced Manufacturing

Springer
London
Berlin
Heidelberg
New York
Barcelona
Hong Kong
Milan
Paris
Santa Clara
Singapore
Tokyo

Other titles published in this series:

A Systems Approach to AMT Deployment
D.R. Towill and J. Cherrington (Eds)

Human-Intelligence-Based Manufacturing
Y. Ito (Ed.)

Intelligent Manufacturing: Programming Environments for CIM
W.A. Gruver and J.C. Boudreaux (Eds)

Automatic Supervision in Manufacturing
M. Szafarczyk (Ed.)

Modern Manufacturing
M.B. Zaremba and B. Prasad (Eds)

Advanced Fixture Design for FMS
A.Y.C. Nee, K. Whybrew and A. Senthil kumar

Intelligent Quality Systems
D.T. Pham and E. Oztemel

Computer-Assisted Management and Control of Manufacturing Systems
S.G. Tzafestas (Ed.)

The Organisation of Integrated Product Development
V. Paashuis

S.G. Tzafestas

Advances in Manufacturing

Decision, Control and Information Technology

Springer

S.G. Tzafestas, PhD
Department of Electrical and Computer Engineering, National Technical University of Athens, Zographou 15773, Athens, Greece

Series Editor

Professor Duc Truong Pham, PhD, DEng, CEng, FIEE
University of Wales Cardiff
School of Engineering, Systems Division,
P.O. Box 917, Cardiff CF2 1XH, UK

British Library Cataloguing in Publication Data
Advances in manufacturing : decision, control and information technology. - (Advanced manufacturing series)
1.Manufacturing processes 2.Manufacturing processes - Data processing 3.Manufacturing processes - Technological innovations
I.Tzafestas, Spyros G., 1939-
670.2'85

Library of Congress Cataloging-in-Publication Data
Advances in manufacturing : decision, control, and information technology / [edited by] S.G. Tzafestas.
p. cm. -- (Advanced manufacturing)
Selected papers from the Third European Robotics, Intelligent Systems, and Control Conference, held June 22-25, 1998 in Athens, Greece.
Includes bibliographical references and index
ISBN-13: 978-1-4471-1217-4 e-ISBN-13: 978-1-4471-0855-9
DOI: 10.1007/978-1-4471-0855-9
I. Manufacturing processes--Planning. 2. Process control.
3. Decision support systems. I. Tzafestas, S. G., 1939- .
II. European Robotics, Intelligent Systems, and Control Conference (3rd : 1998 : Athens, Greece) III. Series.
TS183.3.A39 1999 98-49602
658.5--dc21 CIP

Reprint of the original edition 1st edition 1999

Typesetting: Camera ready by contributors

69/3830-543210 Printed on acid-free paper

Preface

Manufacturing systems constitute one of the major classes of technological systems that need the synergy of techniques and tools from such disciplines as sensor technology, control engineering, management science, and computer and information technology. New manufacturing systems are continuously being designed, tested and put in operation via these technologies.

This book presents a set of thirty six contributions in the manufacturing field selected from those presented at *the "Third European Robotics, Intelligent Systems and Control Conference (EURISCON'98)"*, held in Athens, Greece (June 22-25, 1998). These contributions cover a wide range of manufacturing problems, and design, control and information technology methodologies and tools that reflect the current research trends in the field, also including several practical how-to-do issues.

The contributions are presented in five parts as follows:

PART I:	*Process planning, scheduling and machine-cell design*
PART II:	*Process monitoring, inspection, diagnosis and maintenance*
PART III:	*System forecasting, optimization and control*
PART IV:	*Robotic and travelling crane systems*
PART V:	*Specific applications*

Part I presents state-of-art-issues and new results in computer-aided process planning, multi-machine scheduling, self-organization in production systems, process plan sequencing of steel structures, welding diesel engines, and machine cell design via mathematical programming.

Part II presents several methodologies and tools for process monitoring in mass production, automated visual inspection of wood boards and printed ceramic dishes, function chart design for control and manufacturing systems, fault-diagnosis in conveyor-belt systems, predictive maintenance in manufacturing systems, geometric calibration of X-ray systems, and fieldbus reliability evaluation in distributed automated system design.

Part III is devoted to forecasting, optimization and control problems and presents a neural network based short-term load forecasting program for power systems, a hybrid knowledge-based and neural network - based forecasting methodology, a process optimization technique for metallurgical plants, a knowledge-based, a min-max and a static and dynamic supervisory control method for discrete event dynamic systems (DEDS).

Part IV presents an overview of force sensing methods in microrobotic systems, a novel neuropredicitve teleoperation scheme employing a predictor model and the neural input, an outline of the robotic assembly and disassembly architectural issues including representative case studies, design issues of arc welding stations aimed at increased productivity, an approach for the integration of the stereo visual and virtual reality parts in intelligent robotic systems, and finally some novel solutions of the collision-free path planning and tracking control problems of automated bridge cranes

Finally *Part V* presents analysis, design, estimation and control issues concerning specific manufacturing systems, namely: teleworking in laser material processing, blind source separation (BSS) based on temperature detection via a double pyroelectric sensor, build-time estimation of stereolithography fast prototyping systems, reconfiguration of a packing production process with flow problems between the various subsystems, computer-based determination of the iron losses in individual magnetic cores, and automated / robotics-based techniques for road construction and maintenance.

Taken together the contributions of the book provide a well-balanced source of information on the recent advances in manufacturing systems design, control and operation. The book can serve as a reference volume for the researcher and practitioner in the field, as well as for the educator and postgraduate student. The editor is indebted to all contributors for their timely contributions and to Springer 's editorial staff members for their assistance in the book preparation.

August 1998 *Spyros G. Tzafestas*

Contents

4. A Study of Self-Organization in a Production Environment
E.S. Tzafestas

5. The Process Plan Sequence in Steel Structures Welding of Great Diesel Engines
N. Nicolich, *P. Persi* and *M. Reini*

6. Design of Part-Machine Cells in Manufacturing Systems Using Mathematical Programming Techniques
S. Sofianopoulou

PART II : *PROCESS MONITORING, INSPECTION, DIAGNOSIS AND MAINTENANCE*

7. Knowledge Based Process Monitoring in Mass Production
G. Petuelli and *G. Blum*

PART IV : *ROBOTIC AND TRAVELING CRANE SYSTEMS*

33. Build-Time Estimation Tools for Rapid Prototyping Stereolithography Systems

J. Giannatsis, *V. Dedoussis* and *L. Laios*

34. An Efficient PC-Based Environment for the Improvement of Magnetic Cores Industrial Process

P.S. Georgilakis, *N.D. Hatziargyriou*, *N.D. Doulamis*, *A.D. Doulamis* and *J.A. Bakopoulos*

35. Analysis and Reconfiguration of a Packing Production Process in an Alimentary Pasta Firm

F. Deperini, *L. Penso* and *D. Pozzeto*

Contributors

Aschemann H.
Meas., Control & Microtech. Dept, Ulm Univ., 89069 Ulm, Germany

Bakirtzis A.
ECE Dept., Thessaloniki Univ., Thessaloniki, Greece

Bakopoulos J.
Schneider Electric AE, Athens, Greece

Benmounah A.
Univ. of Annaba, Annaba, Algeria

Billat A.
LAM, BP 1039, 51687 Reims, Cedex 2, France

Blum G.
Mech. Eng. Dept., Paderborn Univ., D-59494 Soest, Germany

Boerger R.
DMT-Laser Centre Bergkamen, Germany

Budinska I.
Control Inst., Slovak Acad. Sci., 84237 Bratislava, Slovak Rep.

Capkovic F.
Control Inst., Slovak Acad. Sci., 84237 Bratislava, Slovak Rep.

Chan A.
Eng. School, Wales Univ., P.O.Box 917, Cardiff, U.K.

Conrard B.
CRAN, Instrumentation, ESSTIN, Vandoeuvre, France

Danilov C.
RTS Lab., Res. Inst. Informatics, 8-10 Averescu Ave., Bucharest 1, Romania

Dedoussis V.
Ind. Management Dept., Piraeus Univ., 18534 Piraeus, Greece

Deperini F.
DEEI, Trieste Univ. Trieste, Italy

Domales K.
Control Div., Mech. Eng. Dept., NTUA, Athens, Greece

Doulamis A.
DSP Lab., ECE Dept., NTUA, Zografou 15773, Athens, Greece

Doulamis N.
DSP Lab., ECE Dept., NTUA, Zografou 15773, Athens, Greece

Fahlbusch S.
Control &Robotics Inst., Karlsruhe Univ., 76128, Karlsruhe, Germany

Fatikow S.
Control &Robotics Inst., Karlsruhe Univ., 76128, Karlsruhe, Germany

Frankovic B.
Control Inst., Slovak Acad. Sci., 84237 Bratislava, Slovak Rep.

Georgilakis P.
Schneider Electric AE, Athens, Greece

Giannakos P.
Mech. Eng. Dept., NTUA, Athens, Greece

Giannatsis J.
Ind. Managmt. Dept., Piraeus Univ., 18534, Piraeus, Greece.

Gomez Ortega J.
Syst. Auto Eng. Dept., Seville Univ., 41092 Seville, Spain

Grundmann H.-G.
Paderborn Univ., 59474, Soest, Germany

Hatziargyriou N.
ECE Dept., NTUA, 42 Patission Str., Athens, Greece

Horihata S.
Production Eng. Dept., Toyohashi Univ., Toyohashi, 441-8580 Japan

Iung B.
CRAN-GSIP, H. Poincare Univ., BP239, Vandoeurre, France

Haritopoulos M.
LAM, BP 1039, 51687 Relms, Cedex 2, France.

Hartescu F.
RTS Lab., Res. Inst. Informatics, 8-10 Averescu Ave., Bucharest 1, Romania

Harwin W.
Cybern. Dept., Reading Univ., P.O.Box 225, Reading RG6 AY, U.K.

Hofer E.
Meas., Control & Microtech. Dept, Ulm Univ., 89069 Ulm, Germany

Huez R.
LAM, BP 1039, 51687 Reims, Cedex 2, France.

Kallipolites V.
Control Div., Mech. Eng. Dept., NTUA, Athens, Greece

Kiartzis S.
ECE Dept., Thessaloniki Univ., Thessaloniki, Greece

Kitagawa H.
Production Eng. Dept., Toyohashi, 441-8580 Japan

Kontaratos M.
Trends, 9 Kondylaki Str., 11141, Athens, Greece

Kyriakopoulos K.
Control Div., Mech. Eng. Dept., NTUA, Athens, Greece

Lahres S.
Meas., Control & Microtech. Dept, Ulm Univ., 89069 Ulm, Germany

Laios L.
Ind. Management Dept., Piraeus Univ., 18534 Piraeus, Greece

Leger J.-B.
CRAN-GSIP, H. Poincare Univ., BP239, Vandoeurre, France

Lenz M.
Wirtgen GmbH, Hohner Str. 2, 53578 Windhagen, Germany

Liapakis G.
Geotech, 10 Travladoni Str. 15773 Athens, Greece

Mamalis A.
Manufacturing Div., Mech. Eng. Dept., NTUA, Athens, Greece

Maragos P.
Inst. for Language & Speech Processing Artemidos&Epidavrou, Athens, Greece

Mekras N.
IRAL, ECE Dept., NTUA, Zografou 15773 Athens, Greece

Merzak M.
Ferrovial Annaba Co., Annaba, Algeria

Miyoshi T.
Production Eng. Dept., Toyohashi Univ. Tech., Toyohashi, 441-8580 Japan

Naudet Y.
LAM, BP 1039, 51687 Reims, Cedex 2, France

Neunreuther E.
CRAN-GSIP, H. Poincare Univ., BP239, Vandoeuvre, France

Nicolich M.
Energy Dept., Trieste Univ., Trieste, Italy

Oithoff G.
Paderborn Univ., 59474, Soest, Germany

Papadakis S.
ECE Dept., Thessaloniki Univ., Thessaloniki, Greece

Paraskevopoulos P.
Control Lab., ECE Dept., NTUA, Zografou 15773, Athens, Greece

Penso L.
Energy Dept., Trieste Univ., Trieste, Italy

Persi P.
EEI Dept., Trieste Univ., Trieste, Italy

Petridis V.
ECE Dept., Thessaloniki Univ., Thessaloniki, Greece

Petueli G.
Mech. Eng. Dept., Paderborn Univ., D-59494 Soest, Germany

Pham D.
Eng. School, Wales Univ., P.O.Box 917, Cardiff, U.K.

Pozzetto D.
Energy Dept., Trieste Univ., Trieste, Italy

Prokopiou P.
IRAL, ECE Dept., NTUA, Zografou 15773 Athens, Greece

Psarianos B.
Rural & Surveying Dept., NTUA, 15773, Athens, Greece

Reini M.
Energy Dept., Trieste Univ., Trieste, Italy

Rembold U.
Control & Robotics Inst., Karlsruhe Univ., D-76128 Karlsruhe, Germany

Robert, M.
CRAN, Insrtumentation, ESSTIN, Vandoeuvre, France

Rupp T.
Forschungszentrum Informatik, Haid-und-Neu Str. 10-14, 76131 Karlsruhe, Germany

Santa K.
Control & Robotics Inst., Karlsruhe Univ., 76128 Karlsruhe, Germany

Sawodny O.
Meas. Control & Microtech. Dept., Ulm Univ., 89069 Ulm, Germany

Schmidt W.
Paderborn Univ., 59474, Soest, Germany

Shirakawa Y.
EE Eng. Dept., Fukui Univ., Fukui 910-0017, Japan

Sofianopoulou S.
Ind. Managmt Dept. Pireaus Univ., 18534 Piraeus, Greece

Suzuki M.
Production Dept., Toyohashi Univ. of Tech., Toyohashi 441-8580, Japan

Tel F.
Control & IT Dept., TU Budapest, H111 Budapest, Hungary

Terashima K.
Production Dept., Toyohashi Univ. of Tech., Toyohashi 441-8580, Japan

Theocharis J.
ECE Dept., Thessaloniki Univ., Thessaloniki, Greece

Thiriet J.
CRAN, Instrumentation Group, ESSTIN, Vandoeuvre, France

Toth E.
Control & IT Dept., TU Budapest, H111 Budapest, Hungary

Tzafestas C.
Lab. Robotique de Paris, Univ. P.-M. Curie, Velizy, France
(Currently: IRAL, ECE, NTUA, Athens, Greece)

Tzafestas S.
IRAL, ECE Dept., NTUA, 15773 Athens, Greece

Varvatsoulakis M.
Control Lab., ECE Dept., NTUA, Zografou 15773, Athens, Greece

Vargas M.
Syst. Auto. Eng. Dept., Seville Univ., 41092, Seville, Spain

Villermain-Lecolier G.
LAM, Faculte de Sciences, BP1039 Reims, France

Vivas C.
Syst. Auto. Eng. Dept., Seville Univ., 41092, Seville, Spain

Vosniakos G.
Manufacturing Div., Mech. Eng. Dept.,NTUA, 15773 Athens, Greece

Woern H.
Control & Robotics Inst., Karlsruhe Univ., 76128 Karlsruhe, Germany

Xirouchakis P.
LICP-DGM, EPFL, CH-1015 Lausanne, Switzerland

Zaytoon J.
LAM, Faculte de Sciences, BP1039 Reims, France

PART I

PROCESS PLANNING-SCHEDULING AND MACHINE CELL DESIGN

1
Concepts for Integrating Process Planning and Production Scheduling

T. Kis and P. Xirouchakis

1 Introduction

Integration between different phases of the production process is one of today's key issues in manufacturing. This is so for process planning and shop floor control as well. Process planning creates work instructions to manufacture a part and shop floor control schedules and executes these instructions to get the final products on time and with reasonable cost. However, process planning inevitably imposes constraints on the utilization of machining resources. In particular, the routing of the parts and the necessary machining times on the various resources is primarily determined by the process plans attached to them. These constraints are due to the decisions concerning the kinds of processes that should be carried out in an appropriate order to produce a part, which satisfies the original specification.

One possibility to close the gap between process planning and shop floor control is to *postpone* decisions that were taken previously in the process-planning phase. These decisions can be precisely those that affect part routings and machining times. Thus process planners should develop *processing alternatives* that can be subsequently used at shop floor control. One of the most important reasons to do so is to improve responsiveness to the continuously changing demands that the workshop should satisfy. That is, when the product mix changes, the resource requirements may change as well. A smooth and balanced utilization of resources is a *must* that should be aimed at.

Our focus is on scheduling with processing alternatives and in particular how to use state of the art methods in order to prepare schedules and also to provide some feedback information to the process-planning phase. In our discussion we confine ourselves to job-shop scheduling. The processing alternatives of the parts will be represented by means of AND/OR graphs, which capture the essential information needed in the course of scheduling.

In the following we first review some of the work which has been done in the fields of integration and job-shop scheduling. Then we show examples that scheduling can really benefit from processing alternatives. The most technical part of the paper is about applying existing methods to the new problem and also how to gain some useful information for the process planners. We finish the paper with some conclusions.

2 Review of Previous Work

Integration of process planning and shop floor control is not a novel issue, the first ideas appeared in the eighties. Chryssolouris and Chang considered the assignment of resources as the integrator of process planning and scheduling [1]. The selection of an appropriate resource for each operation from among some alternatives was addressed as a multi-criteria decision-making problem and it was supported by a decision making software tool. A more comprehensive approach was taken in the FLEXPLAN and COMPLAN projects [2,3]. The COMPLAN system consisted of different modules including CAPP systems and schedulers. The process planning modules provided *non-linear process plans* to the schedulers, which served as a basis for integration. Subsequently, the scheduler selected one particular path in each of the non-linear process plans attached to the jobs and prepared a schedule. If the result was not satisfactory, then after selecting another path in some of the non-linear process plans the schedule preparation was repeated.

The systems reviewed share the property that the integration is realized through a more flexible process plan description and in the rest of the paper we also confine ourselves to this integration principle. There are various methods to describe processing alternatives. One simple way is to have a linear sequence of macro-operations, where each macro operation comprises a set of chains of micro-operations [4]. The principle behind this representation is to divide the machining of the part into some milestone steps and take care of each step at a lower level of hierarchy. Micro-operations specify machining on work centers, like drilling or milling, and there may be some setup and clamping requirements attached to each chain. For each macro-operation one chain of micro-operation should be selected from the corresponding set, so that a complete sequence of machining instructions is chosen.

The hierarchical view is taken in the ISO 10303 standard as well, where a process plan is defined recursively using the notion of *process plan activity*, which is a *decomposable unit of work* [5]. However, the AND/OR graph representation of the ALPS process specification language provides more general control structures than the ISO standard does while also offering task decomposition [6]. The ALPS language has some notions inherited from concurrent programming like resource allocation and synchronization of activities. The purpose of the language was to bridge the gap between process planning requirements and production control requirements.

However, there is a lack of deeper understanding of scheduling with processing alternatives. Scheduling theory has recently made a break-through in solving job-shop-scheduling problems [7,8,9]. The most promising methods are local search and sophisticated enumeration techniques and their combination. The application and extension of these methods to the more general problem class is an open issue.

3 The Benefit from Processing Alternatives for Scheduling

3.1 Representation of Processing Alternatives by means of AND/OR Graphs

We will consider AND/OR graphs similar to those proposed in ALPS [4], although we use only a subset of the notions there presented. We need only to represent the machining operations, the technological precedences between them, and the control structures AND and OR. There is no notion of decomposition, in other words our process specifications are flat[1]. Consider the example depicted in Figure 1.

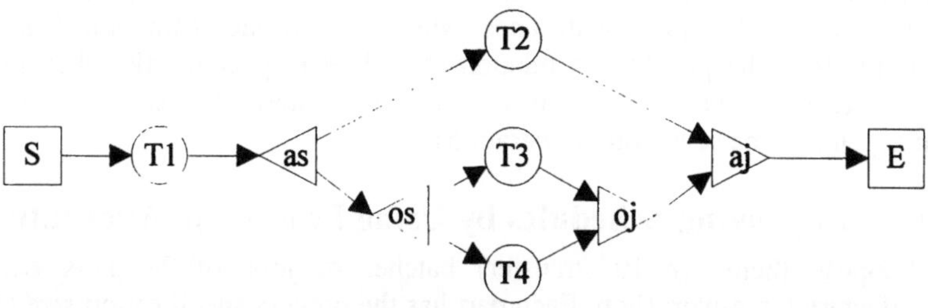

Figure 1. A process specification having AND/OR graph structure.

The depicted process specification features all structural constituents of the representation. It is an acyclic directed graph having three kinds of nodes and directed edges representing precedence relations between the nodes. The graph has a unique starting node labeled *S* that should precede all other nodes, and it has a unique ending node *E* that should be the successor of all other nodes. There are four task nodes *T1*,…,*T4*. Each task node represents a unit of work to be carried out at a given work center in a single run. Thus task nodes have detailed machining instructions, but this is not depicted in the figure. Furthermore, there is an AND-substructure, which is spanned by the pair of nodes AND-split and AND-join, and also there is an OR-substructure determined by the OR-split and OR-join nodes. (AND-split is depicted by a left triangle labeled with *as*, whereas AND-join is depicted by a right triangle labeled with *aj*. The OR-split and OR-join nodes are labeled by *os* and *oj*, respectively.)

For a more precise definition we introduce a simple language which can express the structurally sound graphs. The language consists of task identifiers and three kinds of relations: *seq*($c_1,c_2,\ldots$), *and*($b_1,b_2,\ldots$), *or*($b_1,b_2,\ldots$). The formula expressing the graph in Figure 1 is the following: *seq*(*T1*, *and*(*T2*, *or*(*T3*, *T4*))). The semantics of the language are defined recursively. The *completion of a task* involves the execution of the work instructions attached to the specified machine. The components $c_1,c_2,\ldots$ of a *seq*() relation should be completed from left to right. The branches $b_1,b_2,\ldots$ of an *and*() relation can be completed in any order and in any

[1] We use the term *process specification* instead of process plan, because usually a process plan encapsulates more information than we consider here and also in a process plan machining operations might not be grouped together with respect to resource use.

interleaving of the tasks on *different* branches. Concerning the *or*() relation, one and only one of the branches $b_1, b_2, \ldots$ has to be completed.

In our example assume that we select *T3* in the OR-substructure. The processing should start with task *T1*, then it should proceed with *T2* and *T3* in arbitrary order. Suppose that *T1* is face milling on the appropriate milling machine, *T2* is hole-drilling and *T3* and *T4* are end-milling operations to manufacture a pocket using different work centers. Then the example really makes sense, since hole and pocket making can be sequenced in any order if the volumes of these features do not overlap and the machining accuracy is sufficient with respect to the tolerances.

There are cases when the above graph and language are not expressive enough to describe the technological constraints. One such example is when there should be a precedence relation between a task *T1* in one of the branches of an AND structure and some another task *T2* outside this structure. However, such precedences should be introduced with care. For instance, when *T2* is in one of the branches of an OR structure, then the precedence relation should be respected only when the branch containing *T2* is chosen. As a consequence, there may be *conditional-precedences* that depend on the selection of operations.

3.2 Improving Schedules by Using Processing Alternatives

Suppose there are 10 (transfer) batches or jobs of 20 parts each to be manufactured in a workshop. Each part has the process specification *seq*(*T1*,*T2*,*T3*), where the tasks are the ones specified in Section 3.1. If the processing time of *T3* is dominant, i.e. significantly larger than the processing time of the other tasks, then the work-center to which these tasks are assigned to becomes a bottleneck. By allowing a choice between *T3* and *T4* for each batch the completion time of all of the batches can be significantly reduced. In this case the process specification turns to *seq*(*T1*, *T2*, *or*(*T3*, *T4*)). As we may observe, the reduction of completion time is due to the resource alternatives.

A different way to increase the freedom of schedulers is to relax the strict precedence relation between *T2* and *T3*. That is when making a schedule with the process specification *seq*(*T1*, *and*(*T2*, *T3*)) then after finishing *T1* for each of the parts in a batch, the batch could be sent either to the drilling machine or to the end-milling machine. When there are other batches for production as well, then the freedom in choosing the routings allows achievement of a smooth work flow. This will be justified in the next section after defining a mathematical model for the problem.

When there is a choice in the resources and also in the order of tasks in a process specification, then final parts having the same process specification but following different routes in the workshop may slightly differ from one another in dimensions or surface qualities or in both. These perturbations should be kept within certain limits. One the one hand this is a limitation or boundary condition to this approach. On the other hand process planners should provide process specifications having the property that any selection of tasks and their order satisfying the specification guarantees to yield a final part that meets the requirements.

4 Solving the Scheduling Problem

In this section we first define the mathematical problem which arises when faced with scheduling with processing alternatives. We are going to decompose the problem into two subproblems: (1) resource selection and (2) scheduling partially ordered sets of tasks. However, schedules can be prepared either off-line or on-line. Here we consider off-line scheduling and local search techniques.

4.1 The Resource Selection Problem

This subproblem is related to the load leveling of the workshop. The primary role of OR-substructures in a process specification is to allow selection of resources. However, machine tools may have overlapping technological capabilities. Thus it is possible to do some unit of work either on a single machine or alternatively it can be split between two machines. A fragment of a process specification providing these two alternatives might be *or*(*T1*, *seq*(*T2*, *T3*)), where task *T1* requires a single machine, and tasks *T2* and *T3* require one machine each. Hence, in the presence of many jobs the problem of machine loading is rather non-trivial.

Let a set of jobs be given to be scheduled in the next production period (few days or weeks). We may distinguish between two kinds of problems: (1) feasible selection and (2) minimal selection. The *feasible selection problem* aims at finding a selection of tasks that are sufficient to finish all of the jobs and such that for each machine the amount of work assigned is below its capacity. (The capacity of a machine in a production period is the amount of available or free working hours.) An objective function can be provided, too, expressing the total cost of using the machines. *The minimal selection problem* addresses the selection of tasks such that the weighted-sum of work assigned to each machine is minimal. In this problem there is no explicit limit on using a machine, though overloading of machines could be penalized in the goal function as well. We are concerned only with the former problem, and we discuss a special case, indeed.

We are given a set of jobs. Each job consists of one or more parts having the *same* processing alternatives. If there are more parts in a job, then these parts are processed consecutively and transferred together in the workshop. Consequently, we may think of a single part with the summation of processing times for each task. However, we impose a strong restriction on the structure of the process specifications. Namely, in an OR-substructure there can be no OR-substructure embedded. In our language this means that we do not allow specifications like *or*(seq(*T1*, *or*(*T2*, *T3*)), *T4*). This restriction allows the enumeration of all OR-substructures in a set of jobs regardless of which particular branches are chosen.

We are going to formulate the feasible selection problem as an *integer linear program* involving binary variables. Namely, with each OR-substructure we associate a set of binary variables, i.e. one binary variable for each branch. As an example, consider a fragment of a process specification *or*(*and*(*T1*, *T2*), *T3*, *seq*(*T4*, *T5*)). The OR-substructure has three branches, consequently there are three binary variables x_1, x_2, and x_3 associated with it. We have the trivial constraint $x_1 + x_2 + x_3 = 1$ expressing the necessary choice between branches. Besides these equalities there are *m* inequalities to model the loading constraints of the machines (where *m* is the number of machines). Selection of a particular branch of an OR-substructure loads the machines required by the tasks on the branch. This is modeled by adding terms of the form x_j *proc_time* to the left hand side of the inequalities that limit the load of the machines needed. The linear program has the general form:

$$\min \sum_{jk} x_{jk} c_{jk}$$

$$x_{11}\begin{bmatrix}1\\ \\ \\ a^1_{11}\\ \vdots\\ a^1_{m1}\end{bmatrix}+x_{12}\begin{bmatrix}1\\ \\ \\ a^1_{12}\\ \vdots\\ a^1_{m2}\end{bmatrix}+\cdots+x_{1k}\begin{bmatrix}1\\ \\ \\ a^1_{1k}\\ \vdots\\ a^1_{mk}\end{bmatrix}+\cdots+x_{p1}\begin{bmatrix}\\ \\ 1\\ a^p_{11}\\ \vdots\\ a^p_{m1}\end{bmatrix}+\cdots+x_{pq}\begin{bmatrix}\\ \\ 1\\ a^p_{1q}\\ \vdots\\ a^p_{mq}\end{bmatrix}\begin{matrix}=1\\ \vdots\ 1\\ =1\\ \le C_1\\ \vdots\ C_j\\ \le C_m\end{matrix}$$

$$x_{jk} \in \{0,1\}$$

where a^i_{jk} is the machining time on machine j when selecting the kth branch in the ith OR-substructure (its value equals 0 when machine j is not needed to complete the tasks on the branch), p is the number of OR-substructures and C_j is the capacity (in working hours) of machine j minus the fixed load in the next production period. The goal function is a sum over all decision variables giving the total cost of using the selected machines.

This problem is quite similar to the *MULTIDIMENSIONAL KNAPSACK* problem [10], for which many solution methods have been proposed in the literature [11]. The only difference is that in our linear program we know in advance that exactly p binary variables should be set to 1, i.e. one for each OR-substructure. This additional knowledge can be exploited in a solution algorithm.

It may happen that the problem has no feasible solution. That is, any assignment of binary variables satisfying the equality constraints hurts at least one of the inequality constraints. In this case some of the jobs cannot be completed in the next production period. Finding the right job to put on backorder or subcontract may involve the solution of another optimization problem considering due dates, costs and penalties.

4.2 Scheduling Partially Ordered Sets of Tasks

Suppose that the resource selection problem has been solved: i.e. a selection of tasks is available for each job. Then there still might be some freedom in the order of execution of the tasks of each job. That is, after filtering out the *or*() relations by choosing one particular branch for each of them, the order of tasks in *and*() relations is unspecified. Traditional scheduling methods cannot handle these process specifications adequately, because it might happen that two tasks in *and*() relation could be assigned to overlapping time periods in the course of execution. This case must be prohibited. To this end, we define a disjunctive graph representation of the scheduling problem and show how to apply some well-known methods to find good approximations of time-optimal schedules.

4.2.1 The Disjuntive Graph Representation

Consider a set of jobs, where the process specifications attached to the jobs are restricted to tasks, *seq*() and *and*() relations. The *disjunctive graph* representation of the scheduling problem consists of several *cliques*. The union of all tasks of all process specifications, an extra *starting* and respectively *finishing* node constitute the set of nodes. Each task belongs to precisely two cliques: (1) the tasks of a job form a clique (so called *job-clique*) and (2) the tasks that require the same machine also form a clique (*machine-clique*). All in all, there are $n+m$ cliques, where n is the number of jobs and m is the number of machines.

In each clique there are edges between every pairs of tasks. However, there are two kinds of edges in the graph: *directed edges* and *disjunctive edges*. There is a directed edge between two task nodes if the order of execution is already known. Initially, directed edges may occur only in job-cliques, whenever some tasks are in *seq*() relation. Furthermore, there is a directed edge from the *starting* node to every task node and there is a directed edge from every task node to the *finishing* node.

A disjunctive edge between two tasks expresses the choice in the order of execution. These edges are to be changed to directed edges in the course of scheduling so that a total order of execution is specified in each clique and the graph has no directed cycles. Disjunctive edges can be between tasks of the same job, when the tasks are on different branches of an *and*() relation, or between tasks requiring the same machine. In fact, machine-cliques initially possess only disjunctive edges, while job-cliques may have edges of both kinds.

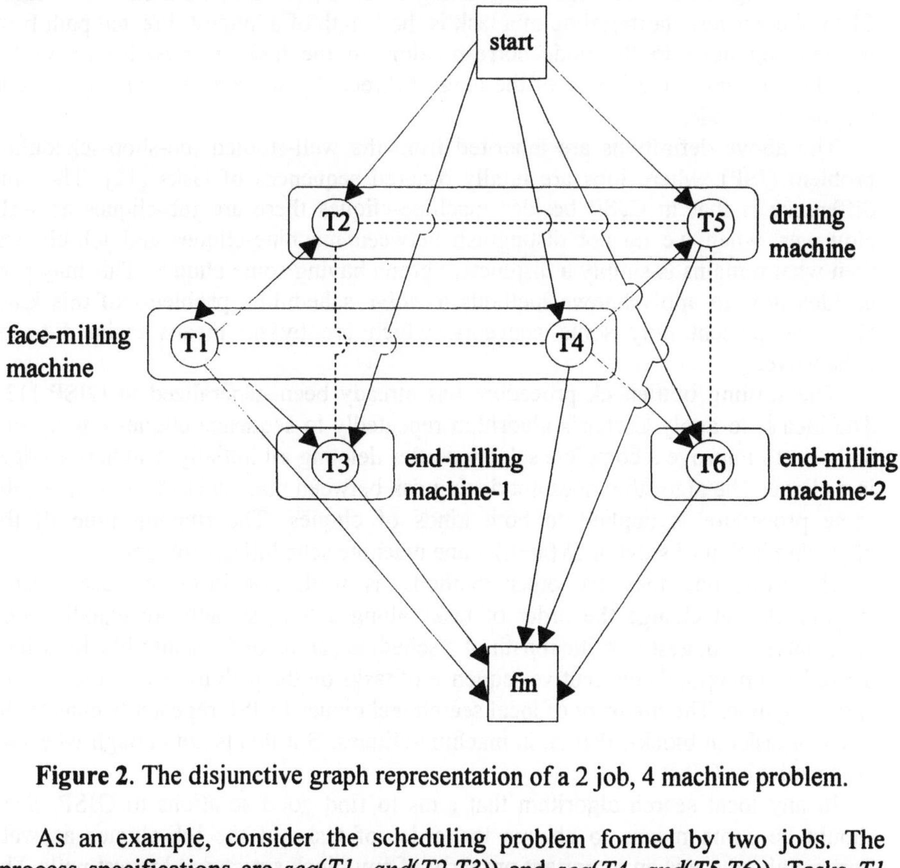

Figure 2. The disjunctive graph representation of a 2 job, 4 machine problem.

As an example, consider the scheduling problem formed by two jobs. The process specifications are *seq*(*T1*, *and*(*T2*,*T3*)) and *seq*(*T4*, *and*(*T5*,*T6*)). Tasks *T1* and *T4* require the face-milling machine, tasks *T2* and *T5* need the drilling machine and the machining of tasks *T3* and *T6* involves different end-milling machines. The disjunctive graph representing the problem is depicted in Figure 2.

In the figure there are 2+4 cliques. Namely, there are two job-cliques and 4 machine-cliques (machine cliques are framed by horizontal bars). Dashed lines indicate disjunctive edges. There is a choice in the order of execution of the pairs of tasks *T2* and *T3*, *T2* and *T5*, *T1* and *T4*, *T5* and *T6*, respectively.

4.2.2 The Generalized Job-Shop Scheduling Problem

The scheduling problem can be either to minimize the maximal task completion time or makespan, or when due-dates are given minimum tardiness, or other standard performance measures. Here we consider only the minimal makespan problem, which we will refer to as GJSP.

A feasible solution to the scheduling problem is a *complete selection* of directed edges which substitutes all of the disjunctive edges of the disjunctive graph representation of the problem such that the graph has no directed cycle. In order to determine the starting time of each task, we assign a weight to each directed edge. Namely, the *weight* of a directed edge is the processing time of the task node from which it originates. The edges originating from the *starting* node have 0 weight. Then, the *earliest* starting time of a task is the length of a longest directed path from the *starting* node to the node corresponding to the task. The *makespan* of the schedule is simply the length of the longest directed path from the *starting* node to the *finishing* node.

The above definitions are inherited from the well-studied job-shop-scheduling problem (JSP), where jobs are totally ordered sequences of tasks [12]. The only difference is that in GJSP besides machine-cliques there are job-cliques as well. However, when we do not distinguish between machine-cliques and job-cliques, then what remains is simply a disjunctive graph having some cliques. This may give an idea how to apply known methods to solve scheduling problems of this kind. Here we present only some concepts, which are by no means claimed to be exhaustive.

The shifting bottleneck procedure has already been generalized to GJSP [13]. The idea is to apply Carleir's algorithm repeatedly to sequence cliques one by one and also to improve a complete selection after deleting all initially disjunctive edges in a clique. The algorithm does not distinguish between machine and job-cliques, the same procedure is applied to both kinds of cliques. The running time of the algorithm is that of solving $O((m+n)^2)$ one machine scheduling problem.

However, one may try other methods as well. For instance, local search techniques that change the order of tasks along a longest path are equally good candidates. A longest or critical path of a schedule can be divided into blocks, where a *block* is a maximal consecutive sequence of tasks on the path to be executed on the same machine. The majority of local search techniques to JSP repeatedly change the order of tasks in blocks, that is, in machine-cliques. But this is not enough when we are faced with GJSP.

In any local search algorithm that aims to find good solutions to GJSP, there should be some means to change the order of tasks in the job-cliques as well. Fortunately, there is an important property of any block structured longest path: The last task of a block and the first task of the following block always belong to the same job. If the directed edge connecting these two tasks substitutes a disjunctive edge, then it can be reversed, so that a different complete selection is obtained. It is easy to show that reversing *any* directed edge along a longest path of a complete selection cannot introduce a directed cycle in the graph (the original statement and proof considered only operations in the *same* block). This fact justifies our statement that we get a complete selection. Combining this with known transformation rules new algorithms can easily be developed.

At this stage we are working on algorithms that use the above principle, but no computational results are yet available. However, there is a lack of common test examples for GJSP either generated randomly or coming from industrial practice. There are already a lot of test examples widely available on the internet for JSP [14] which has proved to be a good basis for comparison of various methods, and perhaps stimulated some work in the field as well.

4.3 Interactions between Resource Selection, GJSP and Process Planning

We have identified two subproblems: first we should solve the feasible selection problem then using the selection found turn to the generalized job-shop scheduling problem. When a satisfactory schedule is determined for the latter, then the schedule certainly has one or more longest paths. The machining times of the tasks on a longest path largely contribute to the length of the path. There are two options to shorten the schedule length: (1) perform these operations faster, or (2) choose other processing alternatives when available.

The first option challenges the process planners. We consider the longest path calculation and analysis as the *backward interface* between production scheduling and process planning. In this way a loop is formed between process planning and scheduling: process planning provides processing alternatives to scheduling and scheduling can point out tasks or processes whose modification may enable better performance in the workshop.

The second option needs efficient algorithms that can select new branches in some OR-substructure of some process specifications. In this way we can iterate between resource selection and scheduling partially ordered sets of tasks.

5 Summary

In the paper we have presented high level concepts for integrating process planning and scheduling by means of processing alternatives. We have meditated about the representation of processing alternatives and discussed the scheduling function in more details. We have also identified two subproblems that inevitably emerge when making production plans in the presence of processing alternatives. The subproblems can be solved in a sequential or iterative manner applying known optimization and scheduling techniques.

Acknowledgements

This work has been supported by Swiss National Research Foundation, grant 2100-049421.96/1.

Bibliography

1. Chryssolouris G, Chan S 1985 An Integrated Approach to Process Planning and Scheduling, *Annals of the CIRP* 34/1:413-415.
2. Tönshoff H K, Kreutzfeld J and Hofschneider, D 1993 Concurrent Process Planning and Workshop Control in Batch Production, *Proceedings of the CIRP Seminars, Manufacturing Systems* 22/3:231-241.
3. COMPLAN, ESPRIT Project 6805, *Public Domain Report*, http://www.mech.kuleuven.ac.be/pma/project/complan/complan.html
4. Brandimarte P, Calderini M 1992 Integrated Process Planning and Shop Scheduling by Local Search Algorithms, In: Pels H J, Wortmann J C (eds) *Integration in Production Management Systems, IFIP*, 257-276.
5. Lee S, Wysk A and Smith J S 1994 Process Planning Interface for a Shop Floor Control Architecture for Computer-Integrated Manufacturing, *Int. J. Production Research* 33/9:2415-2435.
6. Catron B A, Ray B 1991 ALPS: A Language for Process Specification, *Int. J. Computer Integrated Manufacturing* 4/2:105-113.
7. Carlier J, Pinson E 1990 A Practical Use of Jackson's Preemptive Schedule for Solving the Job-Shop Problem, *Annals of Operation Research* 26:269-287.
8. Nowicki E, Smutnicki C 1996 A Fast Tabu Search Algorithm for the Job-Shop Problem, *Management Science* 42/6:797-813.
9. Balas E, Vazacopoulos A 1995 Guided Local Search with Shifting Bottleneck for Job Shop Scheduling, *Management Science Research Report #MSR R-609, Carnegie-Mellon University, Pittsburgh.*
10. Nemhauser G L, Wolsey L A 1988 *Integer and Combinatorial Optimization*, John Wiley&Sons, Inc.
11. Chu P C, Beasley J E 1998 A Genetic Algorithm for the Multidimensional Knapsack Problem, http://mscmga.ms.ic.ac.uk/jeb/jeb.html.
12. Blazewicz J, Domschke W and Pesch E 1996 The Job-Shop Scheduling Problem: Conventional and new Solution Techniques, *European Journal of Operation Research* 93:1-33.
13. Ramudhin A, Marier P 1996 The Generalized Shifting Bottleneck Procedure, *European Journal of Operation Research* 93:34-38.
14. Beasley, J. E., OR-Library, http://mscmga.ms.ic.ac.uk/info.html.

2

Intelligence in Computer-Aided Process Planning for Machining

A.G. Mamalis and G.C. Vosniakos

1 Introduction

Process planning is the systematic specification and sequencing of processes by which a product is to be manufactured economically within a given manufacturing system. A process plan usually contains the route of operations and specification of each operation in the route, i.e. type of operation, operation parameters (e.g. feed, speed etc.), machine, tool(s), jigs/fixtures, standard time, setup time, and additional instructions (e.g. for setup, in-process inspection etc.)

Process planning requires data on production equipment, general knowledge on technological capabilities of manufacturing processes and optimisation skills based on long experience. Therefore it is data-intensive, technological knowledge-intensive and experience-intensive. It is no wonder that manual process planning is dependent on few experts, inconsistent and slow, hence computer aided process planning (CAPP) is increasingly used [1].

Variant CAPP is based on the notion of Group Technology part families. All parts in a part family have similar shape and/or process plans (similar type. sequence and number of operations). When a new process plan needs to be produced, the standard process plan of the appropriate part family is retrieved and modified to tailor the plan to the particular part.

Generative CAPP relies on manufacturing knowledge captured in the system and not on standard process plans. Thus process plans are 'generated' from scratch. Such knowledge requirements are usually product type -specific and company-specific, resulting in dedicated CAPP systems, e.g. for sheet metal parts, forged parts, machined (or even turned) parts etc. [2].

Another distinction is that between interactive and automatic systems. Interactive CAPP systems tend to rely on the user's decision based on presented information. Automatic systems minimise all user interaction. system intelligence, although it appears strange at first, can be manifested equally in either type, although interactive systems by definition rely on human intelligence, too.

In general, the more intelligent the CAPP system, the less the user will be required to think about process planning problems and to suggest solutions or decide on system suggestions. In spite of the wide use of variant systems, in practice these are recognised as being too constrictive. Generative systems are by definition intelligent, but their scope is too wide to render them practical. This work advocates restriction of that scope to specific part types or families.

2 Process planning philosophy

Variant process planning based on rigidly defined standard process plans for part families as in Group Technology composite parts suffers various drawbacks, e.g. the plans need to encompass all possible features to enable editing by removal, and it is hard to accommodate variations that are due to feature interactions [1].

A parametric type of standard process plan is much more flexible and can accommodate variations of presence, dimension and interaction of features. This needs to be based on a language, though. Such a language is ALPS developed by NIST [3]. It provides conditional and Boolean structures as well as grouping mechanisms applied to operations that make up a standard process plan, see Fig. 1.

A standard process plan of that sort can be modified to suit a particular component deemed to be sufficiently represented by it. Strictly speaking, this kind of editing is definitely possible in the interactive mode. Otherwise, automatic editing has to be based on automatic comparison of the features of the particular part and those of the composite part representing the family. The rules according to which editing is performed could be coded into an expert system. However, the complexity added in this way may be too great to justify the *standard plan* approach. Instead, a structured rule set for process planning, i.e. mainly for choosing alternative operations associated to features, would normally be more straightforward.

In either case it is desirable that the process plan derived is fairly open, not committed to specific tools nor referring to processing conditions, but providing

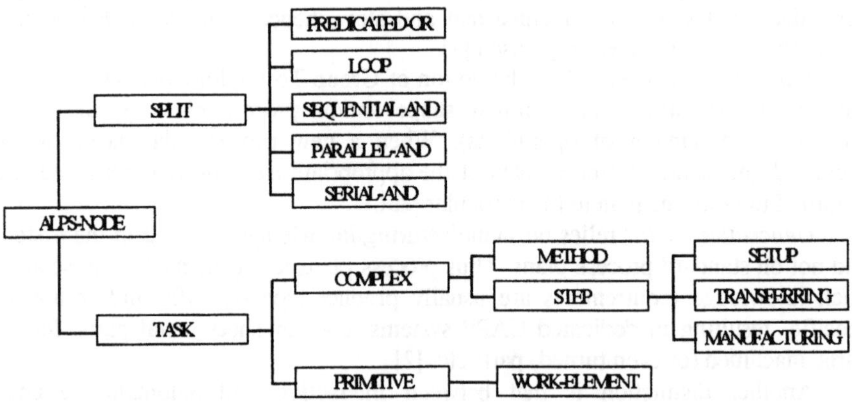

Figure 1 ALPS structure

alternative operations that would accommodate any backtracking, should any of these parameters eventually prove to render an operation unfeasible. This first phase of process planning should be termed 'generic' or 'strategic' planning.

Subsequently this ‘loose’ process plan should be 'optimised' by considering tools, machining and general processing conditions, as well as sequencing criteria. This should be termed 'detailed' or 'operative' planning [4].

A general architecture representing the philosophy outlined above is depicted in Fig. 2 as a synthesis of approaches followed in references [1] and [4].

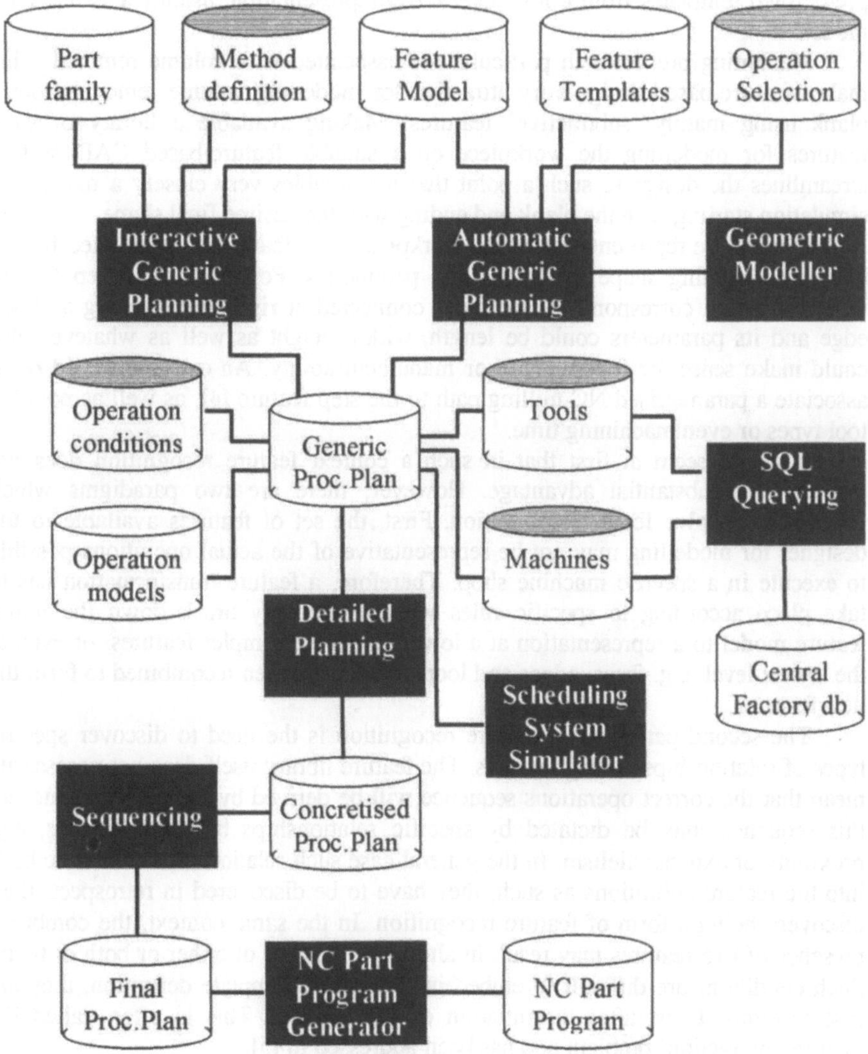

Figure 2 General CAPP system architecture

3 Geometry Interface

Invariably, geometric reasoning in process planning is carried out in terms of features, i.e. shape regions of specific significance in manufacturing method terms. An intelligent CAPP system should not rely on description of these regions by the user, but it should be able to recognise them. The two schools of thought on this issue support either feature based design or feature recognition. The former obliges the designer to model the workpiece as a collection or even a sequence of manufacturing features. The latter extracts a sequence of such features according to pre-defined templates from a low level CAD representation, usually a B-rep solid model.

Machining processes in particular are associated with volume removal. This makes feature-based design very attractive for modelling volume removal from a blank using mainly 'subtractive' features. Making available a library of such features for modelling the workpiece on a suitable feature-based CAD system streamlines the design to such a point that it resembles very closely a machining simulation starting with the blank and ending with the desired final shape.

In a feature representation of the workpiece each feature is instantiated from a template including shape primitives and parameters. For example a step feature template would correspond to two faces connected at right angles along a single edge and its parameters could be length, width, height as well as whatever else could make sense for functionality or manufacturability. An extreme would be to associate a parametrised NC milling path to the step feature [4], as well as possible tool types or even machining time.

It would seem at first that in such a context feature recognition does not provide any substantial advantage. However, there are two paradigms which necessarily involve feature recognition. First, the set of features available to the designer for modelling may not be representative of the actual operations possible to execute in a specific machine shop. Therefore, a feature transformation has to take place according to specific rules which effectively break down the initial feature model to a representation at a lower level, e.g. simpler features, or even at the lowest level, e.g. faces, edges and loops, which are then recombined to form the new features.

The second paradigm of feature recognition is the need to discover specific types of relationships among features. The feature library itself does not necessarily mean that the correct operations sequence will be derived by the designer. Indeed, this sequence may be dictated by specific relationships between features, e.g. proximity or axis parallelism. In the general case such relationships cannot be built into the feature definitions as such; they have to be discovered in retrospect, their discovery being a form of feature recognition. In the same context, the combined presence of two features may result in altering the nature of either or both of them. Such conditions are difficult to embed in the feature template definition; they are best recognised just after instantiation of the features. This is often called the 'feature interaction' problem and has been addressed in [5].

Feature recognition requires extensive geometric reasoning which draws on low level routines such as Boolean operations e.g. for comparing volumes, ray casting, e.g. for determining feature access directions.

Feature recognition in machining is often easier when it starts form the final shape of the workpiece and proceeds towards the shape of the blank by adding instead of removing material. This has the advantage of presenting fewer initial alternatives (features) to choose from.

In either case it is desirable that feature recognition is also done incrementally, i.e. the shape (blank or workpiece) be updated by actually performing the volume operation (subtraction or addition respectively) resulting in a new shape on which the next feature is to be recognised [6].

Feature representations have been easily designed in terms of trees whose leaves are instanced feature templates and whose nodes are operators acting on the features (usually Booleans) [7], see Fig. 3.

Feature recognition is most powerful when based on logic predicates. Expert system type of recognition is possible [1], although relatively slow compared to the so-called adjacency graph method. The latter, however only incorporates edge concavity and adjacency information and its descriptive power is limited. Neural network based recognition is much faster than both and seems to be descriptive

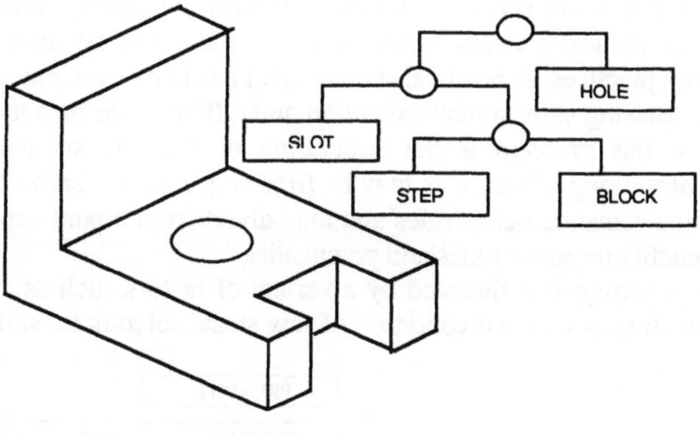

Figure 3 Feature model example

enough [8]. An example of a feature recognition rule is as follows :

```
if feature has a floor block
and the floor block nests no profiles
and the floor block is adjacent to a wall loop
then the feature is pocket
```

The above make for a geometry interface to a process planning system in terms of intelligence and co-operation with the process sequencing module.

4 The planning tasks

4.1 Generic Process Planning

The type of operations that comprise a generic process plan for machining are :

- material removal operations performed with cutting tools

- setup operations consisting of orienting the part or parts (in pallets)
- fixturing operations, consisting of clamping the part(s) on suitable devices
- inspection operations referring to geometrical or surface finish attributes

Each of these operations is associated with a feature (machining, setup, fixturing or inspection) which has to be either selected (in feature-based design mode) or recognised (in feature recognition mode). A feature may and indeed should be associated with alternative manufacturing methods / operations.

Generic process planning performed interactively consists of two phases : First, selecting features from the part family model and associating operations to them leaving room for alternatives and specifying obligatory precedences as necessary. Intelligence in this case is restricted to providing a knowledge base of manufacturing methods pointing to feature types. Second, modifying the part family process plan to suit a particular component within that family. Intelligence in this case can consist of advising the user of editing actions for manufacturing methods according to feature model comparison. Indeed this comparison can be automated, too through an expert system that could use geometric reasoning rules.

Generic process planning performed automatically in reality follows the pure generative approach. Most rules for process selection are largely independent of the specific manufacturing environment, whereas some can be associated with specific company practices. A practical problem that has to be overcome is the large number of rules causing slow system execution and difficult rule base maintenance. One solution to this problem is the structuring of the rule set into sub-sets. Furthermore, paths along which rules may be fired sequentially can be defined [1]. This amounts to a semantic net of rules and may divert from a pure expert system, but certainly benefits response times and practicality.

Operation selection is influenced by a variety of factors such as : part shape and dimensions, dimensions and condition of raw stock, tolerances, surface finish.

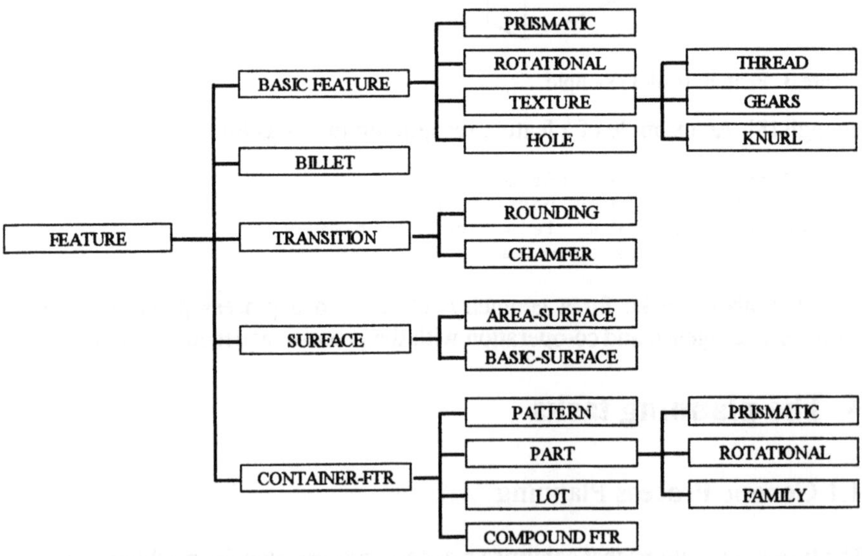

Figure 4 Partial machining feature taxonomy

In addition, essentially any of the four types of operations mentioned above depend on one or more of the others, e.g. machining operations need to know to some extent the fixturing arrangements in order to avoid collisions, and fixturing operations need to know the machining forces in order to keep the part on the table. One way to circumvent this type of chicken-and-egg situation is to assume some sequence for considering selection of these operation types and check for mutual constraint satisfaction afterwards [4].

Machining features are considered first in order to extract all possible machining operations each of which is associated with a setup direction, Fig. 4 depicts a partial machining feature taxonomy.

Setup planning is performed next so as to minimise setups. This is usually a matter of clustering the set of operations into operation access directions in alternative ways.

Fixture planning follows with an aim to check whether the setup orientations are compatible with the fixturing feature types, see Fig.5. Note that every fixturing feature is associated with a number of fixturing elements.

If any discrepancies are found, the setup clusters are changed accordingly; if , even then, no solution is arrived at, alternative operations for the machining features

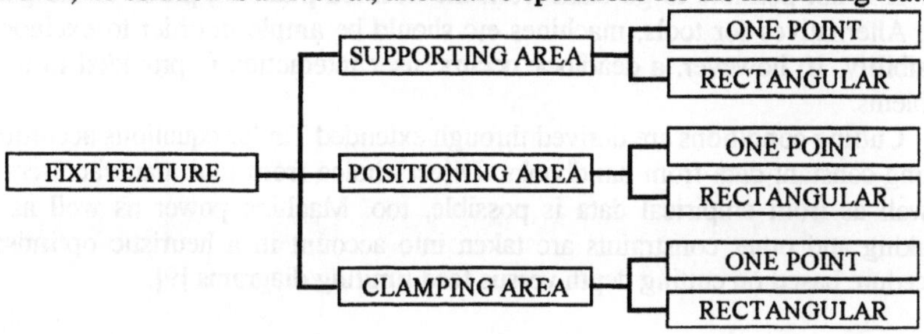

Figure 5 Fixturing feature taxonomy

are chosen.

This philosophy is ideally implemented via a backtracking engine; the Prolog engine has been used for that purpose [1], but coding in a conventional language is possible, too.

Inspection planning is a successor to machining operation planning. Therefore it is performed last, but inspection operations have to be interspersed with the machining operations if in-process inspection is needed.

Deadlocks in backtracking are resolved as a last resort by user interaction, according to system prompts. These usually refer to shortage of alternatives for a particular operation, leading directly to need of augmenting the relevant system knowledge base.

4.2 Detailed Process Planning

The activities of this phase are related to specific factory equipment such as machines, tools and fixtures. The appropriate objects need to be selected at this stage in order to concretise the generic process plan. In addition, cutting conditions need to be determined [2].

It is possible that some of the decisions of generic planning become invalid at this stage, e.g. because some tool type is unavailable, or because some machine is committed to production of other orders. It is, therefore, necessary to provide information on factory equipment as well as on manufacturing order status and lot sizes. The latter are necessary in order to optimise the sequence of operations, should pallets be available.

Machines are selected according to the machine type associated with the operation designated in the generic process plan, on the basis of their power, their ability to perform designated operations as well as their accuracy capabilities. Lot size enters the selection as a discriminating criterion for economic batch quantity calculations.

Tools are selected according to tool type associated with the operation designated in the generic process plan. A simple knowledge base accommodates the experience-based tool selection to cater for manufacturing system idiosyncracies. A simple optimisation algorithm minimises the total number of tools used. This works similarly to the clustering of setups mentioned earlier.

It is not possible to go back from the detailed planning phase to the generic one. Alternatives for tools, machines etc should be ample in order to exclude this possibility. If, however, a deadlock occurs, user interaction is provided to resolve problems.

Cutting conditions are derived through extended Taylor equations according to cutting constant data from handbooks. Direct choice from tool manufacturers data as well as from empirical data is possible, too. Machine power as well as chip breaking and other constraints are taken into account in a heuristic optimisation procedure based on cutting depth versus feed limiting diagrams [9].

4.3 Process Sequencing

Feasible sequencing of operations is straightforward and needs only respect operation precedence constraints. Optimisation or near optimisation of the final sequence, though, needs to follow a criterion. Possibilities are : number of tool changes, number of setups and total machining time. What complicates matters more is the fact that the alternative processing methods (for the same feature) need to be considered, too, in order to find an optimum or near optimum.

Often, the search space is represented as a tree which would need to be traversed in all possible routes in order to calculate the value of the objective

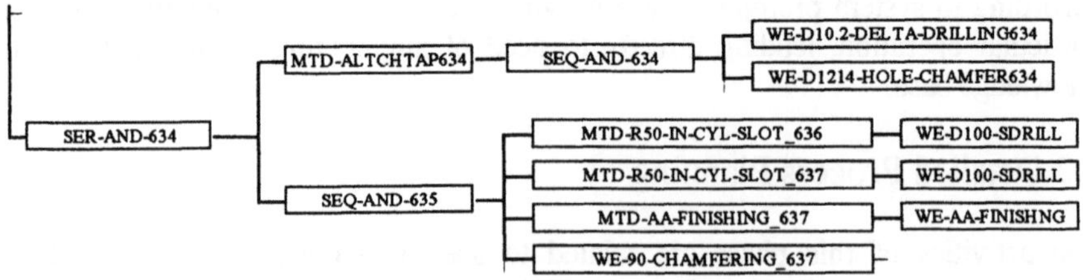

Figure 6 Process planning tree

Work element	Tool	Side	Part	Feature
WE-90X360-AREA-R-MILLING709	SECO-D63-6T30	270	276857	RECT-AREA-709
WE-D215-HOLE-SI-MILLING709	SIDEMILL-D80	0	276857	INCYL-AREA-614
WE-D22-U-DRILLING711	U-DRILL-D22	270	276857	RPTRN-VR-2X2-609
WE-25H7-SINK-SI-MILLING711	RCUT-EML-D20	270	276857	RPTRN-VR-2X2-609

SEQUENCE PRT-PROGRAM SAVE QUIT

Figure 7 Sample process plan

function that needs to be minimised (or maximised), see Fig.6. Such 'brute-force' traversal may be meaningful only in cases where serious restrictions on manufacturing methods are in force; whereas in normal cases the search space becomes too huge and needs some kind of pruning algorithms in order to exclude certain branches of the tree and sequence the remaining ones in order to arrive at a near-optimum in an efficient way. Pruning algorithms are based on heuristic rules. The search follows a breadth first or a depth first pattern depending on the width and height of the tree in conjunction with the effectiveness of the pruning heuristic.

The search space tree corresponds in fact with the ALPS tree that represents the alternative operations of the generic process plan that was concretised during the detailed planning stage. The desired result is a flat structure of purely sequentially arranged operations, see Fig.7. The heuristic tree-pruning algorithm then becomes a transformation of the tree to a list. This list is the actual process plan that has to be suitably formatted before being down-loaded to the shop floor.

A straightforward follow-on function is the transformation from process plan to NC part program. A standard parametric piece of NC code is associated with each operation / feature. The flat structure of the final process plan is scanned sequentially and the appropriate code is substituted with parameters instanced accordingly. Tool changes or pallet rotation statements are also added as necessary, since such events are easily recognised.

5 Data and Knowledge Representation

The programming paradigm followed is object orientation. Within that framework it is possible to accommodate both data and knowledge. Data differs from knowledge in that the latter contains either directly or indirectly procedural information usually expressed as rules. Data and Knowledge was represented in terms of Frames, a compact structure defining an object or a class of objects with their attributes represented by so-called 'slots' and possibly sub-slots called 'facets'. A slot can also contain a pointer to a rule or 'method'. The BEEF frame system in a LISP environment was chosen for implementation of data and knowledge bases [4].

Logic programming implemented with the Prolog language was also tried as the main paradigm with good results. Prolog suitability stemmed from the fact that it not only provides structures for symbolic and pattern-oriented programming, but also an implicit database manipulation language with a powerful 'matching' facility. Prolog also encompasses a backtracking mechanism that is useful in the context of the inherently iterative search associated with the process planning problem.

The main difficulty encountered with both of the above approaches is whenever access to the geometric model is necessary, especially with incremental feature recognition. Geometric manipulation has to be performed within the geometric (solid) modeller environment. Such environments never support languages such as LISP and Prolog and therefore an interface - interpreter for those languages has to be implemented. Alternatively a non-modifiable 'light' geometric model can be deduced from the full model and be made available in a BEEF or Prolog format.

Factory data needed in decision making refers typically to tools, machines and cutting conditions. Tool data, for example, is modelled with the following slots [4]:

```
a-kind-of, stamp, work-element,
geom.dimensions,cutting_angles,
possible_operations,possible_materials,
realtion_of_tool_wear,
TMS_code, TMS_address
```

Knowledge refers to definition of manufacturing methods, selection of operations, tools and machines, cutting condition derivation etc. As an example, definition of hole drilling is as follows [4] :

```
(define_method_alt_chp_tapping
   (feature_type  straight_through_hole)
   (plan_specification
         (alternatives
                  ((thread_unit  metric)
                           (sequential_and
                                   we_d10.2_delta_drilling
                                   we_d12/14_hole_chamfering
                                   we_m12_tapping)))............
```

A sample tool selection rule is as follows :

```
IF feature is through_hole AND
diameter_area is between x and y AND
machine is B&W THEN
tool_holder is K_holder AND
tool is M8
```

6 Data Interfaces

Given that much of the data needed in process planning is also used by other functions in the factory environment it becomes a question of non-redundancy and of data integrity after modifications to link all functions needing this data to a central data repository. This is the factory database. The alternative would be a distributed database environment with some complexity in the distribution model and a need for suitable tools for data query and manipulation.

The factory database in our case is relational and managed with Oracle™. It is queried in SQL by the process planning system (as well as by other functional areas of the factory) to upload mainly tool and machine data needed locally. Although remote queries are possible this should not be confused with distributed data management, since no provisions for concurrency (through locking, time-stamping and similar mechanisms) are in place.

Detailed process planning needs to refer to shop floor schedules both past and predicted in order to finally choose machine tools. Past schedules are accessible through the central database in a way similar to that applicable to the tools. Future schedules are calculated based on a discrete event simulation model with a twofold investigation path. The first one assumes alternative process plans and optimises the choice of machine tools. The second one assumes a rigid decision on machine tools and optimises the timing of production order downloads to the shop floor under the delivery time constraints [10].

7 Conclusions

Process planning involves intelligence at many decision levels. Drawing on major relevant projects in which the authors took part it is recommended to split the planning task into a generic phase, peculiar to part families and independent of the particular manufacturing environment, and a detailed phase, peculiar to a part within a family and allowing optimisation and sequencing of operations. Both are based on features of part families and parts to be manufactured.

Feature based design or feature recognition from a lower level representation essentially build a generic process plan with alternative operations.

A knowledge base defines operations and assigns them to features at the generic level. At detailed level a knowledge base prunes the search tree to optimise - sequence the operations. controlled backtracking allows correction of decisions at previous levels.

Object oriented data and knowledge representation with frames and associated methods is the best paradigm for implementation of this model, preferably in C++.

A central read-only database interface and accompanying scheduling system complement the whole.

References

1. Vosniakos G C 1998 An intelligent software system for the automatic generation of NC programs from wireframe models of 2-1/2-D mechanical parts. *J. Computer Integrated Manufacturing Systems, accepted for publication.*
2. Opas J, Kanerva J and Mantyla M 1994 Automatic process plan generation in an operative process planning system. *Int. J. Production Research*, 34: 332-347.
3. Ray S R and Catron B 1991 ALPS : a language for process specification. *Int. J. Computer Integrated Manufacturing*, 4(2): 105-113.
4. Brite-EuRam Consortium - project 3528 1993 Manufacturing Cell Operator's Expert System (MCOES) Synthesis Report.
5. Laakko T and Mantyla M 1993 Feature modelling by incremental feature recognition. *Computer Aided Design*, 15 (8) : 479-492.

6 Joseph A T and Davies B J 1990 Implementation of a backwards planning technique in the automation of process planning. *Int. J. Advanced Manufacturing Technology*, 5(4) : 321-344.

7. Denzel H and Vosniakos G C 1993 A feature-based design system and its potential to unify CAD and CAM. *IFIP Transactions B: Computer Applications in Technology*, B-10 : 131-144.

8. Nezis C and Vosniakos G C 1997 Recognising 2-1/2-D features using a neural network and heuristics. *Computer-Aided Design*, 29(7) : 523-539.

9. Arecularatne J A, Hinduja S and Barrow G 1992 Optimum cutting conditions for turned components. *Proc. Institution of Mechanical Engineers, Part B*, 206(B1) : 15-31.

10. Mamalis A G, Malagardis J and Kambouris K 1996 On-line integration of a process planning module with production scheduling. *Int. J. Advanced Manufacturing Technology*, 12(5) : 330-338.

3
Single and Multi-Machine Scheduling of Jobs in Production Systems

B. Frankovič and I. Budinská

1 Introduction

The effectiveness of the production is the main task of Flexible Manufacturing System (FMS), which is possible to obtain if the system works reliable, its capacity is utilized by an optimal way so that the cost of production will be minimal and the quality of the products will be at the high level.

One of the tools for obtaining of this goals are the scheduling algorithms which may to improve the efficiency of FMS and its parts, for example, the performance of workshops (W), of production lines (PL) and of shaping, drilling, machining, transport, testing devices, etc.

In this paper a special type of scheduling is considered, where the scheduling algorithm is based on the heuristics approaches and the rescheduling way. The application is focused to the single and multi-machine scheduling in time domain. The problem can be stated as follows: a set of jobs available at time 0 should be scheduled in a workshop created by different types of PL which are in parallel configuration, where the jobs are structured in time. The following assumptions are made: each PL has its own production (technological) plan and their products are used to the completion of the final products. Such system would be, for example, an workshop with an assembly task.

PL contains of different production devices in a given configuration according to the required technological process. It considers that some devices, for example, automatic guided vehicle (AGV) or NC -machines, can serve for more than one PL. From this point of view it is desirable that the separation time between two successive jobs in each PL be defined. The separation time may be lower as constrained processing time.

In case the only transport task executed by an AGV was considered, the problem would belong to the classical single-machine scheduling. That means N jobs are processed on one machine. The machine can realized only one job at the time and each job has to be realized without interruption on the machine. In other words, the arriving job cannot interrupt the realization of the job in the machine and has to wait while the job in a machine be finished.

The solution of such problems is the subject of many publications. In [1] a solution of single-machine scheduling problem is studied. In this problem there are not only chain structured precedence constraints, but also minimal and maximal times separating successive jobs in the same chain. The criterion to be minimized is the makespan. The paper [6] deals with the problem of optimal scheduling of n part-types in a failure-prone machine manufacturing system. In [2] the influence of machine failures on the performances of a scheduling strategy in a job-shop system is appointed out . In [4] such scheduling problem is studied, where the job-shop contains some identical jobs. In this case the disjunctive constraints between the operation can be replaced by conjunctive constraints, some conjunctive constraints can be deleted and the search space can be reduced. The combined Lagrangian relaxation and dynamic programming algorithm for job-shop scheduling to minimize a weighted tardiness penalty function is described in [7]. The papers [5], [3] deal with the economical basis of the scheduling problem.

In this paper Section 2 describes the possible approach to single-machine scheduling of transport jobs and the verification way on the fictitious production systems. In Section 3 the multiple-machine scheduling problem with possible application is described. Finally, the conclusions are presented in Section 4.

2 Formulation of the single-machine scheduling problem

For the formulation of single-machine scheduling problem there is considered a workstation with $p = 1,2,...,m$ number of PL and j_z, $z = 1,2,...,n$ number of jobs in the same PL. Then the total number of jobs is expressed by the equation:

$$N = \sum_{p=1}^{m} \sum_{z=1}^{n} j_{p,z} \tag{1}$$

Let be denote (p,z) the z-th job in p-th PL. Let be denote the processing time of this job as $t > 0$. It assumes that the processing of a job cannot be interrupted. In our formulation there is considered a workshop with the characteristics as follows:

- each workshop consists of various PLs with various type of devices (NC machine, robot, automatic guided vehicle (AGV), testing and control devices etc.)
- various types of workpieces and halfproducts are transported by pallets on the AGV according to the technological process, to the corresponding devices, where the processing time depends on the particular production task
- the speed of transportation to the particular devices may depend on the time needed to perform this job and on the requirement of the processing time beginning on the devices. In other words the move of AGV consists of traveling from one location, where the devices has just finished the task, to the another location where another devices start the task.

In our approach it will be considered the scheduling of the traveling by AGV in time. This is a single machine scheduling problem studied under various assumption and criteria in [1], [5], [3]. Because the main job of AGV is to serve the producing devices the goal may be formulated as follows: to find the time based service model of producing devices and the algorithm of AGV jobs scheduling.

In the such formulated problem there is needful to consider the constraints which are the times among jobs performed. For example, by one AGV in the workshop which serves all producing devices in all PLs. The time between two jobs is the "time window" in which are performed the tasks on the producing devices. By the time window is separated the termination of job (p, z) and the beginning of job (p, z+1) according to the precedence constraints described by PL. It supposes that the time window can be determined by ($k_{p,z}$, $K_{p,z}$), where $k_{p,z} > 0$ and $K_{p,z} \geq k_{p,z}$ respectively and express the minimal and the maximal separation time between job z and $z + 1$.

Let be denote $T_{p,z}(s)$ completion time of job (p, z) in a schedule s on the completion time axle of jobs N in a workshop. The goal of the scheduling is to minimize $T_{p,z}$ $\forall p,z$ for the number of PL, expressed by the relation :

$$Q_p = \sum_{p=1}^{m}\sum_{z=1}^{n} T_{p,z} \rightarrow \min \qquad \text{for } \forall p\, ;\ \forall z\, ;\ \max p \text{ res.};\ \forall z \qquad (2)$$

considering the constraints

$$T_{p,z+1} - T_{p,z} \geq t_{p,z+1} + k_{p,z} \qquad \text{for } \forall z < 1\, ;\ \forall p \qquad (3)$$

where $t_{p,z+1}$ - denotes the required processing time of job z + 1 at p-th PL,

res.

$$T_{p,z} - T_{p,z+1} \geq - t_{p,z+1} - K_{p,z} \qquad \text{for } \forall z < 1\, ;\ \forall p \qquad (4)$$

which ensure that the time window between two successive jobs in the same PL is lower and upper bounded, and

$$T_{p,z} \geq t_{p,z} \qquad for\ \forall z < 1\ ;\ \forall p \tag{5}$$

Constraints (5) says, that no job can be processed before time 0.

The possible construction of a scheduling rescheduling (adaptive) algorithm

For the solution of the problem formulated above it is possible to use the heuristic algorithm introduced in [1], [3], [5]. The role of the adaptive principle is in the rescheduling case if undesirable events appear as, for example, the failure of job or if AGV is unable to perform the required job in available time.

During the computation of scheduling the tasks are:
- to select according to some criterion the PL_p which will be the earliest scheduled. This is the partial schedule $s_l\ (p),\ l = 1$ in PL_p, where are scheduled the successively jobs $(p, 1),\ldots(p,n)$.

It is the heuristic approach . It is also supposed that at each step is computed only one PL.

- after scheduling of first PL it is needful to select the further PL_p from other unscheduled PL for

$$s_{l+1}\,(p)$$

- the role of $s_{l+1}\ (p)$ is to merge the jobs of already scheduled PL_p with the jobs of new PL_{p+1} etc.

- after each computation step it is required to control the constraints.

Algorithm:
Step 1 : choosing the order in which the PLs will be scheduled. First will be scheduled the PL with the highest total processing time (it is the sume of all processing times for all jobs in the PL)

Step 2 : (first schedule) for each job in the first production line to compute the real time of the beginning and the time of the termination of a job with considering of constraints.

```
startT[0][0]=0;
 ( computation of the real time of starting  j-th job in i-th PL)

 endT[i][j]=startT[i][j]+t[i][j];
startT[i][j+1]=endT[i][j]+k[i][j];
```

(computation of the "time window", for each job to compute the earliest (minT) and the latest (maxT) starting time.)

```
minT[i][j]=endT[i][j-1]+k[i][j-1];
maxT[i][j]=endT[i][j-1]+K[i][j-1];
```

Step 3: (s-th schedule) for each job of the i-th PL to compute the beginning and the termination of the job

```
if (t[i][j]<(startT[i-1][j+1]-endT[i-1][j]))
(the processing time of job [i][j] is smaller than the "time window")
        {startT[i][j]=endT[i-1][j];
         endT[i][j]=startT[i][j]+t[i][j];
         minT[i][j]=endT[i][j-1]+k[i][j-1];
         maxT[i][j]=endT[i][j-1]+K[i][j-1];
        }
        else if ((endT[i-1][j]+t[i][j])<maxT[i-1][j+1])
        ( it is possible to shift startT of job[i-1][j] )
                {
                 {startT[i][j]=endT[i-1][j];
                 endT[i][j]=startT[i][j]+t[i][j];
                 startT[i-1][j]=endT[i][j];
                 }
                i=i-1;
         ( computation of new "time windows" for each job in each PL)
                        {startT[i][j]=endT[i][j-1]+k[i][j-1];
                         endT[i][j]=startT[i][j]+t[i][j];
                         minT[i][j]=endT[i][j-1]+k[i][j-1];
                         maxT[i][j]=endT[i][j-1]+K[i][j-1];
                         }
                        }

      else   new schedule with new conditions is required
```

Step 4: repeat step 3 for each PL while there is possible to shift startTime of jobs inside of the "time window" (minT, maxT)

2.1 An illustrative example

A very simple example is used for illustrating the utilization of the algorithm for a single-machine scheduling.

Consider a production system with four parallel production lines, that means four technological tasks. Each of technological tasks contains of 3, 2 res. jobs for AGV. The configuration of the fictitious production system is shown on Fig. 1.
The aim is to propose the sequence of jobs for only automated guided vehicle (AGV), which serves production lines, if it is possible. In case it is not possible, the other AGV is required. Afterwards it is needed to reschedule jobs for two AGV.

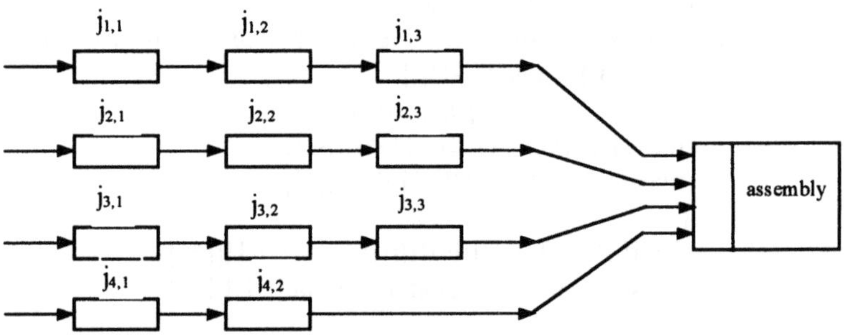

Fig. 1 The fictitious production system with 4 parallel production lines (jobs for AGV)

Table 1 contains the processing times for each of jobs for AGV in production lines and the technological boundaries for them.

Tab. 1. Processing times and constraints for AGV in fictitious production system

ti1	ki1	Ki1	ti2	ki2	Ki2	ti3	ki3	Ki3
3	4	15	2	10	15	6	6	12
5	7	15	2	6	9	3	6	15
6	6	9	4	1	10	5	5	15
4	3	9	4	3	16	0	0	0

t_{ij} - denotes the processing time for j -th job in *i-th* production line
k_{ij}, K_{ij} - denote the technological constraints. That means, the j+1 job in i-th production line cannot start before k_{ij} after K_{ij} after the j-th job is finished. (indexes i, j correspond to the indexes p, z used above)

The case, when the only AGV cannot serve for all PLs in the production system, is shown on Fig.2. The first decision for the solution of that problem is to add new AGV into the production system. Supposing that each AGV should serve independently for two of 4 PLs, the same algorithm could be used. The result is shown on Fig. 3.

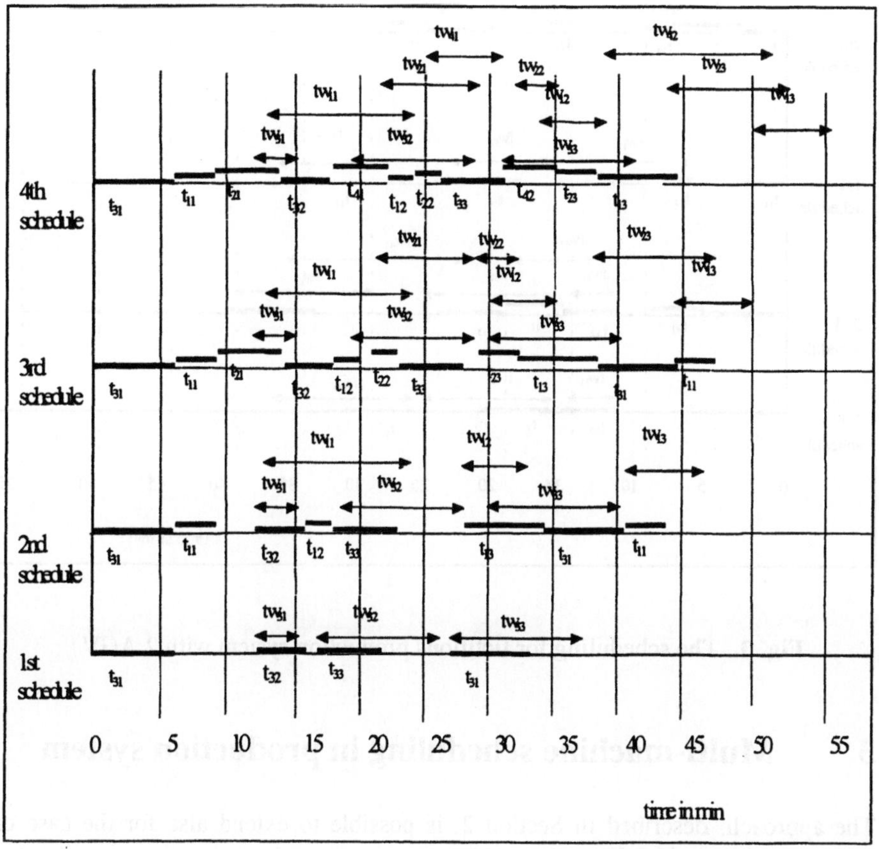

t_{ij} - processing time for *i*-th job in *j*-th production line
tw_{ij} - "time window" for *i*-th job in *j*-th production line

Fig.2. Scheduling for the fictitious PL with one AGV

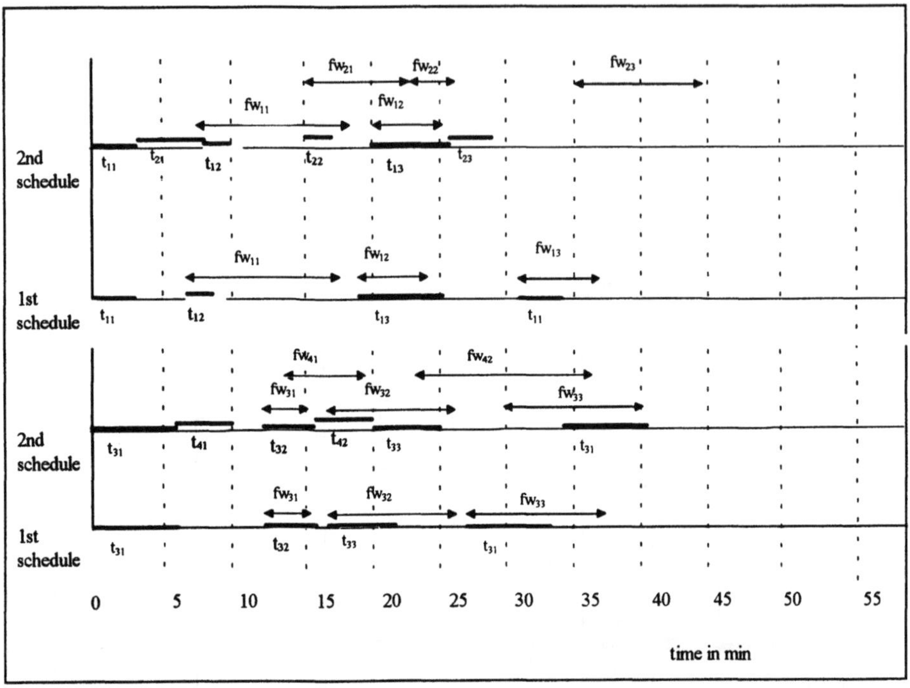

Fig. 3. The scheduling for fictitious production system with 2 AGV

3 Multi-machine scheduling in production system

The approach, described in Section 2, is possible to extend also for the case of multi - machine scheduling.

Formulation of the multi - machine scheduling

Let I_i denotes the different kind of jobs $i = 1,2,...,l$. Each set I_i contains of n jobs $j_z \in I_i$; $p = 1,2,...,n$. Then the total number of jobs with respect of equation (1) can be expressed, for this case, by the equation

$$N = \sum_{i=1}^{l} \sum_{p=1}^{m} I_{i,p} \tag{6}$$

The following assumptions are made :

- the jobs of the same type may be performed at the different machines if they are similarly or from the same type,

- the processing time of each job performed at the machines are different or equal respectively,
- the jobs $j_z \in I_i$ are performed in order with increasing index $c = 1,2,...,$two jobs j_z and j_{z+1} are identical, if their processing times are equal and if they are from the same I_i.

Let the production process (PP) of $h = 1,2,...$ products contains of a set $F_h \subset I_i,$ created from different kind of jobs. Then

$$F_h = \sum F_i, \quad i=1,2, \ldots l \qquad (7)$$

where each set contains of $r_i \in F_i$ number of jobs.
The following constrains are considered :
- constrains of initial conditions of order c is expressed by the relation :

$$t_{i,z,c} \geq 0 \qquad \forall i ; \quad \forall z \in N$$

which denotes the time of beginning of job $j_z \in I_i$ in order c .
- constraints conjunctive :

$$t_{z,i,c} - t_{z-1,i,c-1} \geq T_{z-1,i,c-1} \; ; \quad \forall i \in I_i \; ; \; \forall z \in N$$

This relation express the time between two successive identical jobs. This is the case of single - machine service.
- constraints disjunctive for the case, if the successive jobs $j_z, j_{z+1}, \ldots$, in order c, $c+1$, are from the different sets of I_i . They are expressed by the relation :

$$t_{z,i,c} - t_{z\pm x, i\pm y, c-1} \geq T_{z\pm x, i\pm y, c-1} \qquad ; \forall i \in I_i ; \; \forall z \in N$$

The values of x and y depend on the given technological process and its given configuration.

For the explication of this approach it introduces the fictitious production lines. Let be suppose that the production lines from 4 kinds of devices which are ordered according to the decentralized technological process. It considers, for example, 4 kinds of jobs $i = 1,2,3,4,$ where 1. denotes the transport; 2. the manipulation; 3. machining and 4. testing. Let be the sequences of jobs for the products $h = 1,2,3,4$ in given configuration as follows:

- *for* $h = 1 : i \rightarrow i + 1 \rightarrow i + 2 \rightarrow i + 1 \rightarrow i$
- *for* $h = 2 : i \rightarrow i + 3 \rightarrow i + 1 \rightarrow i$
- *for* $h = 3 : i + 1 \rightarrow i + 2 \rightarrow i + 3 \rightarrow i + 1 \rightarrow i$
- *for* $h = 4 : i + 1 \rightarrow i + 4 \rightarrow i$

On the basis of the such ordered jobs it is possible to deduce the conditions of the beginning and the duration of jobs (considering the given constraints) Fig.4.

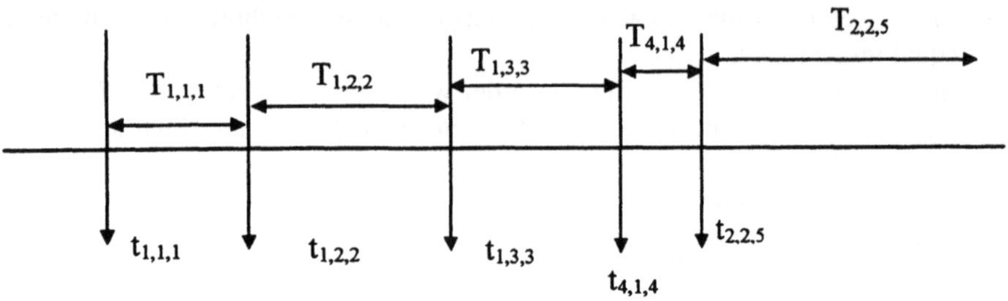

i = 1 (transport)
i = 2 (manipulation)
i = 3 (machining)
i = 4 (testing)
. . . etc.

Fig.4. The conditions of the beginning and the duration of jobs

For each technological process the following assumptions are made:

- each machine may perform in given time only one job
- the interruption of jobs is not considered
- by the technological process the order of jobs transition at the different machines is given . They are fixed in the concrete case.

The aim of the solution is as follows : to determine such configuration of jobs sets F_h performance with verifying all constraints that the total time of performance T will be minimal.

The algorithm of this approach may be described by several steps:

1. the determination of the priority of jobs and PL
2. the determination of the scheduling for each kind of devices in each PL simultaneously by application of the algorithms described for the single machine scheduling
3. the determination of the priority of jobs steps and the minimal total time T
4. the synchronization of the jobs.

Tab.2. Processing time and constraints for NC machine in fictitious production system

i	mi	ti1	ki1	Ki1	ti2	ki2	Ki2
1	2	4	2	20	10	9	20
2	2	7	2	20	6	8	20
3	2	6	4	20	1	11	20
4	1	3	8	20	0	0	0

For the verification of the multi-machine scheduling algorithm will be considered the steps 2 and 4. In this case it is possible to apply the algorithm described and

verified in the Section 2 in such way, that for each kind of devices it is applied simultaneously. After executing of this scheduling it is needed the cooperation of the performance of all I_i. If it is not possible to reach the available synchronization in the first step, it repeats the scheduling and also the synchronization process by considering the constraints for all kinds of devices.

Synchronization of the AGV and the NC-machine is in processing times and constraints. The minimal starting time of the j-th AGV job in i-th PL is the processing time of the j-th NC-machine job in i-th PL.
Suppose that , the rescheduling algorithm could be used independently for AGV and for NC-machine.

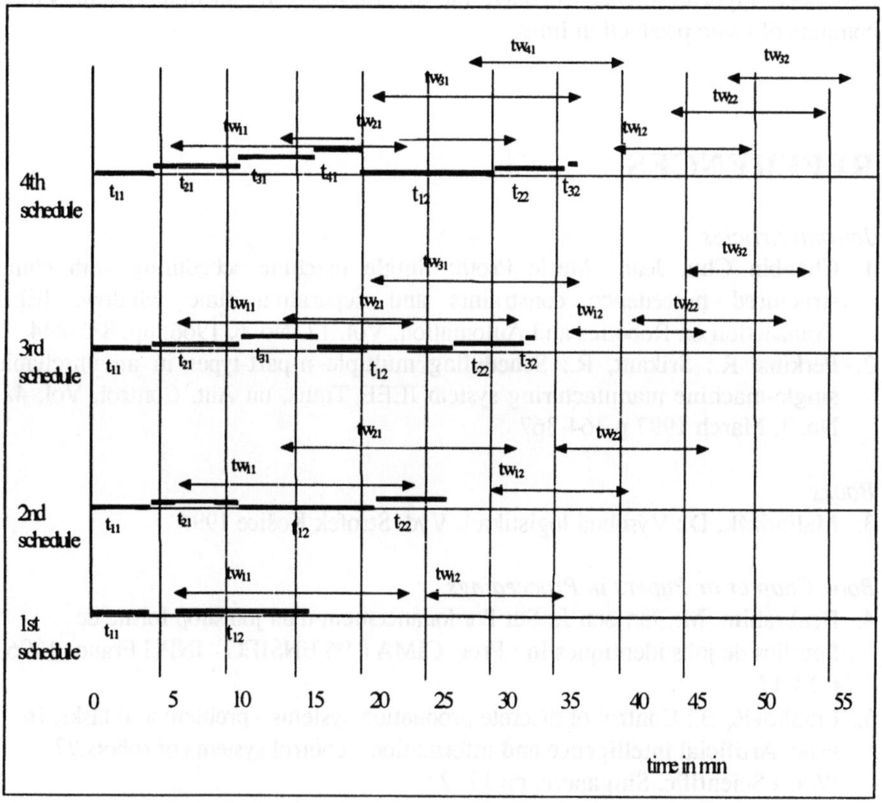

Fig.5. Fictitious production system (jobs for NC-machine)

4 CONCLUSION

The paper features the possibility of the applications of the algorithm for the solution of the single and multi- machine scheduling problem. The algorithms are

based on the heuristic approach and so-called rescheduling way. The application domains of this approach are the production systems, which are built from the production lines with the concrete production task according to the designed technological process and PL configuration.
The approach presented in the paper was used for scheduling of devices performance in the time domain. By the production line structured precedence constraints constitute the special case of jobs sequence execution. The algorithm of single machine scheduling was extended to the multi- machine scheduling in such way, that the production process was divided to the different kinds of production devices and each kind was solved independently. Finally, the so obtained solution was synchronized or rescheduled if the goals was not obtained. The verification of developed algorithm was executed on the fictitious production system which contains of some production lines.

REFERENCES

Journal Articles

1. Chenbin Chu, Jean -Marie Proth: Single machine scheduling with chain structured precedence constraints and separation time window. IEEE Transaction on Robotics and Automation. Vol. 12, No. 6, 1966. pp. 835-844
2. Perkins, R.; Srikant, R.: Scheduling multiple n-part-types in an unreliable single-machine manufacturing system IEEE Trans. on Aut. Control, Vol. 42, No. 3, March 1997 p.364-367

Books

3. Malindžák, D.: Výrobná logistika I. Vyd. Štrofek Košice 1996

Book Chapter or Papers in Proceedings

4. Benbrahim, M.; Janssen J.: Sur l'ordonancement d'un job shop forme de familles de jobs identiques In.: Proc. CIMAT 96 ENSIEG - INPG France 1996, p.14-17
5. Frankovič, B.: Control of discrete production systems - problem and tasks. In Proc. Artificial intelligence and information - control systems of robots 97. World Scientific, Singapore, pp 17 -27
6. Geneste,J.; Domenech, F. Evaluation of job-shop scheduling heuristic in workshop subject to disturbances, In Proc. CIMAT 96, Ensieg - INPG France, 1996, p.9-13
7. Wang P., Luh, B.: A combined Lagrangian relaxation and dynamic programming algorithm for job shop scheduling. In Proc. CIMAT´96 Grenoble pp 3-9

4
A Study of Self-Organization in a Production Environment

E.S. Tzafestas

1 Introduction

Modern cellular manufacturing systems involve distributed environments of robots, machines, materials etc. The principal feature of such a system is that incomplete or imperfect information is given about the input flows and the interactions between parts of the system, so that it has to be adaptive to variations in input and environmental conditions. An additional wish for a cellular manufacturing organization is to be as easily reconfigurable as possible, i.e. as easy to adjust to different environments of the same type as possible [1,2,3,4,5]. For all those purposes, the system has to self-organize in order to respond to the inputs while meeting the operational constraints involved. Self-organization in a distributed environment is achieved through the use of fairly simple local, adaptive behavioral rules that have to be chosen so as to optimize if possible the overall system's operationality, as defined by a set of objectives and/or constraints [6,7,8].

Our long-term goal is to study and understand the nature and specificities of self-organization by examining cellular manufacturing case studies, which, unlike other self-organization environments, define clear goals and operational criteria. In what follows, we will adopt the medium term goal of exploring the notion of behavioral persistence and task specialization in such a case study. The manufacturing environment and the basic behavioral algorithm will be presented in section 2, while the comparative simulation results for persistence and specialization in various environmental setups will be given in section 3. Finally, section 4 will conclude the paper and describe briefly future directions of research.

2 A Case Study

We assume [9,10] an industrial plant that consists of four service units (for instance, assembly cells) that issue demands for tool use. A tool-manager robot goes from unit to unit responding to those demands (Fig. 1). Each service unit increments demands periodically up to an upper limit steady state (which corresponds to a unit demanding maximal tool use). Beyond this point, the unit's demand does not increase anymore (the production of this unit is temporarily stopped). Obviously, this arrival process alone tends to saturate the system (all units tend to maximum demand). Our goal at this stage is to find suitable reactive robot behavioral specifications that will delay this saturation as long as possible, so as to maximize service units' production (assuming that the robot can't be quick enough to service all units at no time). Demands represent service times. The robot stands initially in its rest place in the center; it repeatedly goes to a unit, services it and returns to the center. Each move from the center to a unit or vice versa needs T_M time units, while each service lasts at most T_S. Unless otherwise stated, T_M=10 and T_S=3 and all the results reported are averages over 10 runs, each lasting until demand exhaustion after an initial phase of 1000 time cycles, during which demands are issued.

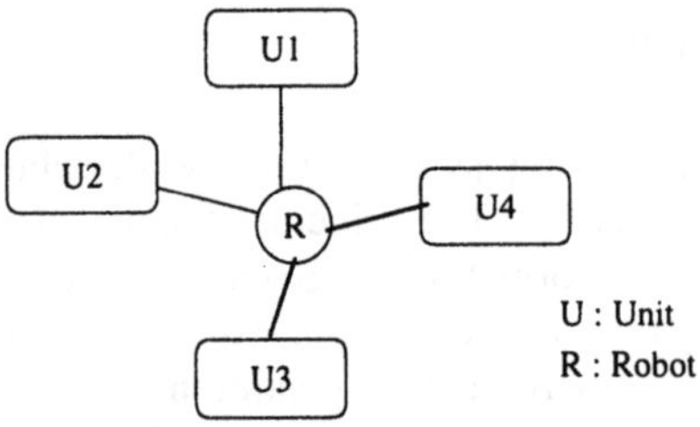

Figure 1. The plant organization

Based on previous experience with reactive robots, we have created one service task for each unit. The tool-manager robot's actuators systems repeatedly evaluate those tasks' motivations and "choose" for execution the one that expresses maximal preference. For this to work, all maximal demands have been normalized to a value d_{max}, which for the present experiments has been set to 40. Each task's motivation is a function of the corresponding unit's demand, such that when demand = 0, motivation = 0. This way, the robot's "nirvana" is when all its motivations are 0; every time its motivations are non-zero the robot tries to act in a way that will bring them back to 0. The presence of the process, that tends to saturate those demands, perturbs the motivational state of the robot and causes it to act/react. The robot control programs are written using a cellular control architecture [10,11], while the actual high-level pseudo-code run by each service task is:

- If the robot is at the corresponding service unit, service it for a computed amount of time.
- If the robot is at center, go to service unit.
- If the robot is elsewhere, go to center.

The tasks' motivations are being evaluated after every such step (move or service). It should be stressed that the navigation and perception systems of the robot have been written once and for all in the beginning and never changed since. Sole intervention in the motivational level of each task has allowed drastic increases in the performance to be observed, as will be shown below.

Note that this is *not* a usual resource allocation problem (see, for example, several papers in [12]): we assume that the individual service units possess a local inventory of tools that they use at will and there is an additional set of tools that need to be shared across units and whose use is sporadic. The tool-manager robot regulates the units' demands for those more "rare" tools, which it is more cost-effective to share rather than possess local instances of them. For this to work, we also assume that individual service units make optimal local use of those tools, for instance by regrouping in time operations that depend on these tools. Such a degree of local intelligence in the service units enables us to afford a tool-manager robot that will decide alone when to retreat from a given unit, in which case it will withdraw all tools previously allocated to it, to make them available to other units. Therefore, the effective operation of the units-manager system depends on the existence of such a cooperative coupling. We can certainly proceed to more complex allocation cases, however, by only defining different motivations for different instances of tools that will be therefore treated and allocated independently.

The best motivational rule found so far, i.e. the motivational rule that optimizes the system's performance, is multiplicative and is applied only to the selected task [9].

$$\text{motivation} = \text{demand} * \text{persistence\_factor}$$

Obviously, the persistence factor should be greater than 1. In the worst case (when all service units are saturated with their maximum demand value d_{max}), it makes sense to have the robot satiate units in turn. In that case, a selected task should persist for as long as its demand is non-zero, hence $\text{persistence\_factor} > d_{max}$. To write the corresponding cellular control program, however, we needed a continuous function for the persistence factor, so we adopted the exponential time-convergent

$$\text{persistence\_factor} = 1 + d_{max}(1 - e^{-t/Tp})$$

which rises from 1 to $d_{max}+1$ (t is the time that has elapsed since selection of the task). Since this rise should happen almost immediately after a task is selected, the T_p constant should be such that $1+d_{max}(1-e^{-1/T_p})>d_{max}$, which gives $T_p<1/\ln(d_{max})$. For the present experiments we used a safe value of T_p approximately equal to $1/2\ln(d_{max})$, $T_p = 0.13$ (since $d_{max} = 40$).

3 Simulation Results

The process has been simulated and studied for multiple robots, in search of self-catalytic models of specialization, that would make each of the robots specialize explicitly (permanently) in a fraction of the service units. Self-catalysis means that the motivation of a task depends on a term that is reinforced on execution of the task. Firstly, a number of experiments have been conducted with additive, multiplicative, multiplicative with an upper bound and self-regulated factors. The second case was far superior to the first, but motivational factors could rise to a very high numeric value; definition of an upper bound solved this problem but resulted in low performance and deadlocks. Self-regulation instead of self-catalysis of the specialization factor has yielded the best results for multiple robots, but still only marginally better than the equivalent case of multiple robots with only persistence. This hardly detectable improvement of the performance with specialization compared with sole persistence can be attributed to the asynchronicity of the system's operation which induces, for example, slightly better regulation of a single-robot 2-unit plant than a 4-unit one that is managed by two robots. The specialization model by self-regulation of the corresponding bias factor is given below :

$$\text{Motivation} = \text{demand} * \text{persistence\_factor} * \text{specialization\_bias}$$

with

$$\text{Bias}(t) = \text{Bias}(t-1) + r * (\text{Bias}_{max} - \text{Bias}(t-1)),$$

$\text{Bias}_{max} = 1$ and all biases are normalized according to $\sum \text{bias}(t)$

We then proceeded to examine the case of interdependent service units where the output of one unit is fed as input demand to one or more other units, so that a flow network is built (such as in Fig. 2). This study was prompted by the observation [9] that specialization as studied above is probably not beneficial because service units are independent and no genuine "load balancing" is necessary in the system. The network structures studied were multi-level structures with full or 1-1 connectivity between levels. To our surprise, and while the self-regulated specialization mechanism has been found to work correctly (for example, in the case of 3 robots and 9 units, each robot quickly specializes to three of the units), there has been no quantitative improvement to the performance of the overall system.

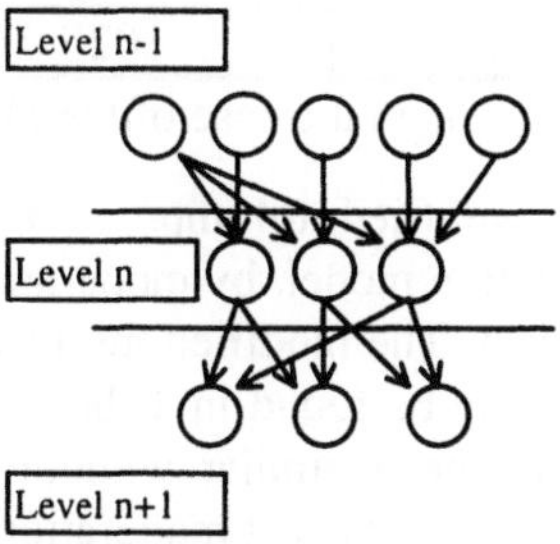

Figure 2. A network of connected units

Next, we studied the case of the initial 4-unit "flat" setup with variable T_M across units and 2 robots. An indicative setup and the corresponding results are given in Tables 1 and 2.

	U_1	U_2	U_3	U_4
T_M	6	8	10	12

Table 1. A variable T_M setup for the 4-unit flat plant

Model	Completion time
Persistence only	1052.25
Specialization	1052.3

Table 2. Results in the case of the plant of table 1

As is obvious from the tables, no significant difference exists between the results without and with specialization. This qualitative property has been verified for various T_M setups and is maintained even is the T_M factor is introduced in the initial persistence law :

```
persistence_factor = (1+dmax(1-e^(-t/Tp))) * lcm(TM, for
                each unit) / TM
```

Next, we studied the standard plant setup (T_M=10, for each unit) with an additional switching cost matrix, in an effort to study the need for specialization whenever switching costs are present and the secondary goal of minimizing the total cost appears. An indicative cost matrix and the corresponding results are given in Tables 3 and 4.

	U_1	U_2	U_3	U_4
U_1	0	1	2	3
U_2	1	0	3	2
U_3	2	3	0	1
U_4	3	2	1	0

Table 3. A symmetric cost matrix for the 4-unit flat plant

Model	Completion time (cost)
Persistence only	1026.0 (136.30)
Specialization	1026.0 (152.15)

Table 4. Results in the case of the plant of table 3

Given these results, we tried to improve the overall operational costs through the same specialization model, by moderating a unit's motivation by the corresponding switching cost. The modified motivation rule is given below, and the corresponding results may be found in Table 5. It is obvious from the table that while the completion time is similar without and with specialization, the overall cost drops dramatically with cost moderation, as was expected.

```
Motivation = demand * persistence_factor
    * (1/cost) [* specialization_bias]
```

Model	Completion time (cost)
Persistence only	1026.6 (78.25)
Specialization	1027.13 (74.475)

Table 5. Results in the case of the plant of table 3 with cost moderation

Finally, an asymmetric cost matrix and the corresponding results without or with cost moderation in both cases of presence or absence of specialization are given in Tables 6 and 7. Again, there is no significant difference between sole persistence and specialization, while cost moderation achieves lower total costs, even if the improvement is not so pronounced as in the case of symmetric cost tables.

	U_1	U_2	U_3	U_4
U_1	0	3	1	2
U_2	2	0	1	3
U_3	2	3	0	1
U_4	1	2	3	0

Table 6. An asymmetric cost matrix for the 4-unit flat plant

Model	Default	Cost moderation
Persistence only	1024.32 (130.925)	1027.25 (101.75)
Specialization	1025.62 (125.225)	1026.2 (106.225)

Table 7. Results in the case of the plant of table 6, without and with cost moderation

The qualitative result from these tables is that in no case is the specialization model superior to the pure persistence one :

- If the units are connected in some kind of network, specialization does not seem necessary, because there is some lower bound in the completion time and no other operational cost exists.
- If the factor T_M varies across units, specialization does not seem necessary, because there is no qualitative difference with the default case (uniform T_M), so that only the absolute value of the completion time's lower bound changes.
- Finally, if there are variable switching costs between units, specialization is indifferent, while moderation of persistence by cost achieves lower overall costs.

In all these cases, it appears that there is no genuine specialization problem, i.e., there is no need for robots to permanently commit themselves to a subset of all service units. Temporary commitment is achieved automatically by the persistence model, which should include cost moderation whenever there are additional switching costs.

4 Conclusion

The above conclusions for our case study are summarized in table 8.

Concept	***Description***
Operationality	Need for a motivational model with persistence
Persistence	• Is based on a worst-case analysis • Incorporates the operational parameters of the agent
Specialization	Useless, in all of the cases : • Units in a network • Variable T_M • Variable switching costs (Symmetric or asymmetric cost matrix) ⇒ There is no specialization problem, or the specialization problem is solved by the persistence model

Table 8. Conclusions

As further work, we plan to study comparatively the symmetric and asymmetric cost matrices, to understand deeper the significant quantitative difference in performance improvement, as shown in tables 4, 5 and 7. We also plan to study adaptive costs, in the sense that the more a task remains active, the less it costs, which captures one if the major features of specialization. Our longer term objective is to study the specialization problem in completely different manufacturing setups and to it transpose to domains other than the industry.

References

1. Jones PF 1992 *CAD/CAM: Features, applications and management*, Macmillan Press Ltd., London
2. Ranky PG 1986 *Computer integrated manufacturing - An introduction with case studies*, Prentice-Hall, Englewood Cliffs, NJ
3. Wright PK, Bourne DA (eds) 1988 *Manufacturing intelligence*, Addison Wesley, Reading, MA
4. Bjørke O 1979 Computer-aided part manufacturing, *Computers in Industry*, (1979):3-9
5. Kusiak A 1990 Manufacturing systems : A knowledge- and optimization-based approach, *Journal of Intelligent and Robotic Systems*, 3:27-50
6. Baldwin KE 1989 Autonomous manufacturing systems, *Proceedings of the 1989 IEEE International Symposium on Intelligent Control*, Albany, NY, September, pp. 214-220
7. Vaario J, Ueda K 1996 Self-Organization in Manufacturing Systems, *Proceedings 1996 Japan-USA Symposium on Flexible Automation*, July, Boston, MA, pp. 1481-84
8. Tzafestas ES 1998 Autonomous agents in cellular manufacturing, *Advances in Intelligent Autonomous Systems*, SG Tzafestas (ed), Kluwer Academic Publishers, pp. 427-438
9. Tzafestas ES 1998 Reactive robots in the service of production management, *Journal of Intelligent and Robotic Systems*, 21(2):179-191
10. Tzafestas ES 1995 *Vers une systémique des agents autonomes : Des cellules, des motivations et des perturbations*, Ph.D. Thesis, Université Pierre et Marie Curie, Paris, December
11. Tzafestas ES 1994 A cellular control architecture for autonomous robots, *Proceedings of the 1994 International Workshop on Intelligent Robotic Systems*, Grenoble, July, pp. 70-79
12. Desrochers AA, Silva M (eds) 1994 *IEEE Transactions on Robotics and Automation, Special Issue on Computer Integrated Manufacturing*, 10(2), April

5
The Process Plan Sequence in Steel Structures Welding of Great Diesel Engines

N. Nicolich, P. Persi and M. Reini

1 Introduction

The structures of the great diesel engines, which are used in ship propulsion and in thermoelectric plants, are produced by electric welding process of elementary components worked out from steel sheets by oxygen cutting. There are four functional parts to be produced in welding department: the monobloc column, the bed plate, the air intake manifold and the gas exhaust manifold. The process plan consists of the following operations: handling and positioning, tack welding, welding and straightening.

Great difficulties arise in the definition and the optimisation of the welding Process Planning (PP): **a**) there are many elementary components (hundreds); **b**) each elementary component is welded to the contiguous one by operations which can in general also be discontinued; **c**) time required for handling and positioning each component, that can be known, depends on its mass and dimensions; but if subassemblies are concerned, the demand in time and resources depends on the former operations in the PP and cannot be known a priori; **d**) in welding process the best position is horizontal: it has to be chosen between handling subassembly for attaining a better position or accepting a more difficult operation; the answer also depends on the former steps in the PP; **e**) numerous constraints have to be considered as: quality, tolerances, technology, accessibility and stability. It is still difficult to approach the PP optimisation problem starting from the precedence-graph [1], which represents all the possible assembling connections, then rejecting all those non compatible with the constraints, because point d) and e) would not be overcome. This approach is actually object of research. In this chapter the authors discuss a first simplified approach to the problem by generating consistent sequences of operations and then selecting the optimal one. In the first place, the attention is pointed out to the structures here discussed, the architecture of which is characterised by a repetition of geometric modules as many times as the cylinders of the engine. This allows a drastic reduction in component number to be treated and the relative mass handling is not difficult. The constraints are given by design specifications which are coded in the drawings. They can be translated in precedence relations between the operations requested by the working set of the module. Other constraints and relative precedence relations, concerning the manufacturing system and the technology, should be added by the planner. The planner compiles a detailed list of all the operations to be performed and a list of the precedences as stated before, that obviously are not enough to identify solely the PP. A precedence matrix or graph is solved and a small number of sequence solutions are expected to be analysed. The selection of the best solution is well accomplished by the help of a relational database, which computes any requested welding process and of an expert planner which answers to the queries about handling and any other set up activities.

1.1. Literature

There are actually three research directions in assembly planning carried out by the authors; i) De Fazio and Witney [1,2], ii) Homem de Mello and Sanderson [3,4,5], iii) Wolter and Chakrabarty [6]. They deal with the representation and the generation of all the assembling sequences of a product on the basis of the connections graph of the components.
The criteria generally adopted are of the following types: stability, easy fixing and setting, low handling and operation time.
De Fazio and Witney [1] describe the liaison diagram of the product being assembled and all possible assembling sequences that can be generated by means of 2L queries to the planner; where L is the number of connections and the queries are: is it true that L_i cannot be done after L_j and L_k have been done? Is it true that L_i cannot be done if L_j and L_k are still undone?
The same authors discuss in [2] an interactive program which solves all the possible assembling sequences starting from the connections diagram of the product. The cut-set method has been adopted generating first all the possible subassemblies and then questions are put to the planner if disassembling is possible. On the base of these information the system generates all the possible sequences and formulates the precedence constraints. The system uses only 5% of the possible queries for each subassembly. As the sequences have been generated, the planner must lessen the number by means of a graphic editing procedure. The results are examined in a more rigorous method considering criteria as stability, fixturing, setting and handling number.
The second research direction, by Homem de Mello and Sanderson, begin with the representation problem of the assembling sequences [3] by means of oriented graph, and/or graph, the set of establishment conditions or two types of precedence constraints. They analyse the reciprocal correlation level, the correctness and the completeness. On the second stage [4], an algorithm is proposed for the generation of all the feasible assembly sequences starting from a relational model of the prod-

uct. The model describes the features, the contacts and the connections. Moreover, the assembling problem is managed by means of the inverse problem of the disassembling. The result is the and/or graph of the product which represents all the feasible sequences. Finally [5], two criteria for evaluating the assembling sequences are presented starting from an and/or graph representation of the product: maximising the number of different feasible sequences and the parallelism level of the assembling operations. Two heuristics are proposed which provide good solutions for each objective they are applied.

The third research group, Chakrabarty and Wolter [6], pointing out the difficulty of the computational solution of the assembling problem, proposes a different approach. The product should be decomposable in a subassembly hierarchy, thus reducing the computational charge for visiting all the feasible solutions. The approach is based on: a) the subassembly decomposition is given; b) a library with information about possible and preferred ways for assembly common structures is prepared; c) the logic of the algorithm is bottom-up; d) the algorithm uses valuation functions for the assembling plans selection, but does not optimise them.

2 The Approach

The aim of this work is not to generate all the possible assembling sequences [1-4], but to introduce a simplified procedure which allows to get a solution by reducing the length of the sequences that have to be evaluated in detail on the basis of productive criteria.

As it has been described before, the welded structure of a diesel engine is produced in separate functional parts or macro-modules, subdivided in modules [5]. This is a consequence of their intrinsic geometry and it depends on the number of the cylinders. Moreover, the module can be composed by elementary-modules due to symmetry.

The first operation hypothesis is to propose the same PP procedure to the different levels of the design; i.e. the elementary-components assembling for obtaining elementary-modules, the elementary-modules assembling for obtaining modules, and so on.

The first difficulty, **a)** in Section 1., has overcome because the elementary components number is strongly reduced.

A second hypothesis has been introduced: the elementary module is assembled by sequential operations (no parallel operations). So if more resources are available, more elementary-modules can be manufactured. The difficulties, described in **c)**, are overcome.

Finally, the constraints, point **e)**, must be considered as precedence relations between operations being sequenced. The constraints are derived from design specifications and from process planner experience.

All the feasible sequences can be found solving the precedence-matrix or the precedence-graph. It is expected that a few number of components and a correct interpretation of all the possible constraints gives a limited number of solutions, about some tenth. On the contrary, it is possible to solve the precedence-graph by

means of heuristics [7] (Fig. 2.1) finding a further reduction of the number of solutions with an acceptable sub-optimisation.
The heuristics, with reference to the precedence-graph, is based on the following rule: proceed along the same branch until it is possible, then pass on a new and free branch. The logic of the rule is evident in several situations: i.e. if the operation a precedes the operation b due to technological constraints (the tack welding must be done before the welding), it is forced to do them in succession because the operative conditions, which are convenient to the first one, are generally convenient to the second. Moreover, if the precedence is an accessibility constraint for the welding process, it is necessary to defer the operation b until a is done, then execute it.

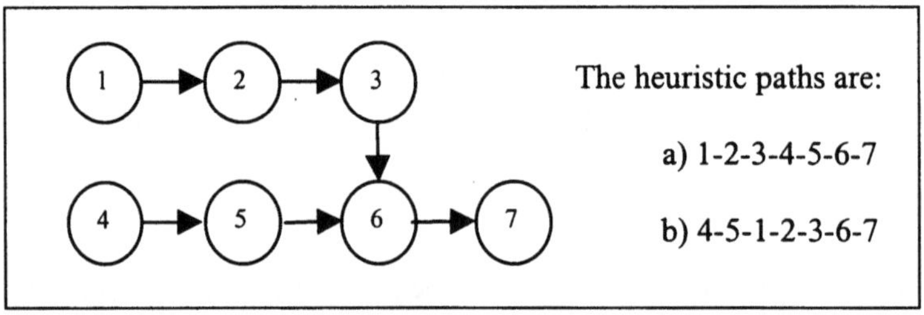

Fig. 2.1 - The heuristic procedure.

The choices of economic kind, described in **d)**, are utilised for evaluating the better sequence among the feasible ones which are obtained as said before. To quantify this final evaluation a relational database has been compiled; it contains data for welding and handling resources and can compute the total processing time, which is required by each operation on the components. The geometry of the components is described too.
An iteration that is characterised by a human expert choice will be introduced.
The heuristic solution of the precedence-graph (or the precedence-matrix) gives a tree of sequences which shows a repetition in more branches of the same bifurcation, at the same operation step. If the last bifurcations, the tails, are solved in favour of one of the two alternatives, it is possible to reach a significant reduction of the output sequences considering a few number of operations.
Reasons which eliminate some tails can be seen as additional constraints to be added to the precedence-graph. Reprocessing the heuristics, the number of sequences is reduced as well as the feasible sequences with respect to the actual constraints system.
Each step of the iteration reduce the solution domain and finally, the evaluation can be carried out taking into account a few number of feasible solutions. The optimum, in the restricted domain, can be attained provided that the new constraints are right and proper for the entire problem. On the other hand, taking into account an heuristic solution only a suboptimal choice is obtained and it is anyway acceptable for the problem with the primary constraints.

2.1. Technological Constraints

The technological constraints are organised in three types as follows.
1) Technological sequences: for each joint a sequence of three distinct operations are ordered: Positioning the edges (L) precedes Tack welding (P), which precedes Welding (S); in symbols: L→P→S ; the elementary components are assumed having edge preparation.
2) Quality of the welding: it is a design specification. The constraints affect the accessibility, the positioning and the operation procedure and, hence, establish the sequence. In the case of high quality in submerged arc welding, the needs are to operating in horizontal position and with the tack welding on the opposite side.

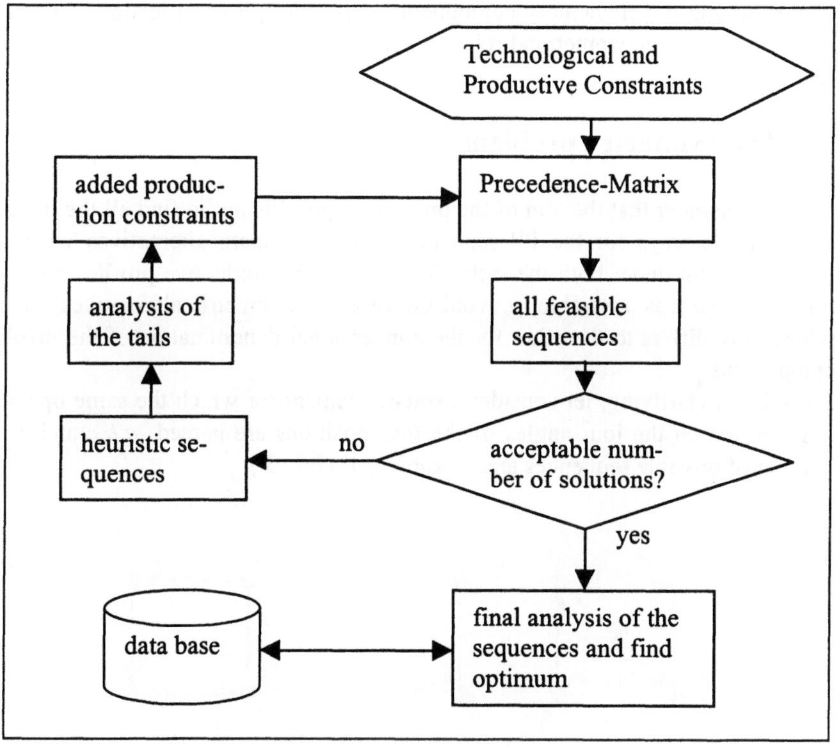

Fig. 2.2 - The proposed approach.

3) Overall assembling quality: this is a constraint which comprehends the two ones just stated before and it must be applied to the whole elementary-module. In particular a local thermal deformation could occur at the thinner or the weaker components. A correct operation sequence can avoid or minimise it; otherwise, the deformation must be eliminated by a thermal or a mechanical straightening operation.

2.2. The Productive Constraints.

This group of constraints is organised in five types as follows.
1) Machine, equipment and fixture which are utilised in welding operations: these constraints depend on the technology (submerged welding arc requests an horizontal position), on withstanding the handling of a subassembly, etc.
2) Continuative use of the resources; the opportunity of grouping the operations which use same resource in order to reducing their flow through the department and preventing their unavailability.
3) Accessibility: in general it concerns hand operated activities because of the dimensions of the equipments.
4) Stability: all the elementary-components must be held in position before joining them with respect of both safety's sake and correct operation execution.
5) Positioning the elementary-components: it is a design specification. Tolerances on distance and on geometrical relation between some components could be specified.

2.3. The Symmetry problem

It is to remember that the aim of the present proposal is not to find all the different and possible ways for the PP sequence, but to compare alternatives which are really different under both the technological and the productive profile. From this point of view it is advisable to avoid the cases of symmetric alternatives, each of which only differs to the other for the conventional denomination of the involved components.
For sake of clarifying, let consider a square element for which the same operation is requested on the four angles. If the four positions are named *a,b,c* and *d* the number of possible sequences are, as known, 4! (Fig. 3.).

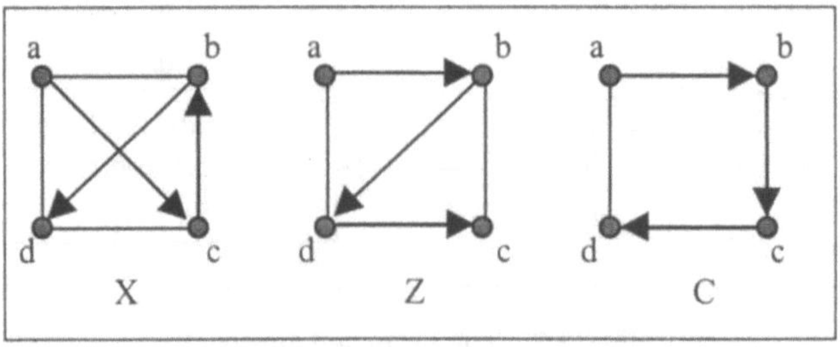

Fig. 2.3 - The symmetric solutions.

The symmetric alternatives occur when:
- The geometry of the sub-assembly (module) is characterised by a symmetry: plane, axial, biaxial, etc.;
- All the operations involved have the same symmetry;

- All technologies and productive constraints, which attain to the above mentioned operations, can be formalised as two or more sets of precedence between operations which respect the same symmetry.
In order to avoid such increase of sequences, it is necessary to break off the symmetry choosing a priori one between the possible alternatives. If the operations are subjected to constraints, the break off can be done in internal way; otherwise, the external way, an auxiliary precedence system is introduced in the precedence-matrix.
In the case of the square element the auxiliary precedence system could be: *a*→*b*, *a*→*c*, b→*d*. Only the sequences starting from *a* can be obtained (Fig. 3.) and the possible solutions are tree: *a,c,b,d* (X), *a,b,d,c* (Z) and *a,b,c,d* (C).

3 Case study

The application has been carried out on an simply elementary-module of the bed plate (Fig. 3.1). Four large plates with *J* edge preparation are jointed to a central element with quality submerged arc welding. The ten small plates request twenty normal arc weldings. If one consider that each joint in general needs three operations (positioning L, tack welding P and welding S), 52 operations are to be expected.
The first step analyses the opportunity of grouping the operations and/or the elementary components. The ten small plates are divided in two groups that are treated as two elementary components. The operations L and P are grouped into one (LP) because in general, and particularly in this case, it is not possible to positioning the element without tack welding. Now it results fourteen operations in total as listed in tab.3.1.

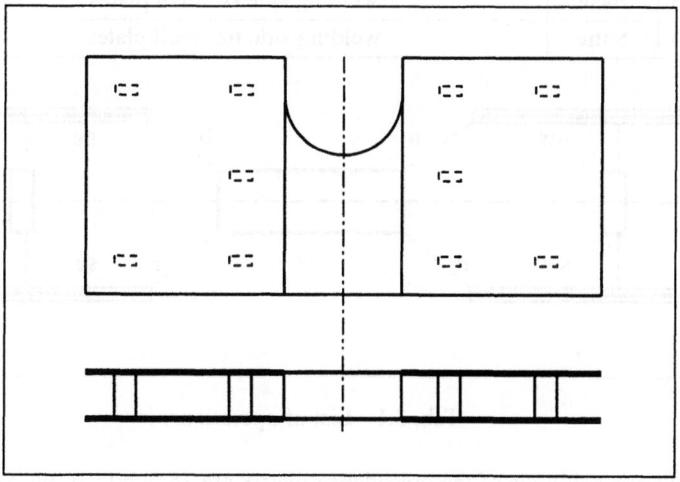

Fig. 3.1 – The case study module.

Technological constraints:

SJa ← LPJa, SJb ← LPJb, SJc ← LPJc, SJd ← LPJd

S4sw ← LP4w, S4nw ← LP4w, S4se ← LP4e, S4ne ← LP4e

The four quality weldings must follow a cross execution:

SJc ← SJa, SJb ← SJc, SJd ← SJb

LP operations must be executed on at least 2 plates before submerged arc welding:

SJa ← LPJc, SJb ← LPJd.

Accessibility constraints for the small plates welding:

LPJa ← LPJb, LPJa ← S4sw, S4nw ← LPJa

LPJd ← S4se, LPJd ← LPJc, S4se ← LPJd

Auxiliary constraints due to the presence of symmetric alternatives;

LPJd ← LPJa, LPJc ← LPJb, LP4e ← LP4w, S4ne ← S4nw, S4se ←S4sw.

1	**LPJa**	positioning and tack welding the plate **nw**
2	**SJa**	submerged arc welding plate **nw**
3	**LPJb**	positioning and tack welding the plate **sw**
4	**SJb**	submerged arc welding plate **sw**
5	**LPJc**	positioning and tack welding the plate **se**
6	**SJc**	submerged arc welding plate **se**
7	**LPJd**	positioning and tack welding the plate **ne**
8	**SJd**	submerged arc welding plate **ne**
9	**LP4w**	positioning and tack welding the small plates **w**
10	**S4sw**	welding side **sw** small plates
11	**S4nw**	welding side **nw** small plates
12	**LP4e**	positioning and tack welding the small plates **e**
13	**S4se**	welding side **se** small plates
14	**S4ne**	welding side **ne** small plates

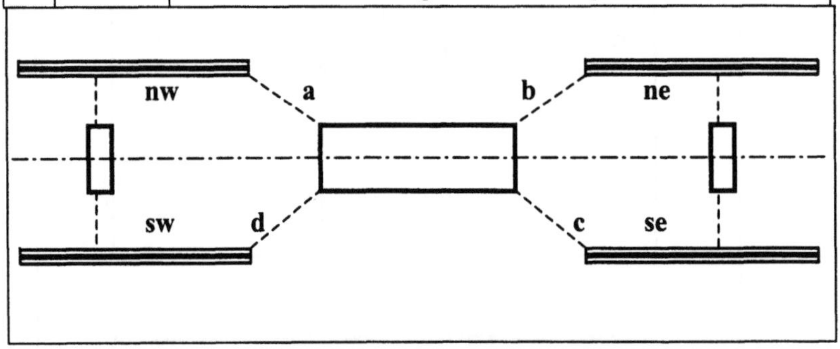

Tab. 3.1 - List of operations.

The heuristic solutions of the precedence-matrix are 75, while all the feasible solutions are 9855. If the auxiliary constraints are not taken into account, the feasible solutions rise up to 50052.

Furthermore, the expert suggests that the four quality weldings (Sja, SJb, SJc and SJd) should be the last operations in the PP. This is an important productive constraint that avoids possible thermal deformation of the four large plate.
A constraints set has to be added to the former one:
SJa ← LPJb, SJa ←SJb, SJa ← SJc, SJa ← LPJd, SJa ← SJd, SJa ←LP4w
SJa ← S4sw, SJa ← S4nw, SJa ← LP4e, SJa ← S4se, SJa ← S4ne.
Adding these new conditions, the procedure provides 18 heuristic solutions (Tab. 2.2) of the precedence-matrix, while all the feasible solutions are 283. If the auxiliary constraints are not taken into account, the feasible solutions rise up to 2268.
In Tab. 3.2 is evidenced that the last five (10-14) operations are the same for all the 18 sequences.
The last bifurcations of the sequence tree are at the 8th operation step. All the operations after these bifurcations are identical:
LPJd, S4nw, S4ne,...; S4nw, LPJd, S4ne,...

Operation step

	1	2	3	4	5	6	7	8	9	10	11	12	13	14
1	LPJb	LPJc	LP4w	S4sw	LPJa	S4nw	LP4e	S4se	LPJd	S4ne	SJa	SJc	SJb	SJd
2				LP4e	S4sw	LPJa	S4nw	S4se	LPJd	S4ne	SJa	SJc	SJb	SJd
3						S4se	LPJa	LPJd	S4nw	S4ne	SJa	SJc	SJb	SJd
4								S4nw	LPJd	S4ne	SJa	SJc	SJb	SJd
5	LP4w	S4sw	LPJb	LPJa	S4nw	LPJc	LP4e	S4se	LPJd	S4ne	SJa	SJc	SJb	SJd
6						LP4e	S4se	LPJc	LPJd	S4ne	SJa	SJc	SJb	SJd
7				LPJc	LPJa	S4nw	LP4e	S4se	LPJd	S4ne	SJa	SJc	SJb	SJd
8					LP4e	S4se	LPJa	LPJd	S4nw	S4ne	SJa	SJc	SJb	SJd
9								S4nw	LPJd	S4ne	SJa	SJc	SJb	SJd
10			LP4e	S4se	LPJb	LPJa	S4nw	LPJc	LPJd	S4ne	SJa	SJc	SJb	SJd
11						LPJc	LPJa	LPJd	S4nw	S4ne	SJa	SJc	SJb	SJd
12								S4nw	LPJd	S4ne	SJa	SJc	SJb	SJd
13		LP4e	LPJb	LPJc	S4sw	LPJa	S4nw	S4se	LPJd	S4ne	SJa	SJc	SJb	SJd
14						S4se	LPJa	LPJd	S4nw	S4ne	SJa	SJc	SJb	SJd
15								S4nw	LPJd	S4ne	SJa	SJc	SJb	SJd
16			S4sw	S4se	LPJb	LPJa	S4nw	LPJc	LPJd	S4ne	SJa	SJc	SJb	SJd
17						LPJc	LPJa	LPJd	S4nw	S4ne	SJa	SJc	SJb	SJd
18								S4nw	LPJd	S4ne	SJa	SJc	SJb	SJd

Tab.3.2 - The 18 heuristic solutions.

Going back on the tree structure, the next bifurcations are found at the 6th operation step with three different solutions.

If bifurcation 8 is solved in favour of one of the two possibilities, 5 sequences are eliminated.
Furthermore, if the preference agrees with the second one (S4nw, LPJd, S4ne,...), the tack welding for the large plate 'ne' and the welding for the small plates 'ne' can be done in sequence with the same equipments.
At this point the planner can introduce a new constraint in the precedence matrix: LPJd ← S4nw.
The heuristics could be reprocessed obtaining a reduced number of feasible sequences. On the other hand, the bifurcations at the 6th operation step can be analysed too. In fact, the planner can consider more convenient to do P and S operations for the small plates on the large plates already tacked on. Such an operation guarantees the planarity of the large plates and the quality of the execution by submerged arc weldings SJa, SJb, SJc and SJd.
The precedence relations to be added to the precedence-matrix are:
LP4e ← LPJc, LP4w ← LPJb, S4ne ← LPJc, S4sw ← LPJb.
The result of the processing of the precedence-matrix is 4 heuristic solutions (Tab.3.3), while all the feasible solutions are 28.

	1	2	3	4	5	6	7	8	9	10	11	12	13	14
1	LPJb	LPJc	LP4w	S4sw	LPJa	S4nw	LP4e	S4se	LPJd	S4ne	SJa	SJc	SJb	SJd
2				LP4e	S4sw	LPJa	S4nw	S4se	LPJd	S4ne	SJa	SJc	SJb	SJd
3						S4se	LPJa	S4nw	LPJd	S4ne	SJa	SJc	SJb	SJd
4		LP4w	S4sw	LPJa	S4nw	LPJc	LP4e	S4se	LPJd	S4ne	SJa	SJc	SJb	SJd

Tab. 3.3 - The heuristic results at the second iteration.

If, finally, the planner thinks to do the operations LP4w and S4sw in succession, a new precedence must be introduced in the precedence-matrix: S4sw ← LP4w.
The result of the processing of the precedence-matrix is 3 heuristic solutions (Tab.3.4), while all the feasible solutions are 22.

	1	2	3	4	5	6	7	8	9	10	11	12	13	14
1	LPJb	LPJc	LP4w	S4sw	LPJa	S4nw	LP4e	S4se	LPJd	S4ne	SJa	SJc	SJb	SJd
2					LP4e	S4se	LPJa	S4nw	LPJd	S4ne	SJa	SJc	SJb	SJd
3		LP4w	S4sw	LPJa	S4nw	LPJc	LP4e	S4se	LPJd	S4ne	SJa	SJc	SJb	SJd

Tab. 3.4 - The heuristic solutions at the third iteration.

The analysis of the 22 PPs in Tab. 3.5 puts in evidence that the majority of the feasible solutions are in contrast with the human expert considerations, as seen before. The main reason of this contrast is the grow of total time, due to a lack in a logic of grouping the operations, compared with the heuristic solutions (no. 1, 6 and 16).Exception is the solution number 22 which characterises the movement of the two large plates, realising in succession all the operations in 'w' and 'e' positions and respecting the accessibility constraints.

Operation step

	1	2	3	4	5	6	7	8	9	10	11	12	13	14
1	LPJb	LPJc	LP4w	S4sw	LPJa	S4nw	LP4e	S4se	LPJd	S4ne	SJa	SJc	SJb	SJd
2						LP4e	S4nw	S4se	LPJd	S4ne	SJa	SJc	SJb	SJd
3							S4se	S4nw	LPJd	S4ne	SJa	SJc	SJb	SJd
4					LP4e	LPJa	S4nw	S4se	LPJd	S4ne	SJa	SJc	SJb	SJd
5							S4se	S4nw	LPJd	S4ne	SJa	SJc	SJb	SJd
6						S4se	LPJa	S4nw	LPJd	S4ne	SJa	SJc	SJb	SJd
7			LPJc	S4sw	LPJa	S4nw	LP4e	S4se	LPJd	S4ne	SJa	SJc	SJb	SJd
8						LP4e	S4nw	S4se	LPJd	S4ne	SJa	SJc	SJb	SJd
9							S4se	S4nw	LPJd	S4ne	SJa	SJc	SJb	SJd
10					LP4e	LPJa	S4nw	S4se	LPJd	S4ne	SJa	SJc	SJb	SJd
11							S4se	S4nw	LPJd	S4ne	SJa	SJc	SJb	SJd
12						S4se	LPJa	S4nw	LPJd	S4ne	SJa	SJc	SJb	SJd
13			S4sw	LPJa	LPJc	S4nw	LP4e	S4se	LPJd	S4ne	SJa	SJc	SJb	SJd
14						LP4e	S4nw	S4se	LPJd	S4ne	SJa	SJc	SJb	SJd
15							S4se	S4nw	LPJd	S4ne	SJa	SJc	SJb	SJd
16					S4nw	LPJc	LP4e	S4se	LPJd	S4ne	SJa	SJc	SJb	SJd
17				LPJc	LPJa	S4nw	LP4e	S4se	LPJd	S4ne	SJa	SJc	SJb	SJd
18						LP4e	S4nw	S4se	LPJd	S4ne	SJa	SJc	SJb	SJd
19							S4se	S4nw	LPJd	S4ne	SJa	SJc	SJb	SJd
20					LP4e	LPJa	S4nw	S4se	LPJd	S4ne	SJa	SJc	SJb	SJd
21							S4se	S4nw	LPJd	S4ne	SJa	SJc	SJb	SJd
22						S4se	LPJa	S4nw	LPJd	S4ne	SJa	SJc	SJb	SJd

Tab. 3.5 - The feasible solutions at the third iteration.

4 Conclusions

The proposed approach to the PP definition in steel structure welding of diesel engines, reminds the work of Chakrabarty and Wolter [6], which characterise the decomposition of the whole structure. The decomposition here discussed allows to face problems of reduced dimensions (the modules) and, moreover, to iterate the procedure at higher levels of assembling operations.

Further characteristic of the procedure concerns the integration of qualitative evaluation by the human expert (analysis of the tails) and precedence-matrix solution algorithm. On a reduced number of attained solutions a quantitative analysis can be carried out by a relational data base. The application to a case study shows the good performance of the method and evidences the opportunity to get a set of very good solutions. The solution set can be directly proposed to the welding department where the final choose will be done with respect to the availability of the resources and to specific situations which cannot be solely expressed by constraints.

AKNOWLEGMENTS

This work has been developed under the support of the grant Ricerca Scientifica 60%- University of Trieste , 1977.
The Authors would like to thank the staff of Grandi Motori SpA (welding Dept.), for their kind availability and the fruitful suggestions.

REFERENCES

[1] De Fazio T L , Witney D E 1987 Simplified Generation of All Mechanical Assembly Sequences. *IEEE Tansactions on Robotics and Automation* vol. RA-3, no 6.

[2] Baldwin D F, Abell T E, Lui M M, De Fazio T L, Witney D E 1991 An Integrated Computer Aid for Generating and Evaluating Assembly Sequences for Mechanical Products. *IEEE Tansactions on Robotics and Automation* vol.7, no.1.

[3] Homem de Mello L S, Sanderson A C 1991 Representation of Mechanical Assembly Sequences. *IEEE Transactons on Robotics and Automation* vol.7, no.2.

[4] Homem de Mello L S, Sanderson A C 1991 A Correct and Complete Algorithm to Assembly Sequence Planning. *IEEE Transactions on Robotics and Automation* vol.7, no.2.

[5] Homem de Mello L S, Sanderson A C 1991 Two Criteria for the selection of Assembly Plans: Maximising the Flexibility of Sequencing the Assembly Task and Assembly Time through Parallel Execution of Assembly Tasks. *IEEE Transactions on Robotics and Automation* vol.7, no.5.

[6] Chakrabarty S, Wolter J 1997 A Structure-Oriented Approach to Assembly Sequence Planning. *IEEE Transactions on Robotics and Automation* vol.13, no.1.

[7] Kusiak A 1990 *Intelligent Manufacturing Systems*. Prentice-Hall, New Jersey, USA.

6
Design of Part-Machine Cells in Manufacturing Systems Using Mathematical Programming Techniques

S. Sofianopoulou

1 Introduction

Group Technology is an innovating manufacturing technique which overcomes most of the usual problems of traditional production systems. The key concept of Group Technology is the grouping of machines into manufacturing cells and of part types into part families. A major part of the design phase of a manufacturing system, consists of identifying a set of independent machine cells, each capable of processing a family of parts. Fully independent machine cells are rare in practice since some parts need to be processed by more than one machine cells, leading to a number of intercellular moves. Transporting parts during their production process complicates the management of cellular manufacturing systems. The efficiency of a cellular manufacturing system depends heavily on the intercellular traffic. The designer of such systems, therefore, tries to allocate machines to cells in such a way so to keep the interaction among cells to the lowest possible level.

Various models have been proposed for solving the cell formation problem. They can be classified into four different categories:

1) A systematic approach, known as Burbridge's Production Flow Analysis [1]. Burbridge's pioneering work is based on parts routing information and consists of an exhaustive analysis of production flow.
2) Matrix rearrangement procedures which obtain solutions from rearrangements of the rows and columns of the part-machine incidence matrix (a 0-1 matrix denoting whether a machine is used by a part or not) with the aim to form several diagonal blocks corresponding to part and machine groups. The ideal situation is one in which all the 1's are in the diagonal blocks and all the 0's are off the diagonal blocks. This seldom happens either because of the type of the data or because of the deficiencies of the algorithms used. Mc Cormick et al. [2] determine cells employing bond energy algorithm. King [3] and King and Nakornchai [4] identify machine cells by rank order clustering of the part-machine incidence matrix, while Chan and Milner [5] apply direct clustering algorithm.

3) Similarity coefficient methods are employed by many researchers to form either machine groups or part groups. In the machine (part) similarity coefficient approach machine cells (part families) are first determined taking into account the similarities between pairs of machines (parts). Machines (parts) with closer similarity coefficients are arranged into the same group following a specific algorithm. Then parts (machines) are assigned to machine cells (part families). The idea of similarity in machine and/or part grouping was introduced by McAuley in his single linkage clustering method [6]. This approach has also been used by Carrie [7]. These methods, however, ignore major practical issues such as the size of the cells and the sequence of machine operations for each part type.
4) Mathematical programming approaches which utilize 0-1 linear or quadratic models. Kusiak [8] presented an integer programming formulation employing a p-median approach. Similarly, integer programming formulations have also been presented by Srinivasan et al. [9], who applied an assignment technique, and Vannelli and Hall [10], who used a graph partitioning approach. An optimal tree search method has been developed by Spiliopoulos and Sofianopoulou [11]. Their method employs two different bounds, one of the quadratic assignment problem and a recently proposed one for the quadratic transportation problem. Other mathematical programming approaches include, among others, the works of Boctor [12], [13], Gunasingh and Lashkari [14], Wei and Gaither [15] and Sofianopoulou [16].

The literature on the manufacturing cell formation problem can also be classified according to the objectives sought. Apart from the minimization of the intercellular moves other objectives sought are:

1) Minimization of exceptional elements (the 1's outside the diagonal blocks of the part-machine incidence matrix).
2) Minimization of the weighted sum of exceptional elements and voids (the 0's in the diagonal blocks).
3) Minimization of the total cell load variation among machines [17].
4) Maximization of the sum of machine similarities within cells [18].

It should be underlined, that although the cell formation problem is addressed by a large number of researchers using a variety of objectives, formulations and solution methods, very seldom a 'complete' solution to the problem is produced, i.e. a solution which determines a partition of *both* the machine set *and* the part set.

The purpose of this chapter is to solve the cell formation problem with the objective of minimizing the amount of intercellular moves subject to cell-size constraints and taking into account the machine operation sequence for each part. Two mathematical programming approaches are proposed. In the first approach the cell formation problem is modelled as a linear integer programming problem. Families of parts, which are to be processed by the corresponding machine cells formed, are identified in a subsequent step. In the second approach a model is presented which tackles *simultaneously* the formation of both machine cells and part families. An novel feature of the proposed models is that there is no need of specifying *a priori* the number of cells to be formed. This is automatically adjusted within the solution procedure. The behaviour of the models is examined through an example of a randomly generated instance of a medium sized production system.

2 Mathematical Formulation

2.1. *Model I*

2.1.1. Machine Cell Formation

The cell formation problem is formulated as follows:

$$\text{minimize} \quad \sum_{i=1}^{M-1} \sum_{j=i+1}^{M} c_{ij}\,(1 - X_{ij}) \tag{1}$$

s.t.

$$\sum_{i=1}^{k-1} X_{ik} + \sum_{j=k+1}^{N} X_{kj} \leq N\text{-}1 \qquad k=1,\ldots,M \tag{2}$$

$$\begin{aligned} &X_{ij} + X_{ik} - X_{jk} \leq 1 \\ &X_{ij} - X_{ik} + X_{jk} \leq 1 \qquad i=1,\ldots,M\text{-}2, \quad j=i+1,\ldots,M\text{-}1, \quad k=j+1,\ldots,M \\ &-X_{ij} + X_{ik} + X_{jk} \leq 1 \end{aligned} \tag{3}$$

$$X_{ij} = \begin{cases} 1, & \text{if machines } i \text{ and } j \text{ are assigned to the same cell} \\ 0, & \text{otherwise}, \qquad i=1,\ldots,M\text{-}1, \quad j=i+1,\ldots,M \end{cases} \tag{4}$$

where M is the number of machines, N is the cell-size capacity and c_{ij} is the total traffic between machines i and j for all parts. In the objective function the term $\Sigma\Sigma c_{ij}(1\text{-}X_{ij})$ represents the total traffic of parts between machines not grouped together, i.e. belonging to different machine cells. Constraint (2) ensures that the total number of machines grouped together with machine k does not exceed N-1. Constraints (3) ensure that machines j and k are grouped together (X_{jk}=1) if machines i and j as well as machines i and k are both allocated to the same cell (X_{ij}=1 and X_{ik}=1 respectively).

The above model can also cater for the case of separation constraints, i.e. constraints which prevent two or more machines to be allocated to the same cell. In the context of the above formulation, these constraints read:

$$X_{kl} = 0 \qquad k < l \tag{5}$$

when machines k and l are not allowed to be placed in the same cell. Similarly, when operational or other reasons require that machines q and t are placed in the same cell then the following constraints are required.

$$X_{qt} = 1 \qquad q < t \tag{6}$$

It is interesting to note that with the machine-to-machine formulation (1), the *a priori* specification of the number of cells to be formed is alleviated. This formulation does not only produce the allocation of machines to cells with the minimum intercellular traffic, but it also gives automatically the number of cells which corresponds to the optimum machine grouping. This feature makes the present formulation distinctly different from the ones used in the literature.

2.1.2. Part Families Formation

Having obtained the assignment of machines to cells, one then seeks to form the corresponding part families. The part families formation problem can be formulated and solved using the following model:

$$\text{maximize} \sum_{i=1}^{M} \sum_{k=1}^{K} \sum_{p=1}^{P} \sum_{s=1}^{S_p} \sum_{\substack{j=1 \\ j \neq i}}^{M} b_{ik}\, d_{psi}\, b_{jk}\, d_{p(s+1)j}\, Z_{pk} \tag{7}$$

s.t.

$$\sum_{k=1}^{K} Z_{pk} = 1 \qquad p=1,\ldots,P \tag{8}$$

$$Z_{pk} = \begin{cases} 1, & \text{if part } p \text{ is assigned to cell } k \\ 0, & \text{otherwise,} \qquad p=1,\ldots,P, \quad k=1,\ldots,K \end{cases} \tag{9}$$

where S_p denotes the number of operations scheduled to be performed on part p, and d_{psi} is a 0-1 coefficient denoting whether the sth operation of part p is performed on machine i or not. K denotes the number of cells formed and b_{ik} denotes the machine-cell allocation matrix (b_{ik}=1 if machine i is allocated to cell k and 0 otherwise). Both K and the matrix b_{ik} are produced by the solution of model (1)-(4). The objective function (7) assigns each part to the cell that has the largest number of machines used in sequence by that part. Each part is allocated to a cell

(8). The model can be further simplified by replacing constraints (9) with constraints $Z_{pk} \geq 0$ resulting thus to a simple LP model.

Detailed description of Model I as well as extensive computational results for random instances of production systems are presented in [16].

2.2. Model II

As it has already been mentioned previously, most of the models proposed in the literature do not produce a configuration which determines a partition of *both* the machine *and* the part set. In this section, a novel formulation of the part-families and machine-cells formation problem is presented, which cures the above mentioned deficiency.

The problem is formulated as follows:

$$\text{minimize} \quad \sum_{p=1}^{P} \sum_{s=1}^{S_p} \sum_{i=1}^{M} \sum_{\substack{j=1 \\ i \neq j}}^{M} d_{psi}\, d_{p(s+1)j}\, (1 - Y_{pi}\, Y_{pj}) \tag{10}$$

s.t.

$$\sum_{i=1}^{M} Y_{pi} \leq N \qquad p=1,\ldots,P \tag{11}$$

$$Y_{pi} + Y_{pk} + Y_{ri} - Y_{rk} \leq 2 \qquad p,r=1,\ldots,P, \quad i,k=1,\ldots,M \tag{12}$$

$$Y_{pi} = \begin{cases} 1, & \text{if part } p \text{ and machine } i \text{ are assigned to the same cell} \\ 0, & \text{otherwise}, \quad p=1,\ldots,P, \;\; i=1,\ldots,M \end{cases} \tag{13}$$

In the above formulation, objective function (10) minimizes the total amount of intercellular traffic which is generated if part p requires processing from machines i and j consecutively ($d_{psi}d_{p(s+1)j}=1$), but it is not assigned to the cell that machines i and j belong ($Y_{pi}Y_{pj}=0$). Note that S_p denotes the number of machines visited by part p. Constraints (11) require that the maximum number of machines which can be assigned to the cell that part j belongs should not exceed N. Constraints (12) ensure that part r and machine k should be grouped together ($Y_{rk}=1$) if machines i and k are assigned to the same cell ($Y_{pi}+Y_{pk}=2$) and part r is assigned to the same cell as machine i ($Y_{ri}=1$), i.e. if $Y_{pi}+Y_{pk}=2$ and $Y_{ri}=1$ then $Y_{rk}=1$. Also constraints (12) implicitly alleviate the need of specifying *a priori* the number of cells to be used.

If certain machines should not be placed in the same cell, separation constraints should be included in the formulation. In the context of the above formulation, these constraints are of the form:

$$Y_{pi} + Y_{pk} \leq 1 \qquad p=1,...,P \tag{14}$$

indicating that machines i and k are not allowed to be placed in the same cell. If, on the other hand machines q and t are to be placed in the same cell then the following set of constraints is required:

$$Y_{pq} - Y_{pt} = 0 \qquad p=1,...,P \tag{15}$$

Formulation (10)-(13) models the complete part-families and machine-cells formation problem. The solution provided by Model II produces *both* the machine cells *and* the part families formed which correspond to the minimum intercellular traffic. Model II is distinctly different from Model I, and most of the models presented in the literature. The latter first assign machines to cells and then parts to part families or the opposite. It is obvious, however, that the amount of intercellular traffic obtained using Model I gives only an upper bound to the optimum objective function value of Model II. This optimum value could had been evaluated through Model I, only if one evaluated the optimum part (machine) assignment for each feasible machine (part) assignment.

3 Illustrative Numerical Example

Optimal solutions to large combinatorial problems like the one presented here are usually computationally expensive even for small instances of the problem. Consequently heuristic methods are more suitable for obtaining solutions to these problems. For the results presented in this section an adapted version of Simulated Annealing heuristic algorithm has been employed. Briefly, the algorithm starts with a random initial feasible solution associating parts to machines. The procedure continues by repeatedly generating new neighbour solutions, which are accepted not only when they improve but also when they worsen the value of the objective function. The latter takes place probabilistically. The acceptance probability is a function of the temperature parameter, which at the beginning is set to a (high) value allowing the acceptance of most of the generated solutions. However, as the procedure evolves, the temperature value is decreased, thus reducing the acceptance probability, while the number of solutions attempted at each temperature level is gradually increased. Details of Simulated Annealing heuristic can be found in [16], [19], [20], [21], [22].

The algorithm was coded in Fortran 77 and run on a 100 MHz Pendium clone PC with Lahey FORTRAN compiler. Results included in this section refer to a randomly generated instance of the part-families and machine-cells formation problem with 25 parts, 15 machines and cell size N=6. The random sequences of machines that were generated to construct the part routes of each one of the 25 parts are presented in Table 1.

	1	2	3	4	5	6	7	8	9	10	11	12	13	14	15
1					1			2			3				
2				3	4		6		2		1				
3	1						2			4	3				
4				5	1		2		3				4		
5		1									4	3		2	
6					5		2		3		1	4			6
7					3						2	1			
8		2	5		4			3				1			
9				1							2			3	
10												1	3	2	
11			2			3									1
12				2	3				4						1
13						3				1			2		
14		5			3	6	2	1				4			
15	3					4				5			1	2	
16			2	3		4	1								
17			1			3	2					4			
18			4			2		1		3		5			
19							1			3				2	
20		1						2				4		3	
21			5	6	3		2	4	1						
22				3			5		1	4			6	2	
23	4					1				2	3	5			
24	2									1		4		3	
25	5		3			2			1	4					

Table 1. Random part routes

For both Model I and II the Simulated Annealing heuristic algorithm was run for 3000 iterations (attempted but not necessarily accepted solutions). Corresponding results are presented in Tables 2 and 3.

Table 2 presents the solution of both the machine cells and part families formation models, which compose Model I. The solution of the cell formation model produces three machine cells. This solution is next fed to the part families formation model which provides us with the members of three part families. The best amount of intercellular moves found is 47.

Table 3 presents the solution of Model II. As it can be seen three machine cells are formed. Two of them are loaded up to their size limit while the remaining machines form a third cell. A part family is associated with each machine cell, the third one being composed of just part 22. The best amount of intercellular moves found is 42.

Cell	Machines	Parts
1	1, 2, 12, 14	5, 8-10, 15, 17, 19, 20, 23, 24
2	3, 4, 6, 10, 13, 15	11-13, 16, 18, 22, 25
3	5, 7, 8, 9, 11	1-4, 6, 7, 14, 21

Table 2. Machine cells and part families formed with Model I

Cell	Machines	Parts
1	1, 6, 10, 11, 12, 14	3, 5, 7, 9-11, 15, 17-20, 23-25
2	3, 4, 5, 7, 8, 9	1, 2, 4, 6, 8, 12-14, 16, 21
3	2, 13, 15	22

Table 3. Machine cells and part families formed with Model II

As expected the amount of intercellular traffic produced by Model II is less, and therefore better, than the one produced by Model I. It is emphasized that the value of 42 can only be achieved through Model I, if one evaluates the best part (machine) assignment for each feasible machine (part) assignment and then choose the one with the minimum intercellular traffic which is the one corresponding to intercellular traffic of 42.

4 Conclusions

In this chapter the part-families and machine-cells formation problem is formulated using two different mathematical programming models subject to constraints pertaining to cell size limitations and taking into account the machine operation sequence for each part. The objective of both models is the minimization of the intercellular traffic. Model I uses a two-step procedure to produce first the machine cells and then the corresponding allocation of parts to part families. Model II on the other hand, uses a more complicated formulation (quadratic) but produces simultaneously *both* the machine cells *and* the part families which correspond to the minimum intercellular traffic. Results for a random test case demonstrate the applicability of both models.

References

1. Burbidge J L 1975 *The Introduction of Group Technology*. Wiley, New York
2. McCormick W T, Schweitzer P J, White T W 1972 Problem decomposition and data organization by clustering technique. *Operations Research* 20:992-1009
3. King J R 1980 Machine-component group formation in production flow analysis: An approach using a rank order clustering algorithm. *International Journal of Production Research* 18:213-232
4. King J R, Nakornchai V 1982 Machine-component group formation in group technology: Review and extension. *International Journal of Production Research* 20:117-133
5. Chan H M, Milner D A 1982 Direct clustering algorithm for group formation in cellular manufacturing. *Journal of Manufacturing Systems* 1:65-74
6. McAuley A 1972 Machine grouping for efficient production. *The Production Engineer* 2:53-57
7. Carrie A S 1973 Numerical taxonomy applied to group technology and plant layout. *International Journal of Production Research* 11:399-415
8. Kusiak A 1987 The generalized group technology concept. *International Journal of Production Research* 25:561-569
9. Srinivasan G, Narendran T T, Mahadevàn M 1990 An assignment model for the part-families problem in group technology. *International Journal of Production Research* 28:145-152
10. Vannelli A, Hall R G 1993 An eigenvector solution methodology for finding part-machine families. *International Journal of Production Research* 31:325-349
11. Spiliopoulos K, Sofianopoulou S 1998 An optimal tree search method for the manufacturing systems cell formation problem. *European Journal of Operational Research* 105:537-551
12. Boctor F F 1991 A linear formulation of the machine-part cell formation problem. *International Journal of Production Research* 29:343-356
13. Boctor F F 1996 The minimum cost, machine-part cell formation problem. *International Journal of Production Research* 34:1045-1063
14. Gunasingh K R, Lashkari R S 1989 The cell formation problem in cellular manufacturing systems-a sequential modelling approach. *Computers and Industrial Engineering* 16:469-476
15. Wei J C, Gaither N 1990 An optimal model for cell formation decisions. *Decision Sciences* 21: 416-433
16. Sofianopoulou S 1997 Application of simulated annealing to a linear model for the formulation of machine cells in group technology. *International Journal of Production Research* 35:501-511
17. Venugopal V, Narendran, T T 1992 Cell formation in manufacturing systems through simulated annealing: An experimental evaluation. *European Journal of Operational Research* 63:409-422

18. Chen W H, Srivastava B 1994 Simulated annealing procedures for forming machine cells in group technology. *European Journal of Operational Research* 75:100-111
19. Kirkpatrick F, Gelatt C D jr, Vecchi M P 1983 Optimization by simulated annealing. *Science* 220:671-680
20. Golden B L, Skiscim C C 1986 Using simulated annealing to solve routing and location problems. *Naval Research Logistics Quarterly* 33:261-279
21. Harhalakis G, Nagi R, Proth J M 1990 An efficient heuristic in manufacturing cell formation for group technology applications. *International Journal of Production Research* 28:185-198
22. Sofianopoulou S 1992 Simulated annealing applied to the process allocation problem. *European Journal of Operational Research* 60:327-334

PART II

PROCESS MONITORING, INSPECTION, DIAGNOSIS AND MAINTENANCE

7
Knowledge Based Process Monitoring in Mass Production

G. Petuelli and G. Blum

1 Introduction

In view of the fact that 60% to 80% of the factory and manufacturing costs are caused by the manufacturing process the need of cost reduction in production becomes obvious. Reliable manufacturing machines and their economic utilisation go together with the increase of product quality. Therefore the monitoring systems are mainly installed in order to increase the availability of production machines, to attain a high rate of utilisation, to assure quality of production and to protect machines against damages. Thus the demand for process monitoring systems in metal cutting processes is obvious [1].

Especially in highly productive mass production it is essential to maintain quality of workpieces and up-time of machines on the highest level possible. On the other hand quality check of each part in mass production, with up to 300 workpieces per minute, is too expensive and time consuming. Consequently the control of product quality in form of random checks very often indicates quality problems too late. Thus the part rejection rate and losses can reach a disturbing level. Furthermore, failed parts must be taken out of production in order to prevent further processing. On the other hand it is very important to protect machines and products against damages caused by tool break. It is therefore a matter of high priority to determine the optimal point and time for an exchange of tools when they have reached the end of their life-span.

To increase service quality of machines and to reduce tooling costs which includes machines, workpiece, fixture and the tool it is essential to monitor the behaviour of tools with respect to tool wear. The systems applied must have a short reaction time and a high selectivity to insure monitoring of tool wear and tool break at the same time. They must be able to react immediately whenever a tool breaks and they must have the capability to detect all other causes of process deteriorations like tool wear or misalignment of workpieces.

2 State of the Art of Set-up and Performance of Process Monitoring Systems

The design of powerful process monitoring systems requires knowledge about typical quantities which can be measured and are related to process and quality data. Research work is being carried out by several institutes to evaluate the behaviour of different processes by applying different sensors to pick-up signals and to extract typical features describing the performance of individual processes [2],[3],[4],[5],[6]. According to these evaluations different aspects have to be considered by the design of dedicated systems. Therefore, monitoring systems can not be standardised but have to be tailored to meet the needs of an individual process to be monitored. The studies especially are carried out to develop monitoring in turning and milling operations. Monitoring systems for drilling and tapping processes are developed to monitor tool wear and tool breakage of tool with diameters bigger than 4 mm only [3],[4]. A system capable to monitor tools with smaller diameters is using coolant to transmit acoustic emission, thus can not be applied in dry machining [7]. The performance of this system is dependent upon the quality of the coolant supply also.

It was found, that for data handling and feature extraction purposes sophisticated software tools are required in order to retrieve detailed information concerning the status of processes respectively of tools [8],[9],[10],[11]. This is especially the case in processes which are influenced by different disturbances simultaneously.

With respect to metal cutting or forming processes typical quantities which can be measured are power consumption, cutting force or torque and acoustic emission. For these purposes off the shelf sensors are available, which can or must be retrofitted into machines next to the process to insure quality of process signals gained. As indicated in fig. 1, the individual components applied must be selected according to the requirements of the monitoring task.

The signals are processed by dedicated monitoring systems. Different components are available on the market, but none of these systems directly can be installed and immediately started [7],[12],[13],[14]. They always must be adapted to the particular process. Engineering work is to be done in order to achieve best monitoring results possible. Appropriate sensors must be selected with respect to their capability to pick-up typical process data with high selectivity. It is of utmost importance that any process deterioration is detected by sensors. Furthermore, the integration of sensors into machines carefully must be evaluated, since quality of information gained to a large amount is influenced by the path of transmission of signals via different components of machines.

	Monitoring			Optimization			Machine Tools						
	Tool Breakage	Tool Wear	Missing Tool	Tool Life	Process	Quality	Transferlines	Revolving Transferlines	Machining Centers	Lathes	Grinding Machines	Bohring Tapping Single Spindel	Bohring Tapping Multiple Spindel
Power Consumption	●	●	●	●	●	●	●	●	●	○	●	●	○
Acoustic Emission	●					○	●	●	●	○	●	●	●
Torque	●	●	●	●	●	●	●	●	●			●	●
Force	●	●	●	●	●	●	○	○		●			
Vision System			●			○	●	●	○				●
Contacting / non-Contacting Probe			●									●	

Performance: ● good ○ poor

Fig. 1: Correspondence between Process Signals and Application

The performance of monitoring systems in total depends on constrains such as reaction time, loading of sensor, signal processing and objective defined by the application and to be reached by aid of the system. Furthermore, monitoring systems available today do not have the capability to indicate causes of disturbances. In general, the only message generated by systems is „machine stop“ due to a faulty process.

Most importantly until now less effort was made to integrate intelligent data processing in order to determine the causes of process deteriorations and to generate messages indicating which actions have to be taken to balance process characteristics. This demand is part of the development done in order to describe the correlation between the measured signal and the quality of the workpiece, which depends on several conditions and is unique to each process and machine [5].

3 Monitoring of Internal Thread Forming

In mass production machines are operated without being monitored permanently by a machine operator. Therefore it is very important to detect any malfunction at the earliest point and time possible in order to reduce down time and the production of parts with limited quality. A typical application is found in the machining of cable connectors. They are produced from strip stock at a production rate of about 300 pieces per minute by aid of special purpose punching and bending machines. Prior to heat treatment, the last step of production is forming of an internal thread sized M2.5 to M6.0. For this purpose a multiple spindles head was integrated into the bending machine. The spindles are accelerated to 8000 rpm for thread forming, decelerated and reversed to the same speed. In this particular application the spindles have been equipped with strain gauges and non-contacting power and data transmission systems to pick-up the torque applied by the thread forming process. Because of limited space tools must be used with two pilot threads (chamfer of a tap) at the most.

A typical time sequence of the torque applied during a forming process is shown in fig. 2. The process start is indicated by an increasing torque to a level of about 0.03 Nm which is caused by the acceleration of the spindle. When the tool touches the workpiece (time T_0) the torque increases rapidly to a maximum value of M_{max} = 0.25 Nm. The maximum torque depends upon different technological data such as coating of tools, forming speed and kind of coolant supplied and is influenced by tool wear also. Approximately at that point and time (T_{Mmax}) the spindle is decelerated, reversed (T_1) and finally stopped. As can be seen, the process last for about 300 ms only.

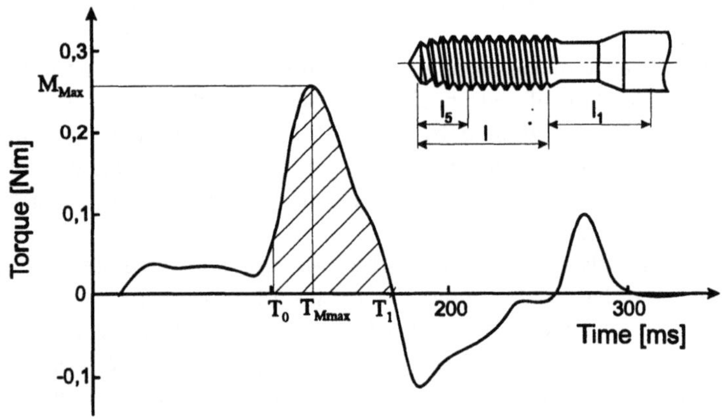

Fig. 2: Time signal of torque in thread forming

The shape of the measured curves changes due to different disturbances that cause degeneration of process stability. Thus, the different characteristics of the signal also vary. Therefore, in the first step of evaluation the different time features mentioned, the maximum value as well as the integral with respect to time of the signal were calculated in order to extract information concerning the interrelationship between process stability and process abnormalities.

3.1 Characteristics of Internal Thread Forming Processes

The most important state of metal cutting and forming processes to be monitored is considered to be tool wear. Progressive tool wear results in increasing torque as well as variation of the time T_0 at which the torque is zero since the spindle is reversed (fig. 3). This shifting of the time is due to an increased number of threads of the tool is needed to form the different threads. Thus the integral also becomes larger.

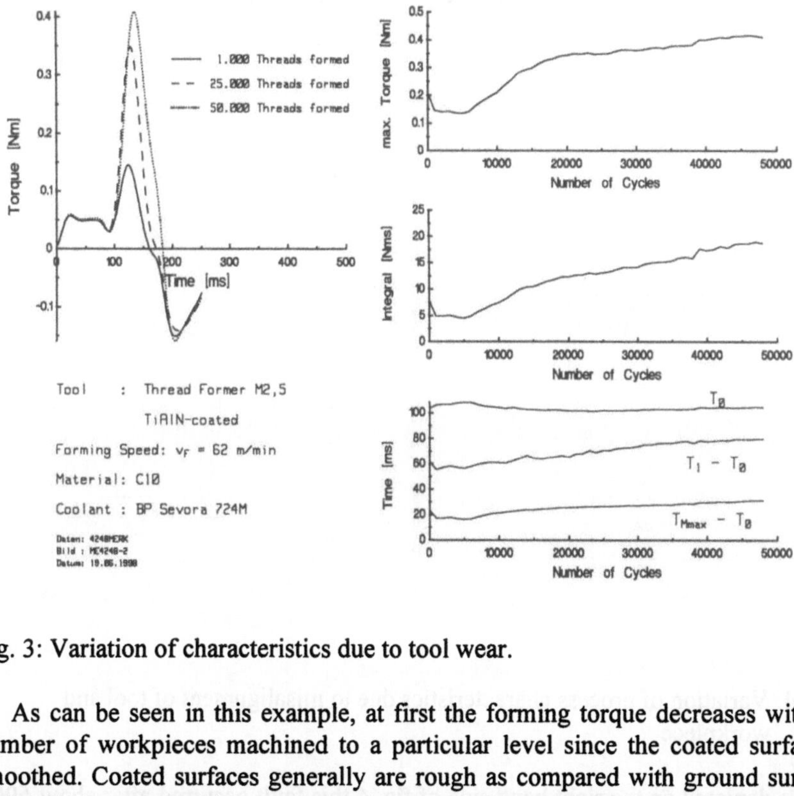

Fig. 3: Variation of characteristics due to tool wear.

As can be seen in this example, at first the forming torque decreases with the number of workpieces machined to a particular level since the coated surface is smoothed. Coated surfaces generally are rough as compared with ground surfaces and will be smoothed by the first processes performed. This does not damage the coating and almost has no influence onto the other characteristics of the process signal. The effect of decreasing torque depends upon the kind of coating applied, in this case a TiAlN-coating was used. The torque almost remains constant until about 6000 parts have been machined. Due to progressive wear the torque increases with a relatively high and constant rate until about 20000 workpieces have been manufactured. From that state on the rate of increase lessens. The increase is caused by progressive adhesive and abrasive wear of the forming edges of the tool. Similar influences can be observed with the integral value, whereas the time characteristics do change to a small amount only. The steadiness of the time related to the process start indicates high stability of the production system. Thus the entity of these characteristics and their relationship is sufficient to monitor tool wear.

On the other hand deviation of position, misalignment an mismatch of tool and workpiece relatively to each other also causes process abnormalities. As can be seen on the left hand side of fig. 4 deviation of position results is a time shift of the signal.

In this case the tool was pushed into the fixture because the clamping obviously was not tight enough. Thus the thread will not be manufactured correctly, the part will be rejected because of bad quality. This in general leads to an exchange of tools even if they did not yet reach the end of their life-span.

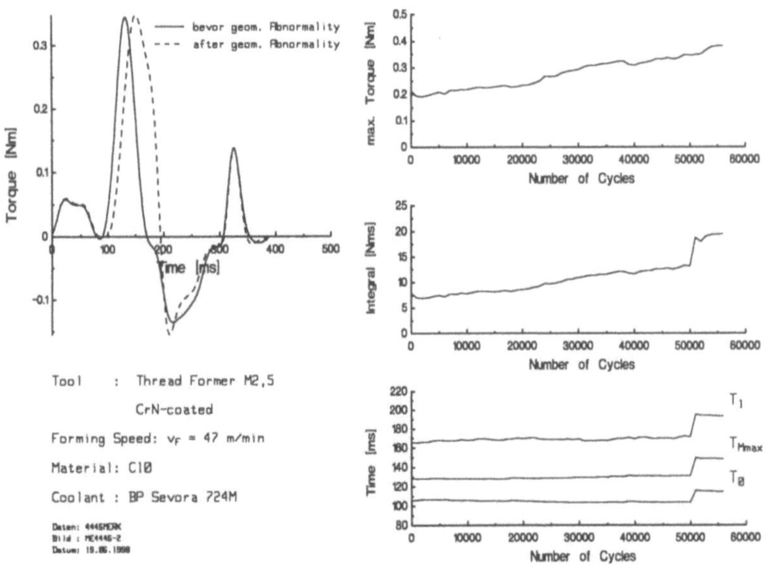

Fig. 4: Variation of process characteristics due to misalignment of tool and workpiece

As depicted on the right hand side of fig. 4 this fault occurred after about 60000 parts were produced. It results in a discontinuity of the curves showing the time features and the integral value of the signal monitored. Contrary to the changes of these features the maximum torque is not affected by the misalignment. The progressive wear of the forming edges of the tool results in an almost linear increase of the torque with respect to the number of parts produced. This behaviour is quite different as compared with the data given in fig. 3 since these data were taken during production with a tool which was coated with TiN.

Taking these features and their changes caused by deviation of position between tool and workpiece thread forming processes can be monitored with respect to misalignment of tool and workpiece.

3.2 Fuzzy Based Process Monitoring

As already mentioned about 2 million thread forming processes were monitored and process signals were measured in order to evaluate characteristics describing process status as well as process degeneration. Special attention was given to the determination of a relationship between causes for process deterioration and their effect onto process signals respectively characteristics extracted. It was found that different characteristics of processes signals are generated and influenced by particular events. Furthermore, it must be taken into account that more than just one cause may occur at a time and is responsible for process disturbances indicated by process characteristics. These different events will influence the variation of features to a different degree. Thus, multiple feature processing is required in order to separate the different influences ant to detect causes for process degeneration as well as the major component responsible for process failures and reduced quality of workpieces. Processing of characteristics requires the introduction and assessment of knowledge which is specific for the process under examination. Since the variations of the features caused by different events are superimposed means have to be applied in order to separate the influences. For this kind of problem fuzzy logic is found to be an appropriate tool.

Based on the above described experience and with the introduction of additional knowledge extracted from the process data, a prototype of a fuzzy based monitoring system was set-up to identify process degeneration caused by tool wear, deviation of position respectively adjustment of tool and workpiece as well as of misalignment of tool which results in an excessive wear and damage of the outer diameter of tools due to contact of tool and fixture. As depicted in fig. 5, the system is of modular design. Thus, provision is made for the extension and integration of additional characteristics to be monitored. Each module is dedicated to a particular cause of process deterioration and contains fuzzy sets as well as membership functions accordingly. Different process characteristics are input into each module, whereas the output of each component is dedicated to the individual feature and process abnormality it is supposed to monitor. As results weighting factors of the individual events monitored are output. They are gained by aid of the mean-of-maximum defuzzification algorithm which proved to give best results as compared with the other standard algorithm.

The system was tested with data acquired in standard production cycles. The results of one test are shown in fig. 6. On the left hand side of the figure a selection of the input data is depicted, whereas on the right hand side the resulting weighting factors with respect to the numbers of parts machined are given. As can be seen, progressive tool wear is indicated by the output shown in the diagram on the upper part of the right hand side. Most importantly, after the production of about 20000 parts a misalignment of the tool with respect to an unidentified part of the machine or fixture occurred which results in a maximum value of the weighting factor indicating this kind of event. Furthermore, deviation of position of the tool due to insufficient clamping could not be detected, the according weighting factor *adjustment* did not vary.

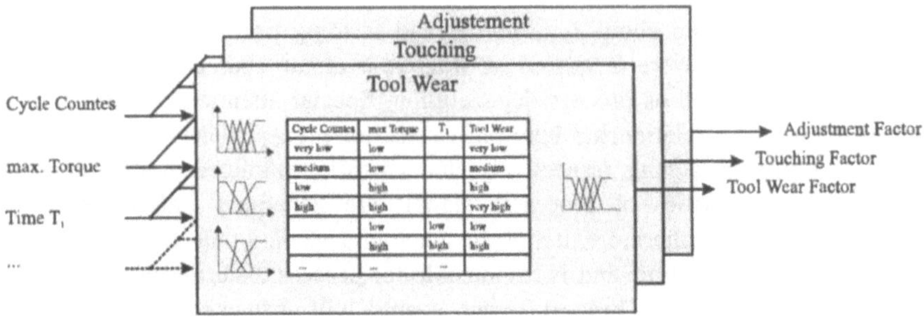

Fig. 5: Set-up of the fuzzy system to monitor thread forming processes

By this application it could be shown that knowledge based process monitoring can be a powerful tool to not just detect process degeneration but the cause of the process failure also.

4 Conclusion

The application of off the shelf process monitoring systems requires engineering work in order to make them work correctly and to detect process disturbances. Nevertheless, these systems do just indicate process degeneration, no means are provided to distinguish different causes of process failure. Therefor, a knowledge based system was developed to monitor internal thread forming processes in mass production with a production rate of about 250 workpieces per minute. Since a multiple spindle head was used for this process strain gauges were applied to measure the torque due to the thread forming. The major objective of this research work was to not just indicate process disturbances but to also generate information about the events that are responsible for the process degeneration and thus the bad quality of workpieces.

Therefor, extensive work was carried out in order to evaluate the relationship between process deterioration and process signal respectively characteristics extracted from these signals. It was demonstrated that different causes of process degeneration do result in variation of torque. The variation depends on the kind of disturbance and generally does influence more than one feature of the signal at a time. Furthermore, more than one event may occur at a time and will cause process degeneration. In order to separate the different influences superimposed in the process characteristics multiple feature processing is necessary. For this purpose fuzzy logiy was found to be an adequate tool. Thus, a system was set-up which is of modular design enabling the extension to monitor the causes of different kind of process deteriorations.

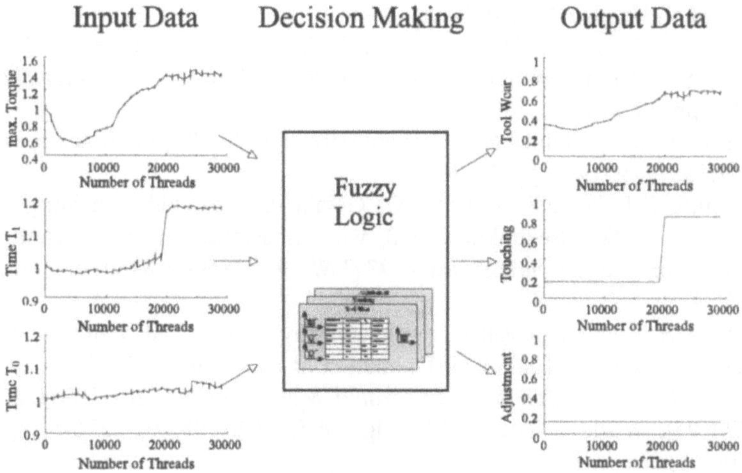

Fig 6: Application of the fuzzy based monitoring system and results gained

The system was applied in production and proved to be a powerful tool to monitor internal thread forming processes and to determine causes of process degeneration.

5 References

1. Weck M, Innovative Fertigungssysteme - Changen und Herausforderungen. "*Wettbewerbsfaktor Zeit in Produktionsunternehmen*", Springer-Verlag, Berlin, 1991
2. Schmitte F-J, Petuelli G, Blum G, Einsatz von Acoustic-Emission zur Prozeßüberwachung beim Bohren und Gewindefertigen. *VDI-Berichte 1255, 1996, ISBN 3-18-091255-3, S. 253-258*
3. Langguth Th, Weber A, Werkzeug-Verschleiß und -Bruchüberwachung beim flexiblen Drehen, Bohren und Fräsen, *Werkstatt und Betrieb 124 (1991)*
4. Beil G, Heider W, Erhöhte Prozeßsicherheit durch Anschnitt- und Werkzeugbruchüberwachung, *mav 10-1993, pp. 10-12*
5. Petuelli G, Blum G, Monitoring of Internal Thread Forming Processes in Mass Production. In: *Proc. of the 3rd International Conference on Automation, Robotics and Computer Vision (ICARCV),* Nanyang Technological University Singapore 1994, ISBN 981-00-5793-8, pp. 788-792.

6. Schmitte F-J, Welkner K, Petuelli G, Blum G, Methoden zur Online-Prozeßüberwachung in der Gewindefertigung, *VDI-VDE Kongress "Meß- und Automatisierungstechnik"* 18.-19.06.98 Ludwigsburg
7. Nordmann K, Werkzeugüberwachung mit neuen Sensortechniken, *VDI-Z Spezial Werkzeuge* Aug. 1994, pp. 42-44
8. Artmeyer M, Lorenz RD, DeVries MF, Process Identification and Modelling Using Structured Topoligies of Artificial Neural Networks, *VDI-Berichte 1179*, 1995, pp. 127-140
9. Dornfeld DA, Monitoring Technologies for Intelligent Machining, *VDI-Berichte 1179*, 1995, pp. 71-90
10. Schmitte F-J, Petuelli G, Blum G, Gemke C, Prozeßüberwachung mit µC-basierter Merkmalsextraktion und wissensbasierter Fuzzy-Auswertung. In: *Proc. of the Fuzzy Neuro-Systeme '97 (FNS '97),* ISBN 3-89601-003-4, pp 269-276
11. Kakade S, Vijayaraghavan L, Krishnamurthy R, Artificial Neural Network for Sensor Fusion and Tool Wear Estimation in Face Milling. In: *Proc. of the 3rd Intern. Conf. On Computer Integrated Manufacturing (ICCIM/AUTOFACT-ASIA'95)*, World Science Publishing 1995, ISBN 981-02-2376-5, pp. 1499-1506
12. NN, Überwachung nach Mass, Perfect Monitoring, Leaflet: *Artis GmbH Bispingen-Behringen*, 1996
13. NN, Brankamp C8060, Leaflet: *Brankamp GmbH, Erkrath* 1996
14. NN: PROMOS, Modulares Monitoring System, Leaflet: *PROMETEC GmbH Aachen*, 1998

8
Recent Developments in Automated Visual Inspection of Wood Boards

D.T. Pham and R.J. Alcock

1 Introduction

Automated Visual Inspection (AVI) is gaining increased interest as a means for controlling the quality of products. AVI gives better accuracy and consistency compared to human inspectors. Applications of AVI systems can be found in various industries including those dealing with electronics, food, metal and textiles. AVI is currently being used in the wood industry, where wood boards are sorted into quality categories based on an assessment of their surface appearance [1]. However, this task has the difficulty that wood is a natural material and so every board is unique. Also, certain defects, such as sound knots, do not differ significantly in brightness from clear wood. Some success has been achieved in this area but much effort still needs to be made to improve the process.

This paper reports on the latest developments in automated wood inspection which have concentrated upon the incorporation of artificial intelligence techniques.

2 Artificial Intelligence

Artificial intelligence involves the development of computer programs which exhibit some form of natural intelligence. The most common artificial intelligence techniques include expert systems, fuzzy logic, genetic algorithms, inductive learning and neural networks [2]. *Expert systems* store knowledge about a particular domain and then are able to solve problems specific to that domain. *Fuzzy logic* can be employed in areas where it is necessary to store or process information of an imprecise nature such as "this is a large defect" rather than where exact information, such as "this is a defect measuring 2x5mm^2", is needed. *Genetic algorithms*, which are based on natural evolution, are able to "evolve" solutions to problems by combining or mutating other possible solutions. Based on the principles of induction, *inductive learning* generates rules automatically from a data set. *Neural networks* mimic the operation, and sometimes the structure, of the human brain in that they have a learning ability and a parallel distributed architecture. Neural networks can be divided into two types: supervised and unsupervised. Supervised networks learn a mapping from inputs to outputs (classes) based on examples. Unsupervised networks group input data according to their similarities and do not need output data during training.

Artificial intelligence techniques have been utilised to solve a wide range of industrial problems including wood board inspection. The AVI process for wood boards involves several stages. First, an image of the board is acquired and enhanced, if necessary, to make it more suitable for later processing stages. Second, the image is segmented into clear wood and defective areas. Third, features are extracted from each area which is found to contain a defect. Fourth, a classifier determines the type of each defective region. Finally, a grade is given to the board based on the number, type and size of defects found. The above-mentioned AI techniques have been applied in all stages except image acquisition and enhancement.

3 Image Acquisition and Enhancement

The first operation which needs to be performed in automated inspection is image acquisition. It is important that a high quality image is obtained to facilitate subsequent processing. Before image acquisition can be carried out, several decisions need to be made. These include the positioning and type of lighting to employ, whether to use grey-scale or colour images and, for wood inspection, whether to utilise an X-ray scanner instead of a camera.

Appropriate lighting is a very important consideration to obtain high quality images. Uncontrolled ambient lighting adds extra difficulties to defect recognition whereas carefully positioned light sources can emphasise defects. For example, inclined illumination can be employed to detect surface roughness.

Grey-scale or colour cameras can be utilised for image acquisition. Colour cameras generate three times as much data as grey-scale images. This increases the computational burden on an automated grading system but gives the potential to improve performance. Recent work on employing colour for wood board inspection has given conflicting results. Silven and Kauppinen [3] found that features from colour images outperformed those from grey-scale images and that the computational overheads of processing colour images can be minimised by ordering the recognition stages based on their complexity. However, Kline *et al.* [4] found that their colour system gave a relatively low performance because the programs identified many clear wood areas as defects. It was suggested that other sensing techniques would need to be employed to find the defects accurately.

It is possible to employ other sources than cameras to obtain data about the board being inspected. X-ray imaging is one alternative. The main advantage of the technique is that it can scan the inside of a log or board. However, the technique generates more data than cameras and is also more expensive. Research into using X-ray images for wood inspection has been carried out for many years and a description of recent work can be found in [5, 6].

To make the acquired image more suitable for later processing, image enhancement techniques may be employed to emphasise the defects. Common image enhancement techniques include morphology and the subtraction of a common background image. Recently, however, researchers have not utilised image enhancement, preferring instead to try to improve the process of image acquisition.

4 Segmentation of Wood Board Images

Thresholding is the most common segmentation method for wood board images because it is fast and simple. However, one major problem is the determination of an appropriate threshold. Kothari *et al.* [7] employed a supervised neural network, the Multi-Layer Perceptron (MLP), to determine thresholds to separate clear wood, defects and background.

Schmoldt *et al.* [6] also employed a MLP network for segmentation but instead of utilising the classifier to determine the optimum threshold, it was used to calculate the type of each pixel (clear wood, bark, decay, knot and split). The MLP had ten inputs, the grey levels of the pixel to be classified and its immediate neighbours and also the distance of the pixel from the centre of the log. The network had five outputs, one for each defect type. Using a MLP for segmentation obviates the need to develop dedicated segmentation algorithms because it learns what constitutes a defect without direct human intervention. Another advantage of this technique is that it can perform segmentation and classification in one single step. However, there are two problems. First, it is not clear what the optimum inputs are for the MLP as just employing the pixel grey level would not be sufficient. Second, the method is very time consuming without dedicated hardware.

The present authors found thresholding to be a useful technique but that no one threshold can be employed to find all defect types [8]. Therefore, they proposed a modular approach for segmentation, consisting of:

- global adaptive thresholding for hard rot. The threshold was based on the number of peaks and valleys in the image histogram and also the mean and standard deviation of grey levels in the image. Any small objects located by this technique were removed because hard rot is a large defect type.
- multi-level thresholding for holes, rotten knots and splits. Pixels with very large or very small grey levels, i.e. nearly black or nearly white pixels, were marked as being defective by this technique.
- row-by-row adaptive thresholding for low contrast defects. The image was split into horizontal lines and pixels were labelled defective if they differed from their neighbours by a significant amount.
- vertical profiling for water streaks. The grey levels of pixels for each column in the image were plotted in a vertical profile. Large valleys in this profile correspond to dark vertical lines generated by water streaks.

Figure 1 shows an image of a wood board. It contains six defects: three pin knots, two rotten knots and a hole. The result of employing the above segmentation techniques on this board is shown in Figure 2.

Recent work on segmentation by the authors has found that adaptive thresholding can be performed using hardware-based functions such as convolutions to increase the speed of segmentation. The use of these standard functions also means that the technique is simpler to understand and implement.

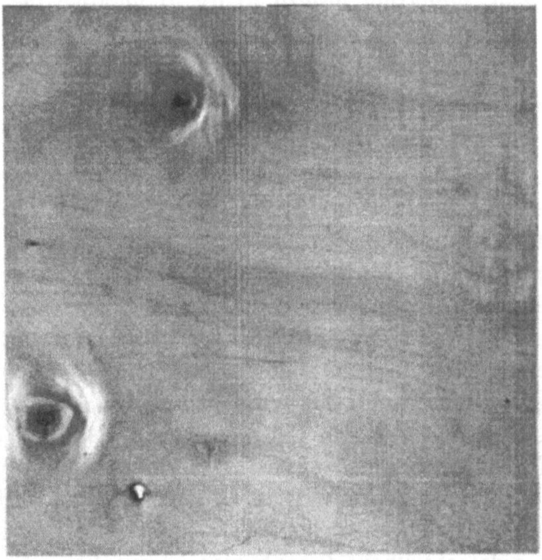

Figure 1 Image of a wood board

Figure 2 Segmentation of image shown in figure 1

As Figure 2 illustrates, after segmentation has been performed, two problems can be experienced. First, many areas, which are clear wood, may be detected as defects. Second, a defect may be represented by more than one segmented object. Two object processing techniques, based on artificial intelligence, have been developed to overcome these problems [9]. The first technique, derived from fuzzy logic, builds up evidences of each pixel representing a defect. Pixels are then grouped into objects and those with a low cumulative evidence are removed. The second technique is inspired by the unsupervised neural network ART-1 [10], shown in Figure 3. With the new technique, objects are sorted into size order before being presented to the network. The network then decides whether to join objects together, based on their size and position, and updates its neurons accordingly.

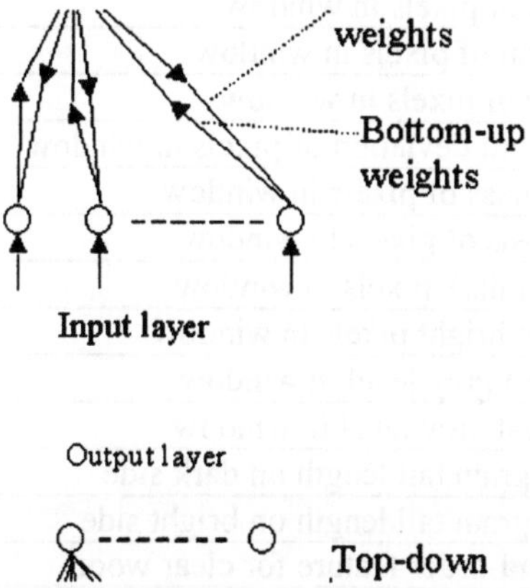

Figure 3 ART-1 network

5 Feature Selection

Currently, for AVI, there is no standard set of features and no method to determine which features to utilise. Much work has been carried out recently on feature selection for wood defect classification.

Work on feature selection has been carried out in the authors' laboratory by Packianather [11]. In his method, statistical measures are calculated to determine how well features differentiate between the classes. These measures were:

- intra-class variation. This determines the amount by which a given feature varies in one type.
- inter-class variation. This shows by how much a given feature varies between separate classes.
- correlation. This determines whether the values of two features are related.

Then, features were rejected if their intra-class variation was above a given threshold or their inter-class variation was below a given threshold. Features were also rejected if they were found to be highly correlated with another feature and were the worst performing of the two features.

Other recent work in feature selection for wood defects has employed genetic algorithms to reduce the number of features [12]. Fifty-three features were selected originally, giving a performance of 83.1% on a MLP. The genetic algorithm found that many of these features were not required and that employing twenty of them gave a similar performance (82.9%).

Feature	Feature name	Window or object feature
F1	Mean of pixels in window	W
F2	Median of pixels in window	W
F3	Mode of pixels in window	W
F4	Standard deviation of pixels in window	W
F5	Skewness of pixels in window	W
F6	Kurtosis of pixels in window	W
F7	No. of dark pixels in window	W
F8	No. of bright pixels in window	W
F9	Lowest grey level in window	W
F10	Highest grey level in window	W
F11	Histogram tail length on dark side	W
F12	Histogram tail length on bright side	W
F13	Second-order feature for clear wood	W
F14	Number of pixels in window with a grey level less than F1-2*F4	W
F15	Second-order feature for dark defects	W
F16	Number of pixels in window with a grey level more than F1+2*F4	W
F17	Second-order feature for bright defects	W
F18	Length of object in X direction	O
F19	Length of object in Y direction	O
F20	Area of all objects in region	O
F21	Elongation of object	O
F22	Circularity of object	O
F23	Proportion of very dark pixels in object	O
F24	Proportion of dark pixels in object	O
F25	Proportion of mid-range pixels in object	O
F26	Proportion of bright pixels in object	O
F27	Proportion of very bright pixels in object	O

Table 1 List of features employed in [14]

To overcome the problem of feature selection, Lampinen and Smolander [13] employed an unsupervised neural network, the self-organising feature map (SOFM) [10]. The pixel values from a window in the image were passed through a Gabor filter. Then, the SOFM was used to cluster the resulting vectors together. Finally, the output of the SOFM was presented to a MLP for classification. One problem with this technique is the very large training sets produced from each image. Also, it remains to be proved whether features obtained using this method produce better results than those chosen conventionally.

In research by the authors, the twenty seven features shown in Table 1 were employed [14]. These consisted of object features, extracted from the located objects, and window features, extracted from a square window with the same centre of gravity as the object. The object features consisted of both shape and shade features. The window features were first-order shade features and second-order shape features. New shape features were developed, based on hardware-based functions, which operate faster than the commonly-used co-occurrence features [15]. As shown in Table 2, when object features were employed together with window features, this gave an improvement in performance of around 10% over using either set individually. Experiments were also carried out by leaving out each of the features in turn and performing classification on the other twenty six. The best results were obtained by just leaving out feature F10. Finally, experiments were carried out on leaving out combinations of the worst-performing features but none of these gave any improvement over just leaving out F10.

	Run 1	Run 2	Run 3	Average
Object features	76.1%	69.3%	75.0%	73.5%
Window features	73.9%	83.0%	69.3%	75.7%
Combined features	86.4%	81.8%	87.5%	85.2%

Table 2 Comparison of object and window features

6 Classification of Wood Defects

The most common classification methods for wood defects have been rule-based classifiers and the MLP. Tests comparing these techniques have revealed that the MLP is the superior method of the two in terms of performance and ease of implementation [16].

The advantage of a rule-based classifier is that it can be seen why the classifier reached a certain decision from the rules which were executed. This cannot be done with a MLP which performs like a black box, mapping inputs onto outputs. The problem with rule-based systems is that it is not simple to generate effective rules. One possible solution to this is inductive learning which can generate rules automatically from examples of features and their associated class. The Rules-3 Plus inductive learning algorithm [17] was tried by the authors to derive rules for wood defects. However, it was found that the generated rules were too specific to the training data. Since then, the Rules-4 algorithm [18] has been developed. It has a superior performance to Rules-3 Plus but has not yet been tried on wood defect classification.

When training a MLP it is necessary to choose the values for certain parameters such as the learning rate, momentum and number of hidden neurons. The authors found that the momentum term was not useful for improving classification performance and that varying the number of hidden neurons also did not affect results significantly [14]. The learning rate was the most critical parameter and low values (around 0.1) were needed for good test set classification accuracy.

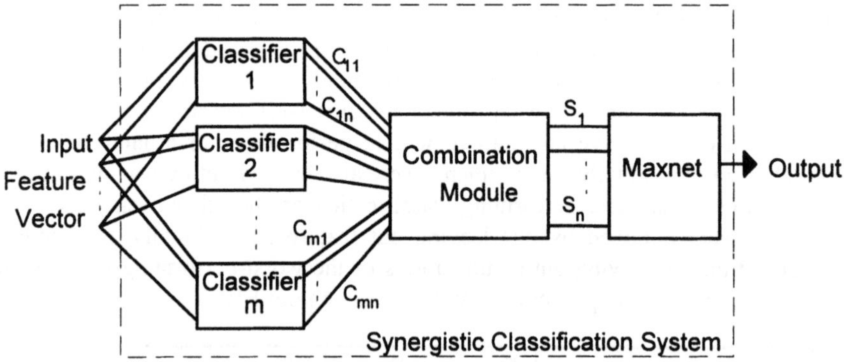

Figure 4 Structure of a Synergistic Classification System

To improve classification accuracies, Synergistic Classification Systems (SCSs), which mix several classifiers into one classification system, can be utilised. Figure 4 shows an example SCS. The system combines together m classifiers, each having n outputs. These are then merged in the Combination module and the largest output of the latter is chosen by the Maxnet to be the overall output. The authors tried varying numbers of MLPs and different combining strategies, finding that combining three MLPs gave an improvement over just one MLP [14]. Employing five classifiers gave the same result as three and more than five classifiers gave progressively worse results.

Research on SCSs has also been carried out in the authors' laboratory by Drake and Packianather [19]. In this work, the MLPs are connected into a tree structure. To classify a given feature vector, first, the vector is passed to an individual MLP. If the MLP determines that the type of the feature vector is one which it classified with 100% accuracy in training, then that type is accepted. If the feature vector is considered to be of another type then the features are passed to a second level classifier. Second-level classifiers only use a subset of the whole feature vector and are designed to differentiate between just two types. Therefore, they are able to concentrate on a smaller area of the feature space.

7 Conclusion

This paper has given details of recent research in AVI of wood boards. Artificial intelligence techniques have been employed at all stages of the AVI process after image acquisition:

- supervised neural networks have been used in segmentation for threshold selection. They have also been employed to classify individual pixels into either clear wood or a defect type;
- fuzzy logic and unsupervised neural networks have been utilised for post-processing of segmented images;
- genetic algorithms and supervised and unsupervised neural networks have been employed for feature selection;
- supervised neural networks have been shown to be superior to other techniques for classification. Recent work has concentrated on combining several such networks together.

The techniques suggested in [8] and [14] gave accuracies of 93% for segmentation and 92% for classification. This shows a considerable improvement over the 55% grading accuracy achievable by humans [20].

References

1 Pham D. T. and Alcock R. J. 1998 Automated Grading and Defect Detection: A Review. *Forest Products Journal.* 48(4), pp 34 - 42.

2 Pham D. T., Pham P. and Alcock R. J. 1998 Intelligent Manufacturing. in *Novel Intelligent Automation and Control Systems, Vol. I.* ed. J. Pfeiffer. Papierflieger, Clausthal-Zellerfeld. 16p.

3 Silven O. and Kauppinen H. 1996. Recent Developments in Wood Inspection. *Int. J. of Pattern Recognition and Artificial Intelligence.* 10(1), pp 83 - 95.

4 Kline D. E., Widoyko A., Wiedenbeck J. K. and Araman P. A. 1998. Performance of Color Camera Machine Vision in Automated Furniture Rough Mill Systems. *Forest Products Journal.* 48(3), pp 38 - 45.

5 Ohman M. 1998. Grade Prediction of Pinus Sylvestris Logs with the Aid of a Radiographic Image of the Log. *Scandinavian Journal of Forest Research.* 13(1), pp 83 - 89.

6 Schmoldt D. L., Li P. and Abbott A. L. 1997 Machine Vision Using Artificial Neural Networks with Local 3D Neighborhoods. *Computers and Electronics in Agriculture*, 16(3), pp 255 - 271.

7 Kothari R., Klinkhachorn P. and Huber H. A. 1991 A Neural Network Based Histogramic Procedure for Fast Image Segmentation. *Proc. 23rd Sym. on System Theory*, Columbia, SC, pp 203 - 207.

8 Pham D. T. and Alcock R. J. 1996 Automatic Detection of Defects on Birch Wood Boards. *Proc. Instn. Mech. Engrs., Part E, J. of Process Mechanical Engineering*, Vol. 210, pp 45 - 52.

9 Pham D. T. and Alcock R. J. 1998 Artificial Intelligence Techniques for Processing Segmented Images of Wood Boards. *Proc. Instn. Mech. Engrs., Part E, J. of Process Mechanical Engineering*, Vol. 212, pp 119 - 129.

10 Pham D. T. and Chan A. B. 1998 Unsupervised Neural Networks for Control Chart Pattern Recognition. *CIRP International Seminar on Intelligent Computation in Manufacturing Engineering - ICME 98*, Capri (Naples), Italy, pp 343 - 350.

11 Packianather M. S. 1997 *Design and Optimisation of Neural Network Classifiers for Automatic Visual Inspection of Wood.* PhD Thesis, School of Engineering, Cardiff University.

12 Estévez P. A. 1998 Combination of Neural Networks and Genetic Algorithms for Classification and Prediction Tasks. in *Novel Intelligent Automation and Control Systems, Vol. I.* ed. J. Pfeiffer. Papierflieger, Clausthal-Zellerfeld. 6p.

13 Lampinen J. and Smolander S. 1996 Self-Organising Feature Extraction in Recognition of Wood Surface Defects and Color Images. *Int. J. Pattern Recognition and Artificial Intelligence.* 10(2), pp 97 - 113.

14 Alcock R. J. 1996 *Techniques for Automated Visual Inspection of Birch Wood Boards.* PhD Thesis, School of Engineering, Cardiff University.

15 Pham D. T. and Çetiner B. G. 1996. A New Method for Describing Texture. *Proc. IWISP '96, 3rd Int. Workshop on Image and Signal Processing on Advances in Computational Intelligence.* Manchester. eds. B. G. Mertzios and P. Liatsis. Elsevier Science. pp 187 - 190.

16 Cho T. H., Conners R. W. and Araman P. A. 1991 A Comparison of Rule-Based, K-Nearest Neighbour and Neural Net Classifiers for Automated Industrial Inspection. *Proc. IEEE/ACM Int. Conf. on Developing and Managing Expert System Programs*, Washington, DC, pp 202 - 209.

17 Pham D. T. and Dimov S. S. 1997 An Algorithm for Incremental Inductive Learning. *Proc. Instn. Mech. Engrs.* Vol. 211, Part B, pp 239 - 249.

18 Pham D. T. and Dimov S. S. 1997 The RULES-4 Incremental Inductive Learning Algorithm. *Applications of Artificial Intelligence in Engineering XII*, R. A. Adey, G. Rzevski and R. Teti (eds) Computational Mechanics Publications, Southampton, Boston, pp 163-166, ISBN 1 85312 471 0.

19 Drake P. R. and Packianather M. S. 1998 A Decision Tree of Neural Networks for Classifying Images of Wood Veneer. *Int. J. of Advanced Manufacturing Technology.* Vol. 14, pp 280 - 285.

20 Polzleitner W. and Schwingshakl G. 1992 Real-time Surface Grading of Profiled Wooden Boards. *Industrial Metrology, 2*, pp 283 - 298.

9
Automated Visual Quality Inspection of Printed Ceramic Dishes

C. Vivas, J. Gómez Ortega and M. Vargas

1 Introduction

The application of machine vision techniques permits automation of many industrial tasks that short time ago needed human supervision. One of such tasks is visual automated inspection of industrial products, since it is often related to complex recognition problems. Most automated visual inspection systems used today attempt to identify defects by comparing the image of the object to be inspected to a non-defective pattern. Shape and textural models are usually needed to describe similarity measures between images, so the system can evaluate how close the inspected object is to the reference pattern.

This chapter is concerned with the problem of automated quality inspection of printed ceramic dishes. In this work some computer vision techniques that allow identification of certain kind of defects on the decorative printings of dishes are presented.

The inspection task to be accomplished is developed through a comparison of images. The dish image to be tested (work dish image) is compared to a non-defective reference pattern. Then a similarity measure between both images is computed, so the algorithm can decide if the dish image is defective or not.

The main problem that arises in the design of this inspection system, is the selection of a proper comparison technique capable of identifying defective pieces while allowing slight differences between correct ones. For instance, small relative deformations and subtle scale variations between the work image (image to be tested) and the reference image pattern should be allowed, while more severe distortions should be taken as a defect. We have to obtain an effective similarity measure between grey scale images. On the other hand, the algorithm should yield sufficiently good results to justify the development of an inspection system that could be practically used.

In figure 1, an example of typical dishes to be inspected by the system described in this chapter can be seen.

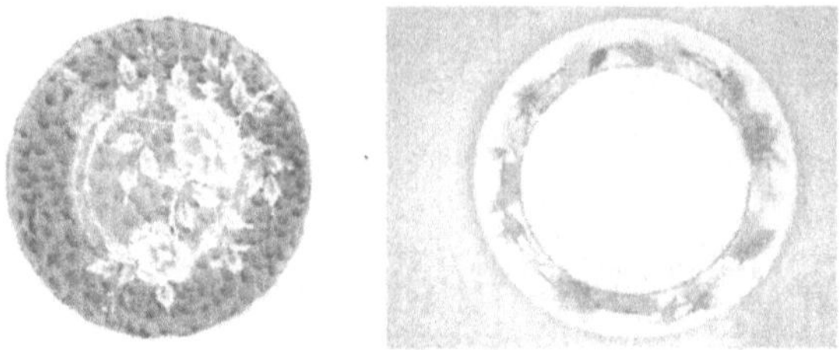

Fig. 1.- Typical dishes to be inspected.

Before the comparison process, some image pre-processing is needed, such as image filtering, location and orientation.

The inspection task is developed in four stages (see figure 2). The first one is image acquisition and filtering. As will be seen below, a technique based on the averaging of multiple static images has been employed to obtain a quality image of dish printings.

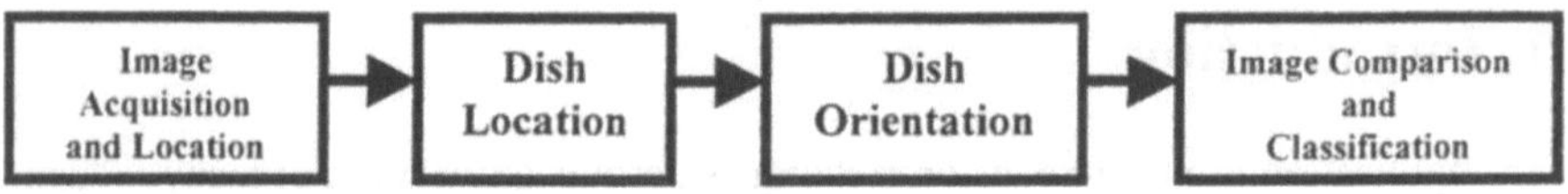

Fig. 2. - Scheme of Process Stages

The second and third stages consist of locating and orientating the image of the dish within the camera frame. These tasks are needed when the position and heading of the dish images within the camera reference frame are not precisely known. This could happen for instance, when the feeding device is a conveyor belt, where the dishes are laid from a feeder. A precise location and orientation of the dish within the camera frame are needed in order to allow a correct matching of the work dish image and the reference non-defective image. Anyway, if the feeding and dish printing devices automatically locate and orientate the dishes in a known way, these stages can be greatly speeded up. This fact is due to the reduction of uncertainty in the position and orientation of dishes, which allow reducing search intervals of the algorithms (section 3) for a precise computation of these parameters.

The fourth stage is image comparison and classification. The comparison is developed in the Fourier domain employing a percentage of the low frequency components of the image spectrum. Since the comparison of images must allow slight differences between them, low frequency components seem to be suitable to characterize images in the comparison task. As will be explained below, these components hold information related to general shapes in images, while detail information is held on high frequency components.

A two hidden-layered Perceptron neural network carries out comparison and classification of images. In the training phase a number of defective and non-defective dish images are presented to the network. This process is carried out off-line. As will be explained, once the network has converged, the system is ready to perform classification of dishes that were never presented to the network in the training phase.

The system was implemented on DSP's network based on the TMS320C40 processor from Texas Instruments. All the algorithms were designed to get advantage of the parallel execution abilities of the network, so the execution time could be minimized.

2 Image Acquisition and Filtering

The first step in a visual quality inspection system is to take a clear picture of the working piece. Because classification is carried out using a single image of the dish, it is very important to minimize the noise in this picture. The image capturing technique adopted here is based on the averaging of multiple static pictures of the dish (see [1]). The method consists of taking several images of the dish, averaging grey-scale values of corresponding pixels on each image. This process attenuates impulsive noise in images, while preserving edge information, which is not the case of other mask-based algorithms (Median filter mask, Gaussian filter, etc). Correlated noise keeps unaffected by this averaging process, although discarding high frequency components in a later step will attenuate it. The proposed technique is fast enough and demonstrated to be sufficiently effective to requirements of the system.

3 Location and Orientation Phase

3.1. Dish Position Estimation

Locating the work dish inside the camera reference frame consists of determining the position of its geometric centre with respect to an image reference frame. This point is defined as the intersection point of the symmetry axis of the dish boundary (for instance, on an ellipse this point corresponds to the intersection of the two major axes).

A very fast and simple iterative algorithm, suitable to estimate the position of the geometric centre of circular and elliptic dishes (the most common shapes) has been used in this work. The algorithm works as follows: First, an edge image of the dish is obtained, applying edge detection and binarization processes to the grey-scale image. The initial point, P_0 (see figure 3), for this algorithm must be within the contour of the dish. Thus, as the dish fulfils almost the whole image, the centre of the image is chosen as P_0.

Then an iterative process is carried out: for each point, P_i, two lines are traced, one horizontal and one vertical, both containing P_i. The intersection points of these lines with the dish contour, A_i, B_i, C_i, D_i (see figure 3), are obtained. Then the next iteration point, P_{i-1}, is computed as the intersection of the midlines of segments Ai-B_i, C_i-D_i.

The algorithm is finished when the distance between two consecutive points is less than a predefined threshold. This threshold avoids the algorithm to get trapped oscillating around some convergence point, when noisy irregular boundaries are present. In practice, no more than five iterations are needed for the algorithm to converge.

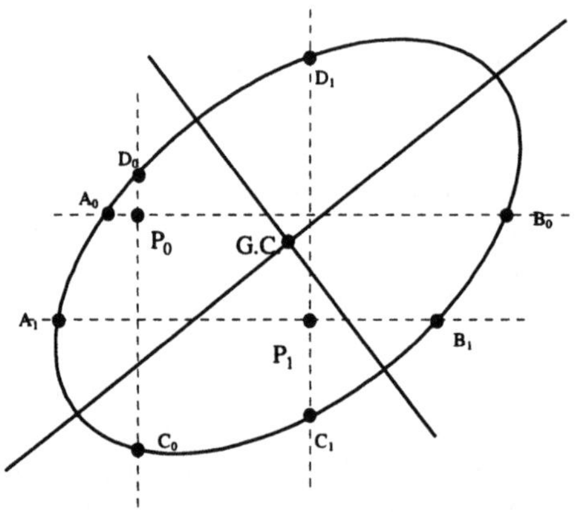

Fig. 3. - Geometric centre estimation scheme

The algorithm works well with continuous dish contours. If gaps are present, the method may fail in computing intersection points. To increase robustness, a set of parallel lines is used to compute each boundary point.

The convergence of this iterative scheme has been mathematically established for a general elliptic shape (for a detailed description, see [2]), and robustness has been experimentally tested for real dishes. Other techniques, such the Hough transform, have been essayed, but with poorer results [2].

3.2. Dish Orientation Estimation

Once the dish geometric centre has been located, it is needed to determine its orientation, that is, to estimate the relative angle between the reference image and the working image. For this purpose, the user will choose three sectors of a circular annulus from the binarized boundary reference image. These sectors will be used as masks to be found in the working boundary image, using the well-known template-matching technique.

Template matching is a search technique that shifts a template over the searching image, in order to find the best matching between the template pattern and a certain region in the searching image, which will be called *the optimum matching*. A measure of the matching degree can be computed through the following equation:

$$c(f_1, f_2) = \sum_x \sum_y f_1(x,y) f_2(x,y) \qquad (1)$$

which is the *binary correlation formula* of two bidimensional functions, f_1 and f_2.

In our case, the matching process is applied by rotating each template (sector) around the estimated geometric centre of the dish, looking for the best match. Thus, three sector orientation estimation angles, θ_i, will be obtained separately. Small radial displacements are also covered in order to take into account the uncertainty in the estimation of the work-dish geometric centre.

A degree of confidence will be associated to each θ_i, from the statistical analysis shown below, in order to find the correct match with the maximum probability. The angle θ_i with the highest confidence value will be used to determine the orientation of the working dish.

The use of three sectors implies some degree of robustness: if the matching of one sector fails, the other two can be used to correctly orientate the dish. These sectors are manually selected, and must be, preferably, non-overlapping and located as far as possible from each other (so defects in one of them do not affect the others). On the other hand, these sectors must contain enough significant information to be found without ambiguity.

3.2.1. Speeding up the Template Matching Algorithm

Systematic search methods, as the template matching approach described above, are inherently slow [3]. Speeding up techniques must be employed to reduce computation time whenever possible.

The approach used to speed up the algorithm consists of limiting the search area by using a previous coarse orientation estimation (*preorientation estimation*), avoiding the need to explore the whole surface. A searching interval around an estimated angle is a priori statistically computed. This interval is considered to be fixed for all the dishes of the same kind to be inspected.

Two different orientation estimation parameters can be used, both starting from the binary edge image [4]:

a) Maximum inertia axis of the image, represented by its angle, θ_a, with respect to the horizontal image frame axis.

b) The vector with origin at the gravity centre and ending at the geometric centre (vector *GRC-GC* in figure 4), with angle θ_b.

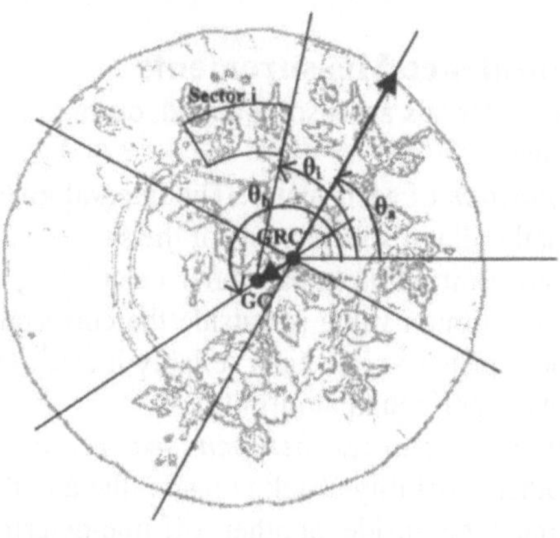

Fig. 4.- Example of binary edge image of a typical dish. Maximum inertia axis orientation (θ_a), *GRC-GC* vector orientation (θ_b) and sector *i* are shown.

A previous, off-line statistical analysis is used to choose the better of these two parameters. In this stage the system is fed with *m* (*m=100*, was experimentally used) dishes of the same kind as the ones used later, on-line. The following pseudocode describes this process:

-For each dish *j* (*1...m*)
  -Calculate θ_a^j and θ_b^j
  -For each sector *i* (*1...3*)
    -Calculate θ_i^j using the template matching process without preorientation.
    -Calculate $\alpha_{ai}^j = \theta_i^j - \theta_a^j$ and $\alpha_{bi}^j = \theta_i^j - \theta_b^j$
-Calculate means and standard deviations:
$\eta(\alpha_{ai}),\ \sigma(\alpha_{bi}),\ \eta(\alpha_{ai}),\ \sigma(\alpha_{bi})$

The parameter chosen is the one with less variance. θ_p will be used to denote the orientation of the selected parameter (that is, $\theta_p = \theta_a$ or $\theta_p = \theta_b$) and its associated mean and standard deviation values will be denoted by: $\eta(\alpha_{pi}),\ \sigma(\alpha_{pi})$.

The following steps describe the way this preorientation parameter is used on-line:

-For each working dish
  -Calculate θ_p
  -For each sector *i* (*1...3*)
    -Apply the template matching process in the angular interval (*confidence interval*):
$\delta_i = [\theta_p + \eta(\alpha_{pi}) - K\sigma(\alpha_{pi}),\ \ \theta_p + \eta(\alpha_{pi}) + K\sigma(\alpha_{pi})]$
    in order to get the optimum orientation, θ_i, for each sector.

(*K* must be chosen, from the statistical analysis, in such a way that θ_i is included in δ_i with a probability of 96%-99%).

Experiments show that typical angular apertures are ±25°. Thus, the use of the preorientation mechanism can reduce the execution time by a factor of 6.

3.3. Other Confidence Measurements

Similar confidence intervals are also obtained, during the off-line analysis, for other significant variables:

-Radial position of each sector. The interval gives the necessary aperture in order to cover small radial displacements of the sector.

-Binary correlation measure. After calculating the optimum matching for each sector, if this optimum value is outside the corresponding confidence interval, the sector must be rejected as not been reliably located (the confidence interval for this variable has its upper bound at infinity).

-Angular interdistances between sectors. For the sectors passing the previous test, another reliability check is made: the angular interdistance for sectors *i* and *j*, θ_i-θ_j , must be inside another off-line-determined confidence interval. Imagine, for instance, that the angular distance between sectors 0 and 1 and between

sectors 0 and 2 is likely to be erroneous, while the angles between sectors 1 and 2 is inside the interval, then it can be assumed that sector 0 has been mislocated.

3.4. Location and Orientation Algorithm

The on-line location and orientation algorithm can be described as follows:

1. - Estimation of the geometric centre of the dish (section 3.1).
2. - Independent estimation of the three chosen sectors (two-stage orientation estimation mechanism, described in section 3.2).
3. - Confidence tests for each sector and for pairs of sectors (section 3.3).
4. - Estimate the orientation of the dish, from the θ_i angle of the sector, among those which passed tests in step 3, with maximum binary correlation. The correlation measures must be normalized with the standard deviation, before the comparison is done:

$$\frac{c_i - \eta(c_i)}{\sigma(c_i)}$$

where c_i is calculated as indicated in Eq. (1).

4 Grey-Scale Image Comparison Test: Quality Test

4.1. Description of the Comparison Procedure

We want to implement a quality control system based on visual data of the pieces to be inspected (test pieces). Quality is measured by comparing the work pieces with a non-defective pattern image. However, this comparison should be flexible enough to allow slight distortion respect to the reference image, as a human being could do. Severe distortions should instead be considered as defects, and the piece must be rejected.

The final objective of this work is to detect significant defects, and not detail defects. In the frequency domain, the detail information of an image is typically associated with high spatial frequency components. Taking advantage of this fact, the proposed algorithm uses a percentage of the low frequency component of the spectrum to carry out the image comparison (see figure 5).

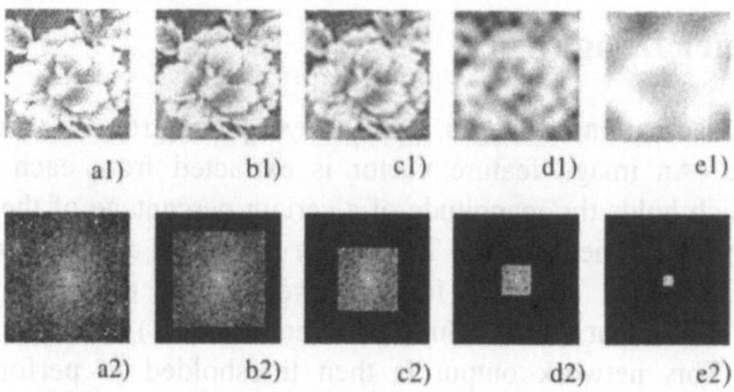

Fig. 5. - Original image of a region of a ceramic dish (a1) and images reconstructed by recovering a percentage of the low frequency spectrum: (b1) 75%, (c1) 50%, (d1) 25%, (e1) 10%. Lower row images represent correspondent spectrums employed in the recovering.

More precisely, comparison is done by using the magnitude of the low frequency components. Phase information is discarded to make use of the translation invariance property of the Fourier transform; that is, for any position in the image, (x_0, y_0), the following equation can be written:

$$F(f(x,y))(u,v) = F(f(x+x_0, y+y_0))(u,v)$$

This property allows comparing images directly in the Fourier domain, with no need of a previous registration of the images to be compared [5]. This preadjust should be necessary if comparison were accomplished in the spatial domain. The reason is that location and orientation algorithms are not so accurate as it would be desirable, so the image windows to be compared may not overlap exactly and comparison will fail.

It can be easily shown that the frequency-domain comparison procedure gives better results if it is made in several small regions separately, instead of doing it in the whole image. Windows of 64x64 pixels size have been used in this work (see figure 6).

Two added advantages to this windows comparison scheme are speed and parallelizing facilities: It's faster to compute n^2, NxN Fourier transforms (imagine n^2 comparison windows) than just one $nNxnN$ transform (the whole image).

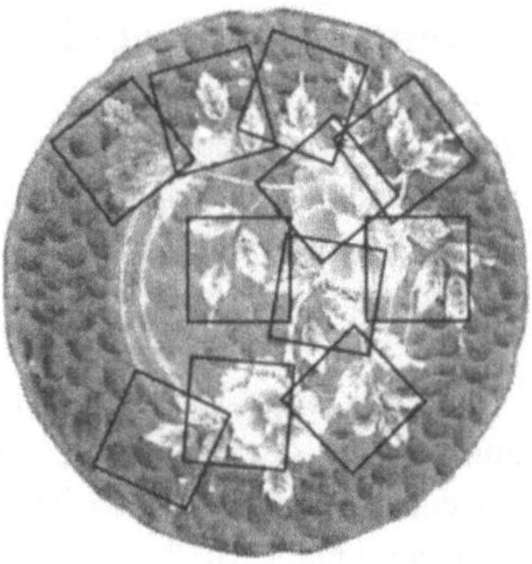

Fig. 6. - Example of eleven 64x64 size, manually selected sectors checked for the quality test.

4.2. Classifier Design

To perform classification of images, a multi-layer perceptron neural network (MLP) was employed. An image feature vector is extracted from each window to be compared, which holds the magnitude of a certain percentage of the low frequency components in the Fourier domain. This vector is used as a natural input to the first layer of the network, so after forward propagating this input we obtain a classification coefficient (with values between 0 and 1) from the single neuron output layer. This network output is then thresholded to perform the desired classification into two classes: defective and non-defective patterns.

The network was trained by using the well-known backpropagation algorithm. A momentum was also added to speed up convergence.

Well established results about the discriminatory power of these networks [6], permit us to assure that 2 or more hidden layers are necessary to perform the classification in the problem at hand. This is because of the presumably non-defective patterns distribution on the *n*-dimensional feature space (where *n* is the number of feature vector components or the number of input neurons). Non-defective images should be similar enough to each other, to provide a well-localized feature vector distribution in the feature space. That is, a single closed n-dimensional surface in the feature space should content most of all possible non-defective patterns, letting all other defective ones outside this 'classification hypersurface'.

This kind of complex classification hypersurfaces can only be implemented on a MLP neural network by using 2 or more hidden layers. In theory, 2 hidden layers suffice to perform any classification problem, no matter how complex it can be. The number of neurons on each layer only limits the complexity. Some other theoretical results may give us some hints about the optimum number of hidden units. The number of neurons in the first hidden layer is related to the number hyperplanes which compose the classification hypersurface, and neurons in the second hidden layers are related to the number of non-connected hyperregions of single pattern type. This last result permit us assure that only 2 neurons in the second hidden layer will be probably necessary to accomplish classification. In this case, the number of hidden neurons was experimentally optimized, as we will see later.

5 Experimental Results

Some experiments on real dishes were made to test performance of the proposed algorithm. A two hidden-layer neural network was used and different percentages of the low frequency spectrum, as well as a variable number of neurons on each hidden layer were tested. Also different threshold values were tried.

The training pattern set was built as follows: one 64x64 pixels window was manually selected from a non-defective reference dish image. This window was located on every image of a set of 100 pictures of the same dish at different positions and orientations on the camera frame, by using the algorithms already described. This way, we got 100 64x64 pixels windows of the same region of a dish void of defects, which were logically used as non-defective training patterns.

Defective patterns were more difficult to obtain, so we decided to generate them by distorting non-defective ones. Two different distortion models were used: the first consists of a certain random non-linear deformation of images in a way they can not be further considered as acceptable images. The second one consists of some kind of random dotting or occlusion of small zones of the images, simulating normal defects on the dishes we are working on. Figure 7 shows examples of these two models of distortion. This way, 100 defective patterns were generated, one for each non-defective image, so finally 200 patterns were used to train the network. A different 200-pattern set for the same dish region was also generated to test network performance.

In the experiments, defective training patterns were generated using both models of distortion, which is the most interesting case for practical purposes. A second hidden layer with 2 neurons was tested. Figure 8 shows the percentage of correct pattern classification with different number of neurons in the first hidden layer. The best classifier is obtained with 90 hidden neurons and using 50% of the low frequency components in the comparison. With this selection, the optimum classification value was of 91%.

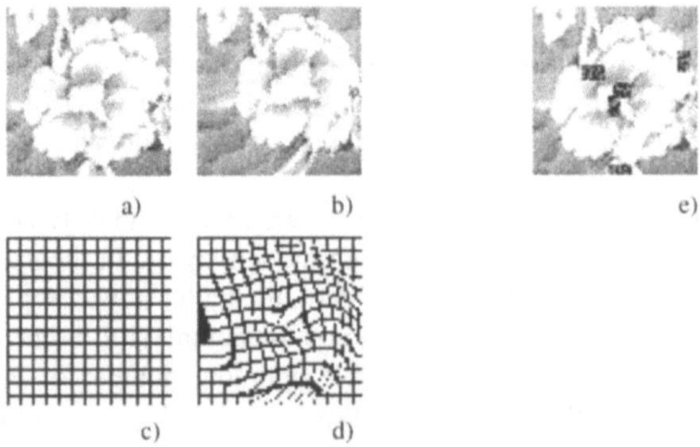

Fig. 7. - Example of a region of a non-defective dish (a), and two kinds of distortions over it: non-linear deformation (b), and random dotting (e). Figure (d) shows the same non-linear deformation in (b), but applied over a regular square pattern (c).

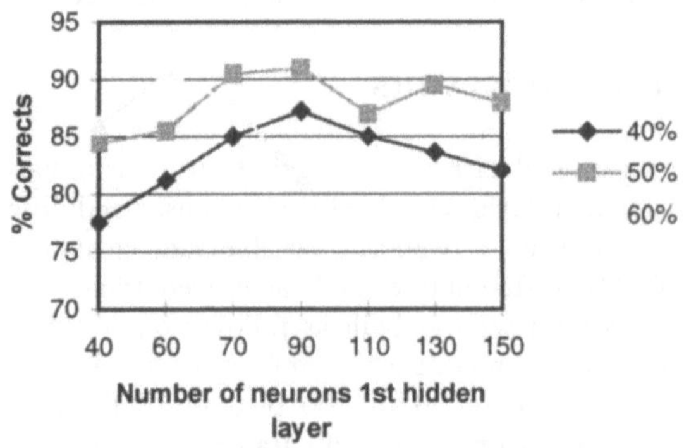

Fig. 8. - Percentage of correctly classified patterns using 2 neurons on the second hidden layer.

Figure 9 shows the percentage of patterns correctly classified as a function of the spectrum threshold value. It can be observed that the best threshold is between 0.2 and 0.3 and that the classification results are relatively insensitive to this threshold value around its maximum. This is a very good property since classification threshold does not become a critical parameter for the system performance.

It is also worth noting that classification performance presents a maximum for a certain number of hidden units on the first layer. This agrees with some theoretical results, which demonstrate that an over-dimensioned hidden layer reduces generalization abilities of these kind of neural networks.

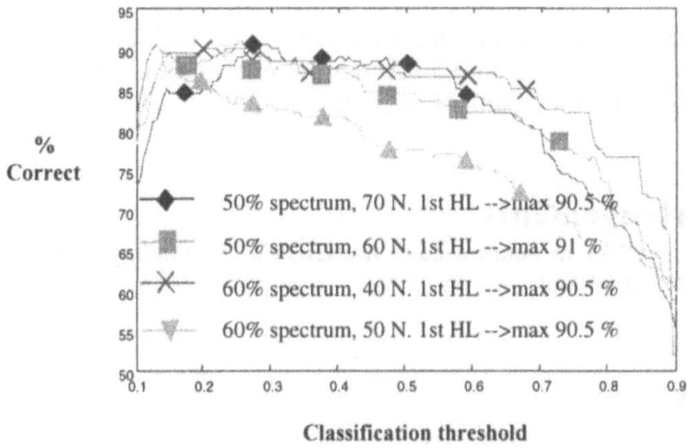

Fig. 9. - Classification curves for the best topologies with 2 neurons on the 2nd hidden layer.

6 Implementation

All the proposed algorithms were tested on a system based on the TMS320C40 DSP from Texas Instruments. The system included an 8-bits grey-scale image acquisition card controlled by a C40 DSP, and 4 other DSP's which provided parallel execution abilities. The whole system, mounted over two HEPC2-M motherboards [7], was connected to the expansion bus of a PC/AT. A uniform lighting system was also provided consisting of several fluorescent lamps, and 400x300 grey scale images of dishes were used to test the system performance.

Algorithms were parallelized by using a star topology where a central node controls executions and information exchange between processors. External processors behave like slaves of the central control node and just process the information transmitted from it. A complete cycle time between 2 and 3.5 seconds, depending on the complexity of the dish to be inspected, were obtained with this architecture (see table I). Approximately, up to 30 different comparison windows can be inspected without increasing execution time. If more comparison windows are needed, execution time increases proportionally.

Process	Execution Time (ms)
Image acquisition + filtering + binarization	300
Geometric centre + principal inertia axis computation	119
Window FFT + Neural classification	110
Template matching orientation	1000-2500

7 Conclusions

A system for automated quality inspection of ceramic dishes is presented. A neural network based classifier has been trained obtaining an optimum performance of more than 90% correctly classified dishes.

The system has been parallelized on a commercial DSP based system to get reasonable execution times (2 to 3.5 seconds for each piece). The system has an acceptable performance for medium-low complexity pieces.

8 Acknowledgements

The authors would like to acknowledge the CICYT for funding this work under TAP95-307 and TAP96-884.

References

[1] González R.C. and Wintz P. 1987 *Digital Image Processing*, Addison Wesley.

[2] Vivas C. and Gómez-Ortega J. 1997 *Automatic Visual Inspection System for Ceramic-Dish Quality Control. Implementation on a DSP-based Hardware*, ISA Internal Report (in Spanish), Seville, July.

[3] Goshtasby A. 1985 *Template Matching in Rotated Images. IEEE Trans. Pattern Anal. Machine Intell.*, vol. PAMI-7, May.

[4] Liao S.X., Pawlak M. 1996 *On Image Analysis by Moments, IEEE Trans. Pattern Anal. Machine Intell.*, vol. PAMI-18, March.

[5] de Castro E., Morandi C. 1987 *Registration of Translated and Rotated ImagesUsingFinite Fourier Transforms. IEEE Trans. Pattern Anal. Machine Intell.*, vol. PAMI-9, pp 700-703, Sept.

[6] Haykin, Simon 1994. *Neural Networks. A Comprehensive Foundation*, Macmillan.

[7] Traquair Data Systems, *HEPC2-M TIM40 Mother-Board User's Manual*, 1995.

10
Grafcet : Methodological and Formal Issues

J. Zaytoon, G. Villermain-Lecolier

1 Introduction

Grafcet or function charts for control systems is an international standard used for the specification and the implementation of logic controllers in manufacturing systems [1], [2]. This model is the basis of the Sequential Function Charts (SFC) international standard used for the implementation of logic controllers. Throughout the twenty years that have passed since it was defined [3], Grafcet is becoming widely used in the industry [4] and in education [5]. The main contribution of Grafcet, which uses a Petri-net like formalism, is that it allows a clear modelling of inputs and outputs and of their relations. It also allows modelling of concurrency and synchronisation. This makes Programmable Logic Controllers (PLCs) more tractable and simplifies the simulation of the control logic of the system. Many PLC builders today use the Grafcet as a specification and/or as a programming language. Among the large companies using it widely or recognising it as an internal standard are: Siemens, Renault, Peugeot, Michelin, and others. However, and in spite of its advantages, Grafcet has long been criticised because it is not supported by a formal foundation that allows to insure correctness and safety requirements on the one hand, and because it lacks adequate methodology that allows an efficient development of high quality models in the case of complex systems on the other.

The objective of this contribution is to review the current active research works undertaken recently to overcome the above problems. These works include:

syntactic and semantic extensions to Grafcet (section 2), development methodologies (section 3), verification (section 4), and control synthesis (section 5). They benefit from the recent advances in software engineering, in manufacturing-systems engineering, in discrete-event systems theory, and in formal methods for real-time systems. The authors' contribution will be particularly emphasised.

2 Syntax and semantics

Grafcet is a discrete-event modelling tool that integrates the ability of Petri nets for concurrent modelling, and the use of variables and Boolean functions to represent complex decisions. These features, together with its simplicity to represent the behaviour of control systems and its normative character, explain its wide industrial implementation. Grafcet consists in describing parallel and synchronised sequences of elementary operations applied to the plant with due consideration to plant's response. The basic concepts of this model are quite clear and simple : the *step*, the *action*, the *transition* and its associated *receptivity*. Five rules were originally used to describe the evolution of a given Grafcet [2].

Many extension have also been introduced in the literature to consolidate the modelling power of Grafcet. These extensions mainly include the use of different types of actions and the introduction of macro-steps and partial Grafcets with forcing orders hierarchy [6]. Recent extensions have also been developed to provided Grafcet with hierarchical structure, abstraction levels, and reuse facilities [7], [8]. These facilities allowed, for example, to extend Grafcet application scope to the control of batch systems.

The application of the five evolution rules of Grafcet results in changing a Grafcet situation into another situation. This new situation may be unstable (if some transitions are firable in this situation) and must change again before the occurrence of a new input event. Issues related to reactivity, determinism and stability have therefore been raised and led to the definition of a consistent temporal framework in which the modelling universe of Grafcet is partitioned into an internal and an external time scales that have no common measure [6], [9]. At the external time scale, all changes of the values of the receptivities associated to the transitions are considered as soon as they occur and their consequential reactions (the actions associated with the next stable situation) are perceived as occurring at the same time instant; this ensures the reactivity of the model. At the internal time scale, a reaction involves a number of consecutive internal evolutions whose durations are as small as necessary; the actions associated with intermediate unstable situations do not affect the controlled plant. This behaviour guarantees the determinism of Grafcet since one and only one output scenario (activation and deactivation of actions) may result from an input scenario.

In spite of their simplicity, Grafcet evolution rules and the above temporal framework are not sufficient to guarantee a unique interpretation of a given Grafcet [10]. To remove persisting ambiguities, the semantics of Grafcet has been completed by an algorithmic "Grafcet player" [11] which organises and explains the interactions between the basic evolution rules and the two time scales of the model. The semi-formal semantics provided by this algorithm insure a deterministic interpretation of Grafcet and reinforces its synchronous and reactive nature.

3 Development methodologies

The existing complex relationships between a manufacturing system and its environment and the necessity of meeting the increasing quality and dependability constraints require a modelling framework which is able to link the design of these systems and the development of their logic controllers. Few methodologies have therefore been proposed to provide a framework for the development of Grafcet starting from a high level specification of the MS. Three of these methodologies [12], [13], [14] are functionally based, and two others [15], [16] are object oriented. The two methodologies proposed by the authors are briefly presented below.

3.1 A functionally based methodology

The methodology proposed in [14], which is represented by the self-explaining SADT diagram of Fig. 1, deals with the design of the control system according to three viewpoints:

1- Functional specification to describe the organisation of the manufacturing activities and flows using structured analysis and design technique "SADT" [17]. This technique is applied using top-down hierarchical decomposition which relaxes system complexity. To distinguish the role of the different activities and production flows without ambiguity, a temporal extension of SADT is proposed. This extension is based on the annotation of SADT arrows into material arrows, information arrows, and synchronisation arrows. Synchronisation arrows are further classified according to the interval logic relations [18] to describe the temporal relationships between related SADT activities.

2- High-level co-ordination and execution of SADT activities using Petri nets [19]. Mapping rules are provided to generate the Petri nets from the temporal extension of SADT. These Petri nets are used to simulate and to validate the SADT model since they allow to guarantee a number of safety and liveness properties, and to detect conflicts. They are also used to ensure a modular development of the individual control tasks.

3- Specification of the local control tasks using partial Grafcets corresponding to the local control tasks for individual machines. These tasks need only a partial view of the system state and their execution is co-ordinated by the Petri net.

A co-operation framework is used to ensure a consistent shift from the external behaviour specification point of view to the specification and the implementation of local controllers tasks. This framework is based on the use of a generic Grafcet skeleton for each leaf activity of SADT on the one hand, and the use of a pre-

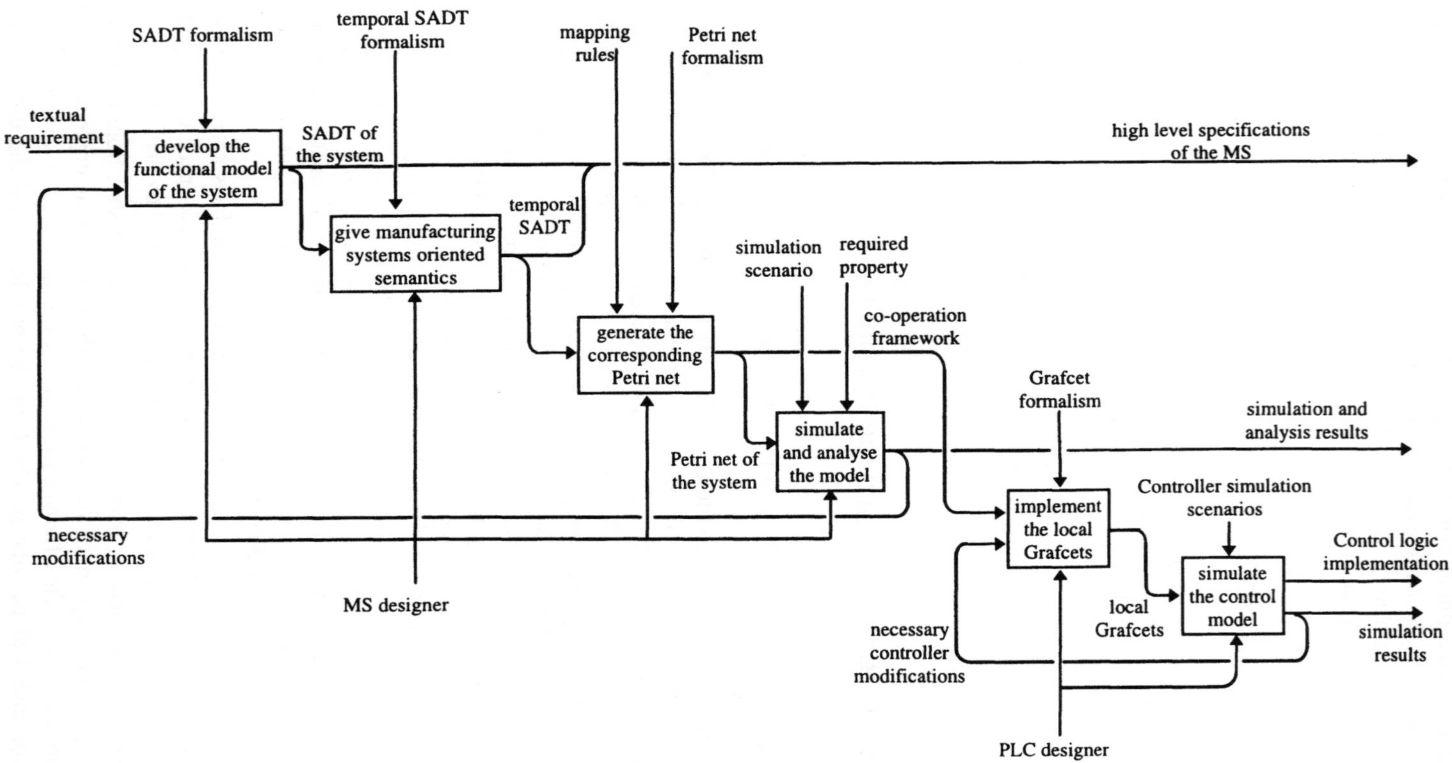

Figure 1. The functional methodology given in terms of SADT.

established communication and synchronisation protocol between a Grafcet and the corresponding Petri sub-net on the other hand. In this way, the PLC designer implements the local control tasks separately; the consistency of the global behaviour of the system can be guaranteed thanks to the formal verification possibilities of Petri-nets. The application of this methodology in a number of manufacturing systems has shown that it provides a useful and practical guide to the development of specific designs of manufacturing-system controllers when the requirements are well identified and the development cycle is relatively short.

3.2 An object-oriented methodology

The methodology proposed in [16] is based on an extension of the object modelling technique "OMT" [20]. This well-established software-engineering technique uses three complementary models (object, functional, and dynamic) well adapted to represent the different views of manufacturing systems. The extension proposed is based on the definition of a generic functional layer on top of the three models of OMT. This layer uses the following mechanisms that are equivalent to those of the object model: functional generalisation, functional aggregation, sequential aggregation and iteration. The use of these mechanisms, both in the object and the functional method of OMT, provides a uniform means to establish a generic functional scheme and a generic structural scheme for each specific manufacturing sector. Starting from such generic schemes, the following steps (sketched in Fig. 2) are used to develop the sequential controller for a given manufacturing system.

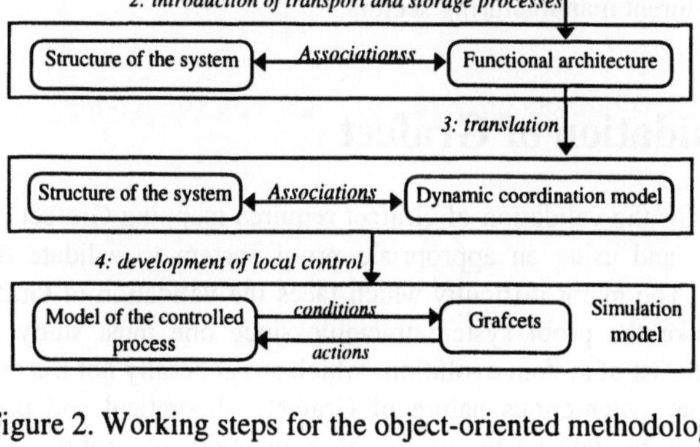

Figure 2. Working steps for the object-oriented methodology.

1- The generic functional scheme related to the appropriate manufacturing sector is instanciated, by graphical selection, to obtain a particular functional scheme corresponding to the sequence of processes required to carry out the target application. One or many objects of the generic structural scheme of the appropriate manufacturing sector must be associated with each of the processes of the lowest decomposition level of the particular functional scheme.

2- Transport and storage processes are introduced, in order to remove the discontinuities between the transformation processes of the particular functional scheme. This results in the definition of the functional architecture of the system. Each of the transport and storage processes must be associated with one or several objects which are instanciated from the generic structural scheme.

3- The dynamic model is automatically generated by means of dedicated translation rules. These rules go through the functional architecture and replace the processes related to an aggregation, a generalisation or an iteration, by a corresponding modular structure. The generated dynamic model corresponds to the specification of the control system at the co-ordination level. The objects controlled by these tasks are determined by the associations that have been established between the generic structural scheme and the functional architecture.

4- The control engineer should next develop the local controller for each control task by designing the partial Grafcet to be executed when the associated state in the dynamic model is active. Each partial Grafcet is used to control the operations and the interactions of the object(s) associated with its control task. The execution context of a partial Grafcet is given by the attributes of the associated objects. The values of input attributes of these objects condition the evolution of the partial Grafcet which in turn updates the output attributes according to its active actions.

This methodology is well adapted to manufacturing systems when the requirements are not well identified in advance or when the designs must be easily adaptable to meet evolving requirements. It was validated using a number of real-sized systems belonging to the manufacturing sector of printed circuit boards. However for a wide use of this methodology, considerable effort will have to be undertaken to establish certified libraries for the generic structural and functional shames for different manufacturing sectors.

4 Validation of Grafcet

A framework for the validation of Grafcet requires mapping Grafcet into a suitable formal model, and using an appropriate proof system to validate the equivalent formal model. The major difficulty which faces the validation of Grafcet is related to how to make the proof system tractable since one must study a state space representing the set of system evolutions which are generally not tractable due to the parallel and the synchronous nature of Grafcet. Theoretical and practical means should therefore be devised to handle the complexity of proofs. Active research work has been carried out recently to establish formal support for the verification (of internal consistency, stability, reinitialisation, deadlock-free) and the validation (of safety, liveness and timeliness properties) of Grafcet.

The approach proposed in [21] is based on the use of Petri nets which are graphically close to Grafcet and hence they are suitable for the mapping of structural aspects of Grafcet. This approach allows to check freedom from deadlock and to establish some invariants for Grafcet. However, it is limited to the validation of a very restricted sub-class of Grafcet in which the receptivities of transitions are limited to single events without any possibility for simultaneous firing of transitions. Generally speaking, Petri nets are not well adapted to validate Grafcet because they are based on asynchronous semantics whereas the Grafcet is a synchronous model. Furthermore, the basic verification procedures of Petri nets no longer apply to interpreted Petri nets which represent the Grafcet-like extension of Petri nets [22].

Synchronous languages such as SIGNAL and ESTEREL have also been used as a support for the validation of Grafcet [23], [24]. In this case, dynamic equations [25] or automata calculations [26] can be used to check stability, freedom from deadlock, as well as a number of liveness properties for synchronous-language programs which are equivalent to Grafcet.

Roussel and Lesage [27] have proposed an efficient algorithm for mapping Grafcet into an equivalent automaton whose states represent the stable situations of Grafcet as perceived at the external time scale. This automaton, which provides semantics to Grafcet in accordance with the interpretation defined in [11], can be checked by the software tool MEC [28] in order to validate safety and some liveness properties related to activation and deactivation of Grafcet steps.

The approach developed in [29] extends validations to timeliness (also called quantitative) properties. In the first place, Grafcet is mapped into temporal automata [30] describing the evolution of a system in terms of a number of clocks. The state space of the resulting automaton is potentially infinite because the clocks range over the real numbers. Symbolic model checking [31], [32], which allows to represent state sets (so-called regions) symbolically rather than enumeratively, is used to verify TCTL-properties of the generated temporal automaton.

Two approaches have also been developed to validate Grafcet together with the plant it controls. The first approach [33], which does not address the semantics of Grafcet properly, is based on the use of TCCS (Timed Calculus of Communicating Systems) / TML (Timed Modal Logic) framework [34]. The second approach [35], which will be presented below, is based on an extension of the TTM (Timed Transition Model) / RTTL (Real-Time Temporal Logic) framework [36]. In these approaches, the integration of a model of the controlled plant is motivated by the fact that the plant represents the primary goal of the automation, whereas the control is only a means to achieve this goal and to ensure that the plant behaves safely and reliably. Therefore, the properties to be verified must primarily refer to states and behaviour of the plant. Furthermore, the use of a realistic model of the plant limits the evolution possibilities of Grafcet to those that can really be reachable when the plant is on-line controlled by Grafcet, and this limits the complexity of proofs.

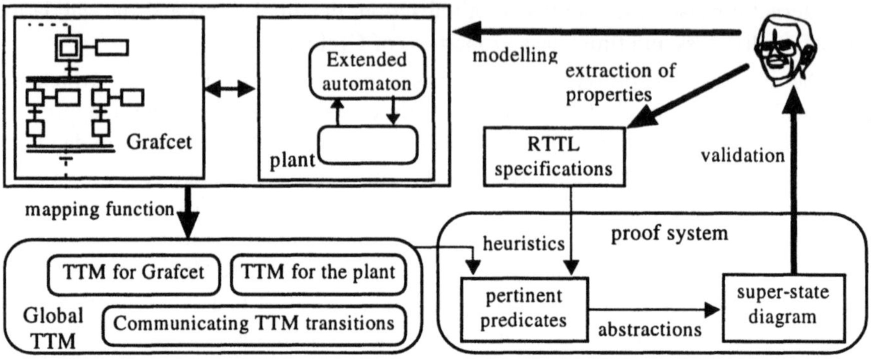

Figure 3. Validation of Grafcet controlled systems.

The approach proposed in [35] is illustrated in Fig. 3. In this approach, the control-system designer starts by modelling the controller using Grafcet, and the plant using extended automaton [37]. A mapping function [38] is used to translate the resulting model into an equivalent TTM, and a dedicated proof system [39] is used for validations. This proof system exploits some intuitive structural properties related to Grafcet and to its interactions with the plant in such a way as to focus only on system executions which are directly concerned with a specified property.

The mapping function insures a correct temporal behaviour of Grafcet in terms of synchronism, determinism and reactivity; it represents a formal definition of the temporal behaviour of Grafcet [11] in terms of the TTM. It also reflects the asynchronism and the non-determinism of the controlled plant. The TTM was chosen because of its flexibility to integrate synchronous and asynchronous behaviours within the same modelling framework. Furthermore, the TTM is based on the use of temporal intervals which are quite useful for the modelling of plant evolution times, since these times cannot be precisely known during the specification phase.

The dedicated proof system is based on the verification of a concise super-state diagram representing an abstraction of system behaviour in terms of a number of pertinent predicates for the property under study. These predicates allow to preserve causality and certain temporal relationships between system evolutions related to the property. A pertinent predicate corresponds to an aggregation either of states of the plant, or of steps of Grafcet. Such a predicate is determined by heuristics which are based on the structure of the modelled system and on plant-controller interactions. Safety properties can be proved as well as a class of liveness and timeliness properties. These properties are expressed using the RTTL. This approach has also been extended to verify hybrid systems whose discrete controllers are specified using Grafcet [40], and the plant continuous dynamics is approximated by means of timed transition models.

5 Control synthesis

The large success of Grafcet has contributed to extending its use from the specification to the implementation of logic control. However, the technological constraints related to the software and hardware characteristics of the target PLCs or PCs entail a behavioural difference between the specification model and the corresponding implementation. Formal synthesis procedures are therefore required to obtain a correct and optimal control implementation for a given Grafcet.

Two approaches, which uses Grafcet as a control model in the framework of the supervisory control theory [41], have been proposed [42], [43]. These approaches are based on the synthesis of a control implementation that represents the minimal possible restriction of the behaviour of a given Grafcet, in a way to satisfy the given safety and liveness specifications and to assure non blocking behaviour of the closed loop system. The difficulty that faces these approaches is mainly related to the semantic difference between Grafcet model (based on conditions, events, logic operators, double time scale interpretation, synchronism, reactivity, possibility of simultaneous actions and simultaneous transition firings), and the formal model of the supervisory control theory (based on asynchronism, particular events interpretation, controller-plant interaction).

The first approach [42], which has been successfully applied to a real size industrial system, can only handle a sub-class of Grafcet in which the logical expressions of Grafcet transitions are limited to single events, representing the edges of input variables. This approach does not consider the constraints induced by the controlled plant.

The other approach [43], [44] attempts to take into account all the features of Grafcet as defined in the [2]. This approach (Fig. 4) is based on six steps which are described below.

1- Modelling the plant behaviour and the safety and liveness user requirements, using automata. Then, modelling the required control behaviour using Grafcet. The automata representing the plant behaviour correspond to spontaneous event generators [41]. Controllable events Σ_c are associated to the activation and deactivation of Grafcet actions. These events are controllable because it is possible to prevent their occurrence by an appropriate conditioning of Grafcet actions. Uncontrollable events Σ_u are initiated by the plant. These events, which cannot be disabled by control action, are associated to the rising and falling edges of Grafcet inputs.
2- Extraction of the graph of reachable situations (GRS) of Grafcet by applying the algorithm given in [45]. The GRS is a uniform and completely specified Mealy machine whose states correspond to the stable situations of Grafcet to which the active outputs are associated. The transitions of GRS are given by logical expressions combining events and conditions.
3- The supremal language of the supervised plant is next obtained according to the classical synthesis algorithm [46]. The supervisor realisation which is used here is that of a discrete event system S, in which the enabling/disabling action of the

supervisor is implicit in the transition structure of S. Therefore, the transition structure of S corresponds to the maximum non-blocking allowable behaviour of the controlled plant with respect to the given safety and liveness specifications.

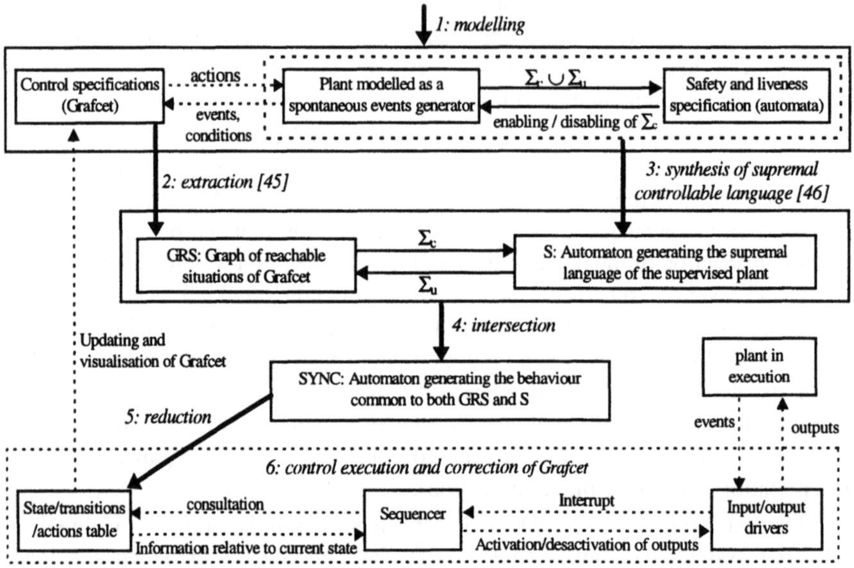

Figure 4. Synthesis of an optimal control implementation

4- The sequences of events which can be generated both by GRS and by S are extracted by intersection of the two corresponding languages. This language intersection is achieved by a dedicated synchronisation operator [43] that goes through the automata GRS and S to construct the automaton SYNC. A transition labelled with a controllable event is included in this automaton when S accepts this event, provided that the event is not in contradiction with the current output state of GRS. Transitions labelled with uncontrollable events are included whenever S can generate these events, irrespective of whether GRS can accept them or not.
5- The fifth step is used to generate the maximum non-blocking execution controller by removing all the blocking situations from SYNC as well as the evolutions which may lead to these situations. The resulting automaton represents the most permissive sub-set of the behaviour of Grafcet which is both non-blocking, and satisfies the supervisory specifications. This automaton is implemented as a state/transition/actions table.
6- The last step consists in executing the generated optimal control by means of an interrupt driven sequencer evolving according to the state/transition/actions table. During execution, the information relative to the current entry of the table is used to update the Grafcet's situation and to highlight the corresponding actions that are disabled in the table. These desactivated actions represent the corrections (or the restrictions) induced by the synthesis procedure. The

resulting control implementation is event driven: when an interrupt occurs, only a limited number of events corresponding to the current state of the table are tested to determine the next state of the controller. A maximum level of reactivity is guaranteed in this way because no further calculations are needed to determine control evolutions,

The developed Software tool includes an on-line part corresponding to step 6 and an off-line part implementing the other steps. The off-line part integrates edition facilities, the software tool AGGLAE [45] for the generation of GRS, and other modules developed using C++ to implement the synthesis, intersection and reduction procedures.

6 Conclusion

Active research work has been carried out recently to provide solutions to specific problems related to the practice of Grafcet which is a widely used model for the specification of sequential controllers. Syntactic and semantic extensions have been proposed in the literature to improve the modelling power of Grafcet, and to provide formal support that guarantees the synchrony, reactivity and determinism of Grafcet. Methodologies have also been developed to link the design of manufacturing systems and the development of their controllers given by Grafcet. This allows to derive the control specifications starting from high level functional and/or structural system models. Once these specifications are established, it is important to validate their consistency and their conformity to user requirements. Various formal methods have therefore been proposed for the validation of different types of safety, liveness and timeliness properties related to Grafcet. A more challenging problem is that of providing a formal framework for the automatic synthesis of an optimal control implementation, starting from a given Grafcet and a number of user requirements. Such a framework has been recently established.

An important effort remains to be undertaken to develop an integrated framework that captures the benefits of the works carried out separately alongside each of these four areas, and to develop an integrated software tool that supports the implementation of such a framework.

7 References

1. David R 1995 Grafcet : A powerful tool for specification of logic controllers. *IEEE Trans Control Sys Technology* 3: 253-268
2. IEC 1988 *Preparation of function charts for control systems.* International Electrotechnical Commission: Publication 848.

3. AFCET 1977 Normalisation de la représentation du cahier des charges d'un automatisme logique. *Final Report of AFCET Commission, J Automatique et Informatique Industrielle*. no. 61-62.
4. Baracos P 1995 Automation engineering in north America. *In: Proc. 2nd. Int Conf Indust Automation*. Nancy, France, pp. 3-8.
5. Kheir N A, Aström K J, Auslander D, Cheok KC, Franklin GF, Masten M, Rabins M 1996 Control systems engineering education. *Automatica* 32: 147-166.
6. UTE. 1993 *Function charts GRAFCET - extension of basic principles*. Union technique d'Electricité : Publication UTE C03-191.
7. Arzén KE 1994 Grafcet for intelligent supervisory control applications. *Automatica* 30: 1513-1525.
8. Duméry J J, Faure J M, Frachet J P, Lampérière S, Louni F 1996 A tool for the structured modelling of discrete events systems behaviour : The Hypergrafcet. *In Proc. CESA'96 IMACS/IEEE Symposium on Discrete Events and Manufacturing Systems.* Lille, France, pp. 519-524.
9. Frachet J P, Colombari G 1993 Elements for a semantics of the time in Grafcet and dynamic systems using non-standard analysis. *A.P.I.I.* 27: 107-125.
10. Lhoste P, Panetto H, Roesch M 1993 Grafcet: from syntax to semantics. *A.P.I.I.* 27: 95-105.
11. Lhoste P, Faure J M, Lesage J J, Zaytoon J 1997 Comportement temporel du Grafcet. *European Journal of Automation* 31: 672-689.
12. Chapurlat V, Giaccone T, Monneret G, Pereyrol F, Prunet F, Simottel D 1993 A structured model for specification of discrete events systems: The ACSY model. *A.P.I.I.* 27: 65-80.
13. Morel G, Lhoste P 1993 Outline for discrete part manufacturing engineering. *In: Proc. 7th. Annual European Computer Conf.* Paris, France, pp. 146-155.
14. Zaytoon J 1996 Specification and design of logic controllers for automated manufacturing systems. *Robotics & CIM* 12: 353-366.
15. Wilczynski B K, Wallace B K 1992 OOPS in real-time control applications. In: S. Adiga, (ed) *Object oriented software for manufacturing systems.* Chapman & hall, London, pp.194-229.
16. Zaytoon J, Villermain-Lecolier G 1997. Two methods for the engineering of manufacturing systems. *Control Engineering Practice* 5: 185-198.
17 Ross D T, Schoman K E 1977 Structured Analysis for requirements definition. *IEEE trans. S/W. Eng*. 33: 86-95.
18. Allen J F 1983 Maintaining knowledge about temporal intervals. *Com. ACM.* 26: 832-843.
19. Peterson J L 1981 *Petri Net Theory and the modelling of systems.* Prentice Hall, London.
20 Rumbaugh J, Blaha M, Premerlani W, Eddy F, Lorensen W 199. *Object-oriented modeling and design*. Prentice Hall, Englewood Cliffs, New Jersey.
21. Aygalinc P, Denat J P 1993 Validation of functional Grafct models and performance evaluation of the associated system using Petri nets. *A.P.I.I.* 27: 81-93.
22. David R, Alla H 1994 Petri nets for modelling of dynamic systems. *Automatica* 30: 175-202.

23. André C, Gaffé D 1996 Proving properties of Grafcet with synchronous tools. *In: Proc. CESA'96 IMACS/IEEE Symposium on Discrete Events and Manufacturing Systems.* Lille, France, pp. 777-782.
24. Marcé L, Le Parc P 1993 Defining the semantics of languages for programmable controllers with synchronous processes. *Control Engineering Practice* 1: 79-84.
25. Le Borgne M, Le Guernic P, Benveniste A 1989 Polynomial ideal theory methods in discrete events and hybrid dynamical systems. *In Proc. 28th. Conf. Decision Control.* pp. 2695-2700.
26. Boudol G, De Simone R, Roy V, Vergamini D 1989 Automatic verification methods for finite state systems. *In: LNCS.* vol 407, Springer, Berlin.
27. Roussel J M, Lesage J J 1996 Validation and verification of Grafcets using finite state machine. *In: Proc. CESA'96 IMACS/IEEE Symposium on Discrete Events and Manufacturing Systems.* Lille, France, pp. 765-770.
28. Arnold A, Bégay D, Crubillé P 1994 *Construction and analysis of transition systems with MEC.* World Scientific Publishing.
29. Marcé L, L'Her D, Le Parc P 1996 Modelling and verification of temporized Grafcet. *In: Proc. CESA'96 IMACS/IEEE Symposium on Discrete Events and Manufacturing Systems.* Lille, France, pp. 783-788.
30. Alur R, Courcoubetis C, Dill D L 1990 Model checking for real-times. *In: Proc. 5th. Annual IEEE Symposium on Logics in Computer Science.* Philadelphia.
31. Alur R, Henzinger T A, Ho P H 1996 Automatic symbolic verification of embedded systems. *IEEE Trans. Software Engineering* 22: 181-201.
32. Henzinger T A, Nicollin X, Sifakis J, Yovine S 1994 Symbolic model checking for real-time systems. *Information and computation* 111: 193-244.
33. Kristensen C H, Andersen J H, Skou A 1995 Specification and automated verification of real-time behaviour - a case study. *In: Proc. 3rd. IFAC/IFIP Workshop AARTC.* Ostend, Belgium, pp. 613-628.
34. Cerans K, Godskesen J C, Larsen K G 1993 Timed modal specifications - Theory and tools. *In: Lecture Notes in Computer Science.* vol 697. Springer, Berlin.
35. De Loor P 1996 Du TTM/RTTL pour la validation des systèmes commandés par Grafcet. PhD thesis, LAM-University of Reims, France.
36. Ostroff J S, Wonham W M 1990 A framework for real-time discrete event control. *IEEE Trans Auto Cont.* 35: 386-397.
37. Ostroff J S 1989 *Temporal logic for real time systems.* Wiley, London.
38. Zaytoon J, De Loor P, Villermain-Lecolier G 1995 Using a real-time framework to verify the properties of GRAFCET. *In: Proc. 3rd. IFAC/IFIP Workshop AARTC'95.* Ostend, Belgium, pp. 233-238.
39. De Loor P, Zaytoon J, Villermain-Lecolier G 1997 Abstractions and heuristics for the validation of Grafcet controlled systems. *European Journal of Automation* 31: 561-580.
40. Zaytoon J, De Loor P, Goeldel C., Villermain-Lecolier 1996 Modélisation et vérification des systèmes hybrides à l'aide du TTM/RTTL. *European Journal of Automation* 30:491-518.
41. Ramadge P J, Wonham W M 1989 The control of discrete event systems. *Proceedings of the IEEE* 77: 81-98.

42. Charbonnier F, Alla H, David R 1995 The supervised control of discrete event dynamic systems: a new approach. *In: Proc. 34th. Conf. Decision and Control.* New-Orleans, USA.
43. Zaytoon J, Ndjab C, Roussel J M 1997 On the supremal controllable Grafcet of a given Grafcet. *In: Proc. 2nd IMACS MATHMOD Conf.* Vienna, Austria, pp. 371-376.
44. Zaytoon J, Ndjab C, Carré-Ménétrier V 1997 On the synthesis of Grafcet using the supervisory control theory. *In: Proc. of IFAC Conf. CIS'97.* Belfort, France, pp. 391-396.
45. Roussel J M 1994 Analyse de Grafcets par génération logique de l'automate à états. Ph.D. thesis, LURPA, ENS-Cachan, France.
46. Wonham W M, Ramadge P J 1987 On the supremal controllable sublanguage of a given language. *SIAM J. Control Optimization* 25: 637-659.

11
A Synergistic Self-Organizing System for Control Chart Pattern Recognition

D.T. Pham and A.B. Chan

1 Introduction

This paper presents a synergistic system which combines Adaptive Resonance Theory (ART2) and the Kohonen Self-Organising Map for control chart pattern recognition. There are 6 principal types of patterns which can be observed in the control charts used in Statistical process Control (SPC). They are Normal, Cyclic, Increasing Trend, Decreasing Trend , Upward Shift and Downward Shift [1][2][3]. All these patterns indicate abnormalities in the process that must be corrected, except Normal. ART2 and the Kohonen Self-Organising Map are two unsupervised neural network paradigms suitable for pattern recognition applications. Though they offer many advantages, they also suffer from a number of shortcomings. The architecture of the synergistic system proposed in this paper is aimed at combining the advantages offered by the two separate paradigms and avoiding their respective shortcomings. Firstly, an introduction is given to the definition of the 6 principal types of control chart patterns mentioned earlier and the various neural networks which have been applied for their recognition in the past. Secondly, a brief review of ART2 and the Kohonen Self-Organising Map is given. The architecture, training and test strategies of the proposed synergistic system and some simulation results are then presented.

Statistical Process Control (SPC) makes use of control charts to assess whether a particular process is functioning correctly. Critical process variables are periodically sampled and the average values of each sample plotted on a "mean" control chart. Abnormalities in the process will manifest themselves as unusual developments in the chart. Effective control is achieved by early detection of such developments and taking corrective actions.

A control chart can exhibit six main types of patterns: *Normal*, *Cyclic*, *Increasing Trend*, *Decreasing Trend*, *Upward Shift* and *Downward Shift*. Figures 1 depicts these 6 pattern types. Correct identification of these patterns is important to achieving early detection of potential problems and maintaining the quality of the process under observation. When a process is operating normally, that is its control chart shows a *Normal* pattern, its variables fluctuate within a range which is consistent with natural noise and randomness. All the other patterns indicate that the process is abnormal and remedial actions are required in order to bring the process back to its normal state.

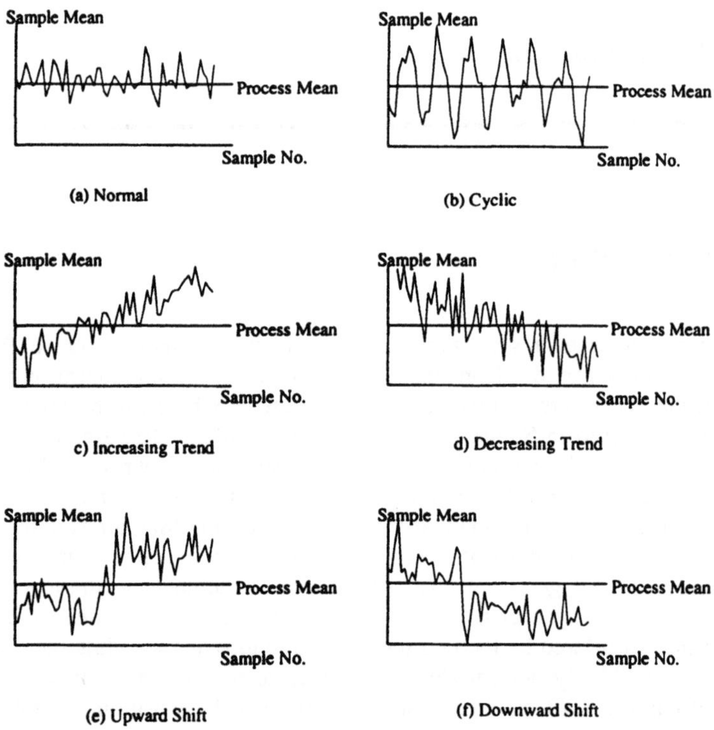

Fig 1.Main types of control chart patterns.

Many neural-network-based schemes have been devised for automatic identification of control chart patterns. Supervised neural networks [4] were successfully employed in [1, 5-8]. However, a disadvantage of using this type of neural networks is that they cannot be easily re-trained to identify new patterns. Due to the inherent randomness in control charts, it is possible that there are patterns not covered by the original data employed for neural network training. In this case, a supervised network would have to be trained again from the beginning using both the original training data and the new patterns to be learnt. This is the classical *stability-and-plasticity* dilemma [9].

The ART2 architecture can, by design, overcome the stability-and-plasticity dilemma. New pattern categories can be learned without affecting or erasing ones which have already been established. This ability relies on a vigilance mechanism which compares the input pattern with the templates of the established classes. If the similarity between the input pattern and a class template is strong enough according to a user-defined criterion, the vigilance parameter, the input pattern will be classified into the class represented by this specific template. The class template will then be updated to incorporate the features of the input vector into its structure. The vigilance parameter safeguards the established class templates from being drastically modified by very dissimilar input patterns, because if such modifications are permitted, the information previously learned by the class templates will be distorted. Although this vigilance mechanism can overcome the stability-and-plasticity dilemma, sometimes it is done at the cost of having a large number of category neurons. This usually happens when the vigilance parameter is set too high. The slightest difference in the input vector will lead to the establishment of a new category. Sometimes, the situation arises in which there exist many established classes but each one of them just accommodates a few patterns. When every input pattern forms a new class, the network gives a trivial solution. It is desirable to have an architecture in which the ability to overcome the stability-and-plasticity dilemma does not result in the creation of trivial class neurons.

The Kohonen self-organising map cannot overcome the stability-and-plasticity dilemma, even though it is also a self-organising neural network. However, the strength of the map lies instead in capturing the topological order embedded in the training data set and hence forming a miniature memory structure of the patterns within it.

In this paper, a synergistic system using a Kohonen self-organising map as a pre-processor and an ART2 network as the end classifier is proposed. The system is given the acronym KART. The KART scheme utilises the memory organisation capability of the Kohonen map to make input vectors more generalised on the basis of the past input vectors learned and hence maintain the accuracy of the ART2 layer without creating many trivial class neurons. Figure 2 shows the structure of a KART network.

2 Training strategy for KART

First, the dimensions of a Kohonen self-organising map have to be specified. This map does not have to be as large as would be required for it to act as a classifier by itself. This map is then trained on a subset of the training data. This subset need not be large but it has to be representative of all the pattern categories in the training data set. The training procedures used for this Kohonen self-organising map are the same as for ordinary Kohonen self-organising maps [10].

Once the weight vectors in the map have become stable, the second stage of training can take place. The training vectors will be presented to the ART2 structure of the KART network through the Kohonen layer. When an input vector $\hat{\underline{\mathbf{x}}}$ is presented, the neuron in the Kohonen layer that contains the weight closest to the input vector in a Euclidean-distance sense will become the firing neuron.

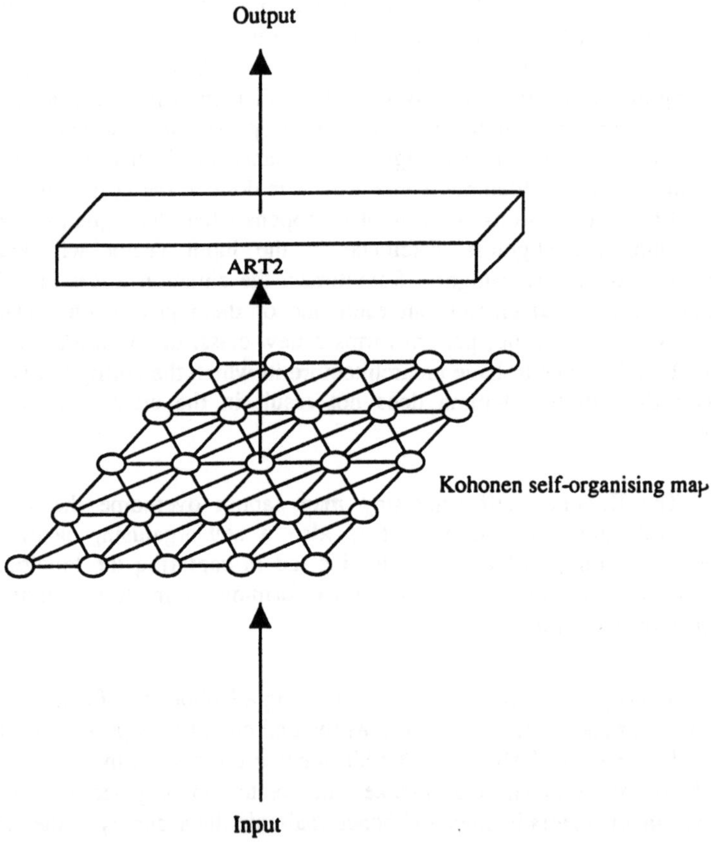

Fig 2. The typical structure of a KART network.

If the weight vector of the firing neuron is $\hat{\underline{\mathbf{m}}}_c$, the input vector $\hat{\underline{\mathbf{x}}}$ will be modified to $\hat{\underline{\mathbf{x}}}'$ according to the following equation before it is passed on to the ART2 layer:

$$\hat{\underline{\mathbf{x}}}' = \frac{\hat{\underline{\mathbf{m}}}_c + \frac{\left\|\hat{\underline{\mathbf{m}}}_c - \hat{\underline{\mathbf{x}}}\right\|}{2}[\hat{\underline{\mathbf{x}}} - \hat{\underline{\mathbf{m}}}_c]}{\left\|\hat{\underline{\mathbf{m}}}_c + \frac{\left\|\hat{\underline{\mathbf{m}}}_c - \hat{\underline{\mathbf{x}}}\right\|}{2}[\hat{\underline{\mathbf{x}}} - \hat{\underline{\mathbf{m}}}_c]\right\|} \tag{1}$$

As previously stated, the Kohonen map serves as a memory. When an input vector is presented, the weight vector in the map that bears the closest resemblance to the input vector will be recalled. This recalled weight vector is not exactly the same as the input vector. It has to be changed into $\hat{\underline{\mathbf{x}}}'$ for inputting to the ART2 layer. This is analogous to when a human is trying to learn something new, he will attempt to relate the new experience to a previous experience. Using equation (1), whenever a new input is applied, it will be related to one of the past inputs learned by the Kohonen self-organising map before it is classified in the ART2 layer. This should reduce the diversity in the training set to avoid trivial classification. The input vector $\hat{\underline{\mathbf{x}}}'$, presented to the ART2 layer, is based on the stored pattern $\hat{\underline{\mathbf{m}}}_c$. If the difference between the input vector $\hat{\underline{\mathbf{x}}}$ and the recalled weight vector $\hat{\underline{\mathbf{m}}}_c$ is marginal, $\hat{\underline{\mathbf{m}}}_c$ will not be modified extensively to produce $\hat{\underline{\mathbf{x}}}'$. On the other hand, if the difference is substantial, then a large amount of modification will be done. This is guaranteed by the updating factor, $\frac{\left\|\hat{\underline{\mathbf{m}}}_c - \hat{\underline{\mathbf{x}}}\right\|}{2}$, in equation (1) which decides by how much $\hat{\underline{\mathbf{m}}}_c$ should be changed (being proportional to the Euclidean distance between $\hat{\underline{\mathbf{m}}}_c$ and $\hat{\underline{\mathbf{x}}}$). Because both $\hat{\underline{\mathbf{m}}}_c$ and $\hat{\underline{\mathbf{x}}}$ are normalised, they lie within a unit hyper-sphere. Therefore, the largest possible Euclidean distance between them is 2. In the extreme case, where $\hat{\underline{\mathbf{m}}}_c$ and $\hat{\underline{\mathbf{x}}}$ are pointing in opposite directions, i.e. they are completely different, the updating factor $\frac{\left\|\hat{\underline{\mathbf{m}}}_c - \hat{\underline{\mathbf{x}}}\right\|}{2}$ will become 1. This results in $\hat{\underline{\mathbf{x}}}' = \hat{\underline{\mathbf{x}}}$ and the input vector $\hat{\underline{\mathbf{x}}}$ will be learned as it is because even the closest weight vector recalled by the memory layer (the Kohonen self-organising map) is still very much different from it. On the other hand, if $\hat{\underline{\mathbf{x}}}$ and $\hat{\underline{\mathbf{m}}}_c$ are not very different, the updating factor will be very close to 0 and $\hat{\underline{\mathbf{x}}}'$ will be very much the same as $\hat{\underline{\mathbf{m}}}_c$. Once $\hat{\underline{\mathbf{x}}}'$ has been formed, learning takes place in the ART2 layer in exactly the same way as in a general ART2 network.

3 Test strategy for KART

The test procedure is the same as the training procedure except that the Kohonen map is not retrained and that no new category neurons will be created in the ART2 layer. An input will be classified into the category neuron to which it is closest as determined by the vigilance test.

4 Simulation results

A set of simulation results with various KART networks are recorded in Table 1.

The Kohonen map for each KART network was trained with a fixed neighbourhood of 3×3. When a firing neuron was decided using the minimum Euclidean distance criterion, the firing neuron and its immediate neighbours would be updated according to the following equation:

$$\underline{\hat{m}}_c(t+1) = \frac{\underline{\hat{m}}_c(t) + \alpha[\underline{\hat{x}}(t) - \underline{\hat{m}}_c(t)]}{\left\| \underline{\hat{m}}_c(t) + \alpha[\underline{\hat{x}}(t) - \underline{\hat{m}}_c(t)] \right\|} \qquad (2)$$

where $\underline{\hat{m}}_c$ is the weight vector of a neuron and $\underline{\hat{x}}$ is the input vector. The updating parameter α followed a linear decay from 1 to 0 with respect to the number of iterations. For each KART network, 10 iterations were used to train its Kohonen front end. A subset of the training data set was used to train the Kohonen parts of the KART networks. From the 366 vectors in the training data set, 90 were chosen for this purpose with 15 vectors taken randomly from each of the 6 pattern classes.

The ART2 layer of the KART networks used the following set of network parameters:

a = 10.0	b = 10.0	c = 0.01	d = 0.99	θ = 0.1

Network	Kohonen map size	Vigilance parameter	ART2 neurons	Classification accuracy	
				trn.dat	tes.dat
1	4x4	0.992	30	72.3%	69.4%
2	4x4	0.996	52	80.2%	77.5%
3	4x4	0.999	68	83.6%	78.4%
4	5x5	0.992	23	79.9%	77.5%
5	5x5	0.996	46	85.2%	80.1%
6	5x5	0.999	60	92.3%	90.7%
7	6x6	0.992	17	78.6%	82.4%
8	6x6	0.996	50	87.8%	84.5%
9	6x6	0.999	52	93.3%	94.1%

trn.dat : classification accuracy when tested with the training data set
tes.dat : classification accuracy when tested with the test data set

Table 1. Classification results for various KART networks.

The results in Table 1 indicate that KART networks generally give good results, both in terms of the number of established category neurons and classification accuracy. The raising of the vigilance parameter in the ART2 layer did improve the classification accuracy regardless of the size of the Kohonen self-organising map. When the latter became larger, the ART2 layer could achieve more accurate classification with a higher degree of generalisation.

5 Conclusions

In this paper, a synergistic system KART, combining the Kohonen self-organising map and ART2, has been proposed. It has been applied to the *control chart pattern recognition* problem and its different properties have been demonstrated.

The KART networks employed the Kohonen self-organising map as the front end and ART2 as the classifier. Inputs were generalised on the basis of past learning before they were classified. KART allowed high vigilance parameters to be used and yet did not lead to excessively many category neurons.

Acknowledgements

The authors would like to acknowledge European Regional Development Fund (ERDF) for their support of the work described in this paper.

References

1. Pham D T, Oztemel E 1996 *Intelligent Quality Systems*, Springer-Verlag, London
2. Pham D T, Chan A B 1997 Control Chart pattern Recognition Using a New Type of Self-Organising Neural Network
3. Chan A B 1996 Control Chart Pattern Recognition Using Self-Organising Neural Networks. PhD thesis, University of Wales Cardiff, UK
4. Haykin S 1994 *Neural Networks: A Comprehensive Foundation.* Macmillan College Publishing Company, Inc.
5 Oztemel E 1992 Integrating Expert Systems and Neural Networks for Intelligent On-line Statistical Process Control. PhD thesis, University of Waels Cardiff, UK
6. Pham D T, Oztemel E 1993 Combining multi-layer perceptrons with heuristics for reliable control chart pattern recognition. *Artificial Intelligence in Engineering VIII*, (eds. Rzeski G, Pastor J and Adey R A), 801-810. Computational Mechanics, Southampton

7. Smith A E, Yazici H 1992 Backpropagation Neural Network Approaches to Process Control: Evaluation and Comparison. Technical Report #90-22-47. Department of Engineering Management, University of Missouri-Rolla.

8. Hwarng H B, Hubele N F 1991 X-Bar chart pattern recognition using neural nets. *45th Annual Quality Congress, American Society for Quality Control*, Milwaukee, 20-22 May 1991, 884-889.

9 Carpenter G A, Grossberg S 1987 ART2: Self-organisation of stable category recognition codes for analog input patterns. *Applied Optics 26(23)*: 4919-30.

10 Chan A B 1996 Self-Organising Neural Networks for Control Chart Pattern Recognition. PhD thesis, University of Wales Cardiff, UK

12

Conveyor-Belt Diagnostic System Using Time-Frequency Analysis

S. Horihata, H. Kitagawa and Y. Shirakawa

1 Introduction

In order to convey coal, ore and other raw materials for iron making, hundreds of conveyor-belts are working in a steel making plant. Because those facilities are weather-beaten, it is very difficult to apply a suitable maintenance service. In addition, conveyor-belts in a steel plant are distributed over a vast outdoor field. Therefore, it is almost impossible to carry out any condition monitoring by operators. In spite of the difficulty in applying a condition monitoring, it is urgently needed to develop an effective method for the detection of conveyor-belt trouble.

In this study, an automatic and centralized diagnostic system of the conveyor-belt is proposed. Since important informations on serious conveyor-belt troubles are included in the electric current of induction motor, which drives all conveyor-belts, the current is measured continuously to detect abnormal fluctuations in it, which are due to serious trouble such as 'slip' and 'meander'. The performance test is carried out at the Kimitsu Works of Nippon Steel co. which is the largest steel making plant in the world.

2 Basic construction of a conveyor-belt

A driving motor of conveyor-belt drives a head pulley. The system is operated by the friction force between pulley and belt, given from a tension weight. Tail and head pulleys support the belt. The make-up of the system is shown in Figure 1.

3 Main troubles of conveyor-belt to be detected

As serious troubles of conveyor-belt are 'slip' and 'meander', those two accidents are selected as main diagnosis items. The breakdown caused by 'slip' occurs, when rain or dust erodes the layer between pulley and belt, and the friction is reduced.

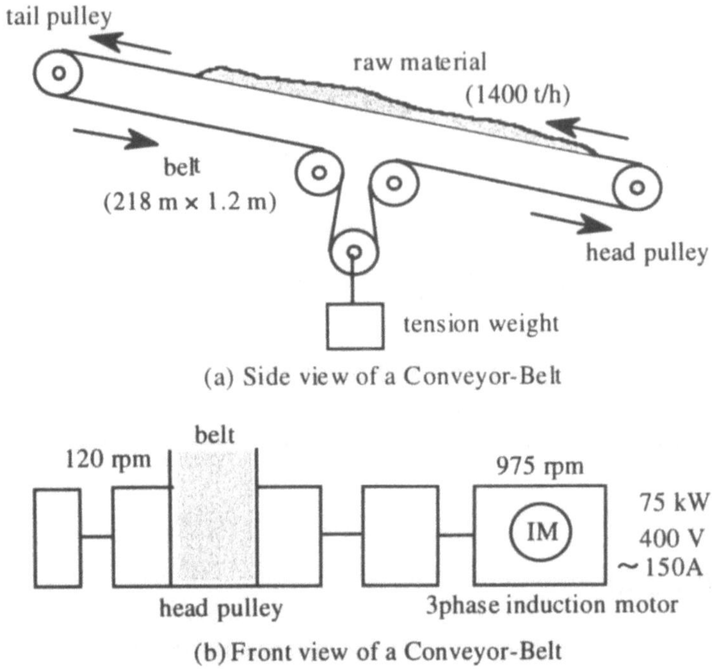

Figure 1: Construction of Conveyor-Belt

This trouble is very dangerous, because a fire accident often brings about soon after the failure occurs. The 'meander' is a phenomenon that a conveyor-belt travels with swinging its load in the transverse direction. It causes such troubles as drop of loads, running off the tracks and breaks in a belt. These troubles must also be found out within a certain time.

4 Basic pattern of electric current of driving motor

The current is measured by a cramp-meter, which is a kind of current transformer, and digitized with the sampling frequency of 10Hz. A variation in current is related to the conveyor-belt operation. By analyzing current signal in detail, a typical pattern of electric current is made clear. Basic patterns are shown in Figure 2.

At the point 'a', a conveyor-belt is started. During starting period of 1-2sec, the current of driving motor flows over 100A. A point 'b' shows no-load state. A point 'c' denotes on-load. A region of the 'd' represents 'meander'. The current fluctuation caused by 'meander' is approximately $\pm$10A and persists for 20-40sec, which is equal to 0.025~0.050Hz. A point 'e' shows overweight condition. A region of the 'f' represents 'slip'. The amplitude of 'slip' signal is over$\pm$25A. At the point 'g', the installation stops.

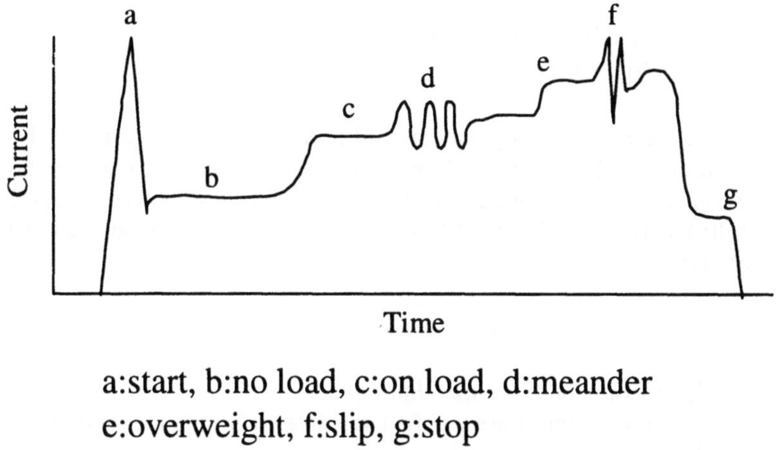

Figure 2: Typical features of motor current

5 Diagnostic system applying the Time-Frequency Analysis

The corollary is composed of two sub-systems. In the pre-conditioning system, the current data that includes both troubles are roughly separated to two categories of the data containing 'meander' and 'slip', respectively. Then the data selected abnormal are send to time-frequency analysis system. In the second system, signals are carefully divided among 'meander','slip' and other phenomena.

5.1 Pre-conditioning system

5.1.1 Classification of presumed 'meander' and 'slip' signals

The signal from 'slip' has very high amplitude, then the presumed 'slip' signal including starting point S(n) is determined by eq.(1) and eq.(2)

$$S(n) \geqq 122 \qquad (1)$$

$$S(n) > S(n-1) \qquad (2)$$

If current signal doesn't satisfy both equations, the signal is assumed to be a presumed 'meander'.

5.1.2 Signal processing of presumed 'meander' signal

As the amplitude of 'meander' signal is over 40A, the signal whose amplitude is lesser than 40A is rejected. Then, to reveal the trend of a current signal more clearly, smoothing takes place by a moving average filter applying eq.(3) and eq.(4).

$$M(n) = \{ S(n-k)+\cdots+S(n)+\cdots+S(n+k) \}/(2k+1) \tag{3}$$

$$d(n)=S(n)-M(n) \tag{4}$$

where S(n) and d(n) are an original data and differential output, respectively. n and k denote discrete values of time. Low frequency components, particularly direct current component, in an original signal are rejected by a pseudo high-pass filter of M(n). If d(n) satisfies eq.(5), the data are sent to second stage system as presumed 'meander' signal

$$10A \leqq |d(n)| \tag{5}$$

5.1.3 Signal processing of presumed 'slip' signal

If following three premises are satisfied, the data are sent to the second system as presumed 'slip' signal

(1) current signal fluctuates over $\pm 25A$

(2) maximum value of current amplitude is over 120A

(3) pattern of the signal is similar to that of start signal

The signals of approximately 120sec period, in which the presumed S(n) is included at the center of the time series, are send to a time-frequency analysis system.

5.2 Time-Frequency Analysis system

In the second system, three kinds of time-frequency analysis are applied.

5.2.1 Time-frequency analysis

A summary of the analysis methods is as follows:

(1) Short Time Fourier Transform, STFT [1]

Spectrogram is defined by eq.(6)

$$ST(\omega, t) = \lim_{T \to \infty} \left| \int_{-T/2}^{T/2} f(\tau) w(t-\tau) e^{-j\omega\tau} d\tau \right|^2 / T \tag{6}$$

where f(t), T and w(t) denote signal, window length and window function, respectively. Spectrogram is the square of STFT. Resolutions of time and frequency are determined by window length T. The higher resolutions of both time and frequency can not be attainable simultaneously. Therefore, a suitable window length must be selected beforehand.

(2) Wigner Distribution, WD [1]

WD is defined by eq.(7)

$$\mathrm{WD}(\omega,t)=\frac{1}{2\pi}\int_{-\infty}^{\infty}f\left(t+\frac{\tau}{2}\right)f^*\left(t-\frac{\tau}{2}\right)e^{-j\omega\tau}d\tau \qquad (7)$$

where $f^*(t)$ denotes the complex conjugate of f(t). WD provides the highest resolution among the three analyses in both time and frequency domains. But application of WD is often complicated by the occurrence of interference terms which are irrelevant to any physical properties of a original signal.

(3) Wavelet Transform, WT [2],[3]

WT of f(t) is given by eq.(8)

$$\mathrm{WT}(a,b)=\int_{-\infty}^{\infty}\psi_{a,b}^*(t)f(t)dt \qquad (8)$$

where $\psi_{a,b}(t)$ is a basic function, called an analyzing wavelet,AW, which is localized in both time and frequency domains. 'a' and 'b' are scale and shift parameter, respectively. The square of WT is equal to scalogram. In this study, Gabor function of eq.(9) is applied as AW, because the concentration of energy in both time and frequency domains is very remarkable.

$$\psi(t)=\pi^{-1/4}\left(\frac{\omega_p}{\gamma}\right)^{1/2}\exp\left\{-\frac{1}{2}\left(t\frac{\omega_p}{\gamma}\right)^2+j\omega_p t\right\} \qquad (9)$$

where ω_p is central frequency of distribution and γ is a parameter which determines the degree of localization in the frequency domain. γ takes the value 5.336 in this analysis.

WT is suitable for the calculation of a certain restricted frequency range of signal, because, for a given scale range, time and frequency analysis can be obtainable by changing the shift parameter. In contrast with WT, both STFT and WD are appropriate to the analysis of a certain restricted time range of signal.

5.2.2 Application of time and frequency analysis

The algorithms of 'meander' and 'slip' detection, applying time-frequency analysis, are summarized.

(1) Detection of 'meander'

From the detailed analysis of 'meander' signal, it become clear that the signal is characterized by the fluctuation pattern in a certain range of frequency. Based on this fact, an algorithm is constructed. To carry out a time-frequency analysis effectively for the restricted frequency range, application of WT is suitable. In the frequency band ranging from 0.025 to 0.050Hz, the instantaneous energy is calculated to classify the failure quantitatively. The instantaneous energy,E(t), which is obtained by integrating the scalogram with respect to the ω, is defined by eq.(10).

$$E(t)=\sum_{\omega=pl}^{ph}P(\omega,t) \tag{10}$$

where P(ω ,t) denotes the scalogram, which is the square of WT, frequency range of integration is between pl and ph .

Noise components are removed from the signal,E(t), by passing through a mediam filter (Figure 3) [4]. A typical pattern of the signal,E(t), which includes both 'meander' and a noise is shown in the upper part of Fig.3. Generally, a noise signal is impulsive and duration time of the signal is much shorter than that of 'meander'. Since a parameter K which defines the window length of the filter is selected empirically to match with the noise signal width, 'meander' signal is distinguish from other signals as shown in the lower part of Figure 3.

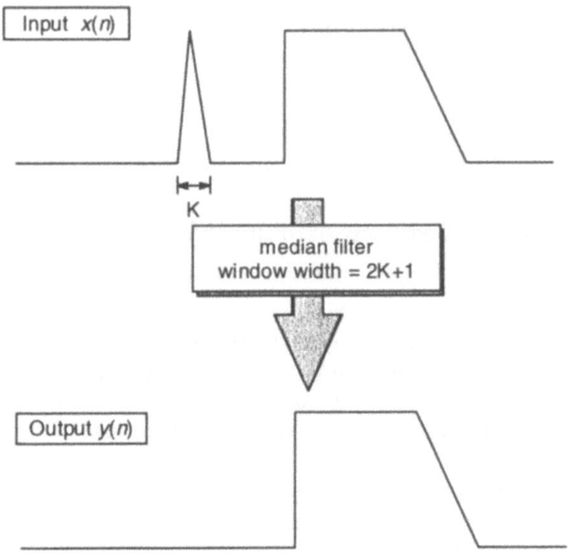

Figure 3: Characteristics of median filter

(2) Detection of 'slip'

As the currents of 'slip' and a start are similar in pattern, start signals are often included in signals which are inferred to be 'slip', and send to the second system. Those signals must distinguish each other.

In the first stage of this system, lower frequency parts in the signals are rejected in order to emphasize high frequency components and to distinguish them more clearly.

By detecting the times of occurrence of both phenomena and by analyzing the difference of energy, or amplitude, of signals, we can draw a clear distinction between 'slip'and the start. Therefore, notice is merely taken of the time-frequency distribution of the neighbor of time of occurrence. Every time the data of window length is taken from an original signal, the signal is analyzed applying either WD or STFT. Therefore both are suitable for the 'slip' detection among three methods. And the calculation takes only a short time. As 'slip' trouble must be detected early, this algorithm is convenient.

5.3 Total system of conveyor-belt monitoring

Conveyor-belt diagnostic system consisting of two sub-system is shown in Figure 4.

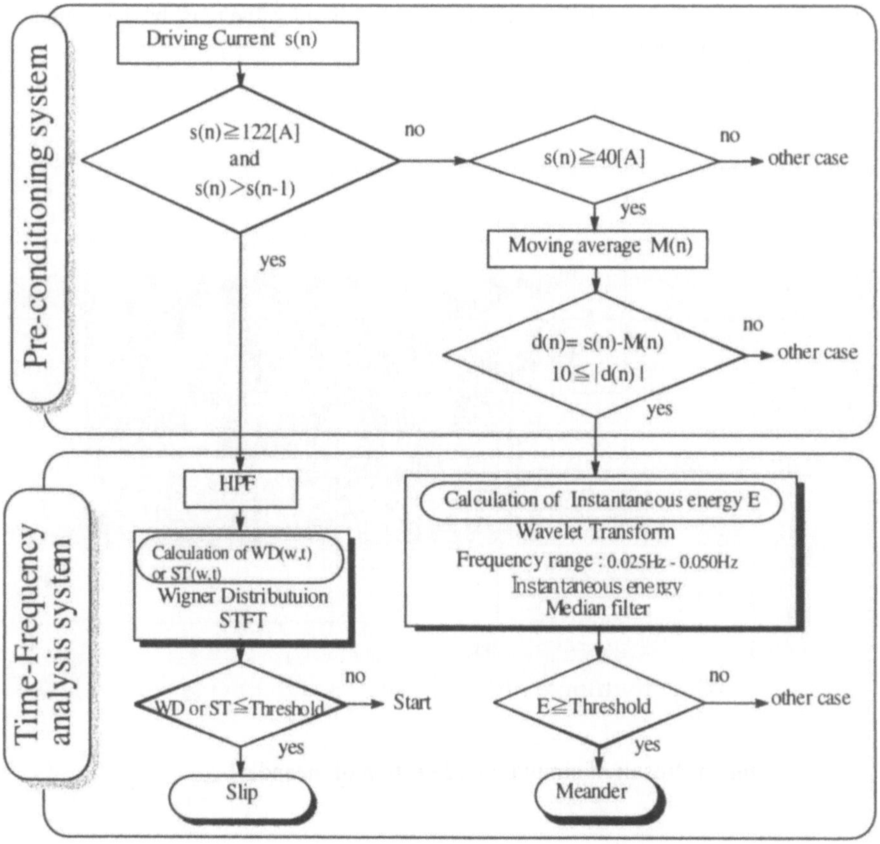

Figure 4: Conveyor-Belt Diagnostic System based on Driving Current Analysis

6 Evaluation of system performance

Field data of approximately 47 hours are measured for several days at the same conveyor-belt and are applied to the analysis.

6.1 Detection of 'meander'

Fig.5 shows output data of mediam filter. Black circles in the figure denote 'meander' and white circles are spurious signals or noise. Threshold values are indicated by cross marks. The marks are connected by a solid line. The threshold value is selected as 95% of minimum width of each 'meander' distribution in this case. This selection leads to an over-estimation level. All 'meander' signals are detected while some spurious signals are also regarded as the accident by the threshold level. When a larger threshold value is selected, information is handled quickly but some 'meander' signals are missed. There is a trade-off between calculation time and signal detectability. The threshold value must be determined by the working condition.

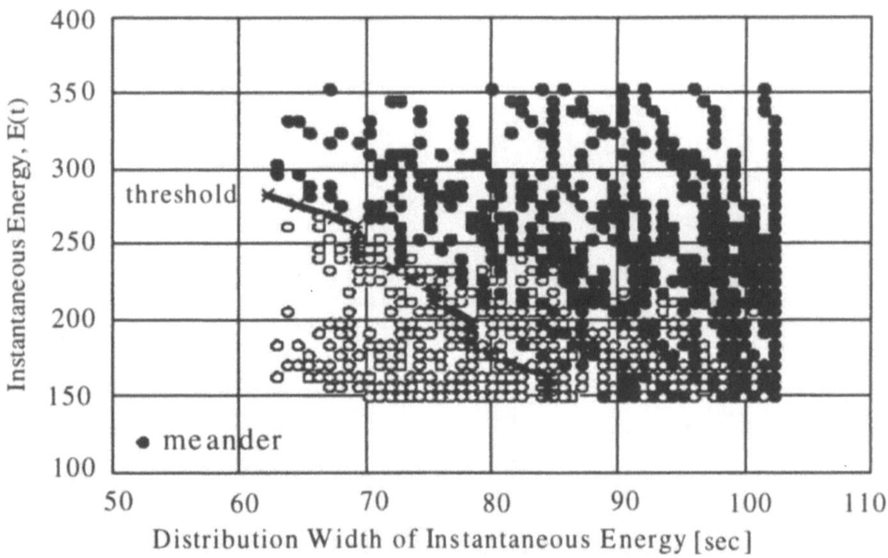

Figure 5: Result of simulation (detection of meander)

6.2 Detection of 'slip'

Fig.6 shows the results obtained by the STFT with window lengths of 0.8sec and1.6sec. Black marks in the figure denote spurious signals or noise, namely star signals and white circles are 'slip'. Because longer window length gives higher frequency resolution in the STFT analysis, a better result is acquired by the result with the window length of 1.6sec. But 'slip's are clearly distinguishable from other spurious signals by the result with the window length of 0.8sec. As a quick detection is desired for 'slip' trouble, a shorter analyzing time is needed. When the window length is selected as 0.8sec, it takes within 0.5sec for total analyzing time.

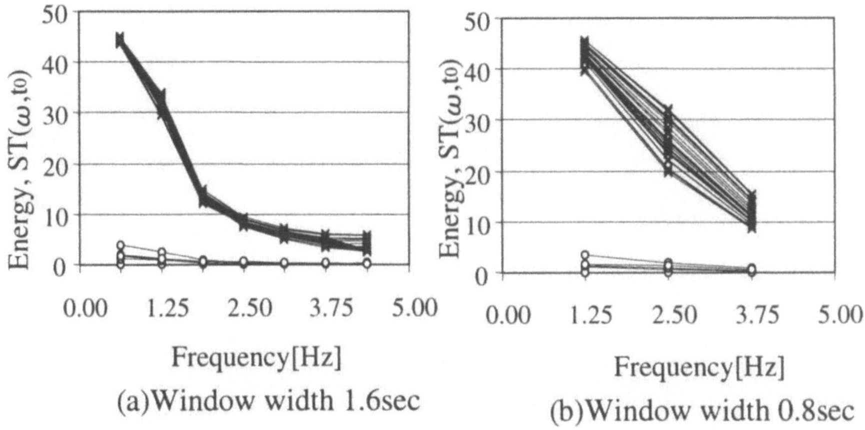

Figure 6: Result of simulation using STFT (detection of slip)

7 Conclusion

An automatic and centralized diagnostic system of conveyor-belt is proposed applying time-frequency analysis for the electric current of driving motor. This system is composed of two sub-system. Those are pre-conditioning system and time-frequency analysis system. In pre-conditioning system, signals from abnormal condition are extracted and those signals are roughly separated into two groups of the presumed 'meander' and the presumed 'slip'. For each group, several pre-conditioning is effectively given and extracted data are sent to the next stage. In time-frequency analysis system, three types of analysis methods, WT, WD and STFT, are used properly, depending on their characteristics. For the detection of 'meander', WT is applied. WD and STFT are adapted to the 'slip' analysis.

It is made clear that the system is very effective to an on-line real-time diagnosis of conveyor-belt operations in steel industries, by applying different time-frequency analysis properly.

REFERENCES

1. Cohen L 1989 Time-Frequency distributions-A review. Proc IEEE 177 7: 941-981.

2. for example, Morlet J, Arens G, Fourgeau I and Giard D 1982 Wave propagation and sampling theory. Geophysics l 47:203-236.
3. for example, Gabor D 1946 Theory of communication. J Inst Elec Eng (London) 93 III : 429-457.
4. for example, Vaseghi S V 1996 *Advanced Signal Processing and Digital Noise Reduction.* WILY TEUBNER, New York.

13 Integration of the Predictive Maintenance in Manufacturing System

J.-B. Léger, E. Neunreuther, B. Iung and G. Morel

1 Introduction

Nowadays maintenance is considered as a key point for the manufacturing system competitiveness because first its cost represents the major part of the operational cost, and second, a system failure can have an important impact on product quality, equipment availability, environment, and operator.

In this context to prevent failure, two main maintenance policies must be applied : the scheduled maintenance and the predictive one. The difference between these two types of maintenance is related to the intervention strategy. Indeed, the strategy of the scheduled maintenance is realised through a planning based on Mean Time To Failure (MTTF) and Mean Time Between Failure (MTBF) whereas the predictive strategy is based on the degradation monitoring, diagnosis and prognosis. The scheduled maintenance generates overcost when the device is replaced while its real life time is not reached, or it generates unavailability of the manufacturing system when the device failed before the theoretical life delay.

One solution to decrease the operational cost and to increase the manufacturing system availability (Figure 1) is to manage continuously all maintenance activities and to control the degradation to move to predictive maintenance [1].

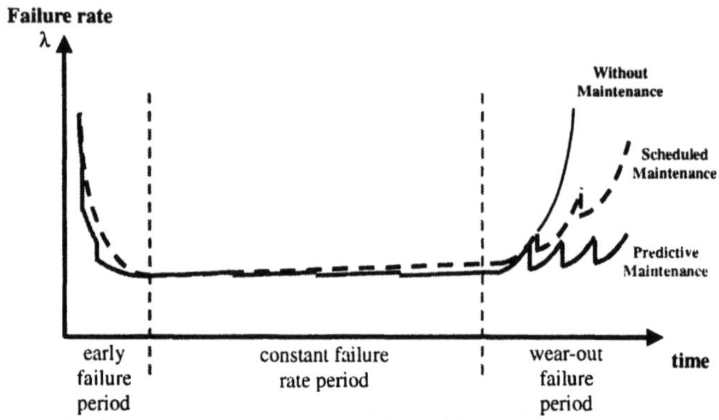

Figure 1: Decreasing of failure rate through predictive maintenance.

2 Predictive maintenance requirements

Predictive maintenance is mainly based on the study of the variations of the system status in terms of pertinent information or data materialising the degradation and the drift behaviour of the system.

The dynamic of degradation describes its past, present and future values, as a control variable, to represent the degradation through time (Figure 2). This representation is required to implement function linked to the predictive maintenance objectives.

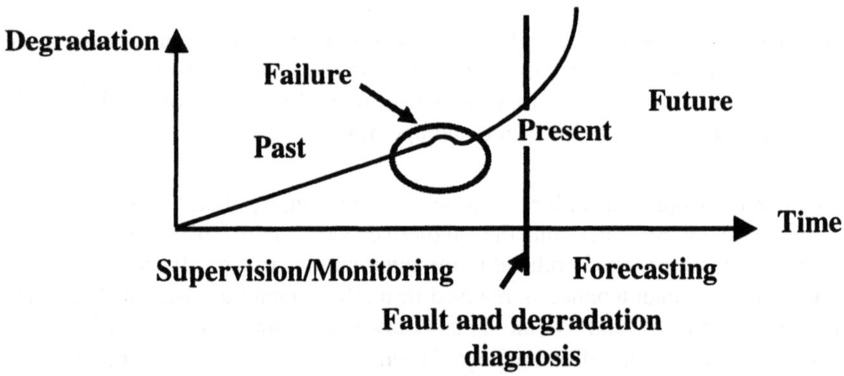

Figure 2: Dynamic of failure.

In this way, predictive maintenance has to control degradation by acting on the degraded system to realise a nominal or a degraded function. This action allows to apply resumption, emergency procedure, reconfiguration, keylocking, redundant functioning...

So, the concept of predictive maintenance is closed to control theory, but by considering that the predictive maintenance goal is to monitor, diagnose, prognosticate online the degradation of manufacturing system to compensate, correct or operate just in time.

To act on the degraded manufacturing system, the predictive maintenance must authorise actions only in total coherence with the manufacturing system behaviour in order to be sure that these actions allow to retrieve a normal state from an abnormal one in safety conditions.

Having a unique and pertinent manufacturing process model integrating functioning and malfunctioning, is therefore a mandatory requirement to better master the dual behaviour of the system (Figure 3).

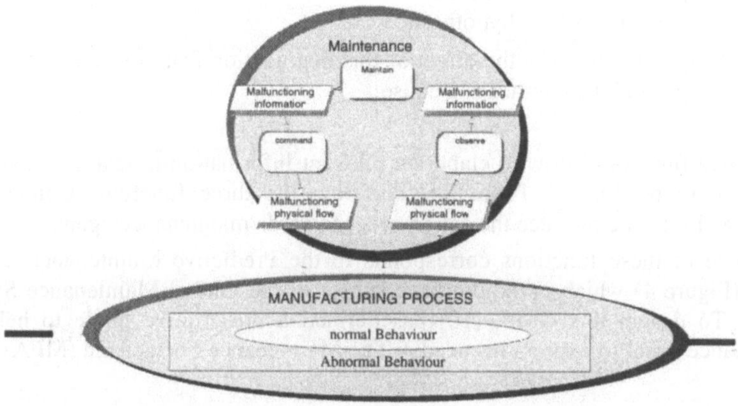

Figure 3: Predictive maintenance automation.

In this way of duality, [2] proposed a theory to formalise the induction of malfunctional behaviour from functional one as explained in the survey of [3] for Manufacturing System. Possible transition between the two normal and abnormal states of a function can be analysed in terms of influence of the function input flow deviation (part, energies, event,...) and of an internal failure on the function failure. Each degradation (drift behaviour) has a specific behaviour resulting from the function input flow value, and each degradation implies difference on the function output flow value (part, energies,...) considered as a symptom of a failure.

3 Predictive Maintenance System for Predictive Maintenance

To support predictive maintenance requirements, the predictive maintenance system must satisfy the whole of previous functionalities related to the dynamic of degradation (monitoring, diagnosis, prognosis, decision, compensation, correction, execution). Although, [4] [5] [6] propose and define the basic concept to integrate the predictive maintenance in maintenance system, there is no reference model today to integrate all the specific predictive maintenance requirements into the maintenance system.

It implies to improve existing scheduled maintenance models (IMMM, IMDS, IMIS, IDS, CIPO, GAMA, VOICE) [7] [8] [9] based on management or corrective maintenance, by integrating the dynamic of degradation related to the past, present and future of degradation value. These three states of degradation are defined from the three following functions :

- ♦ Monitoring produces a degradation value and its behaviour type from observation. The behaviour type of degradation is relevant to understand in what kind of abnormal behaviour the system is.
- ♦ Diagnosis produces the cause of this degradation from its value and type. Diagnosis needs observation to correlate observation and degradation cause hypothesis.
- ♦ Prognosis produces the effect of this degradation from its value, its type, and the degradation cause.

These three functions allow to elaborate relevant information to take a decision on the action to be applied. From this decision, the three functions Compensate, Correct and Execute produce the action order towards maintenance agent.

The whole of these functions corresponds to the Predictive Maintenance System (PMS) (Figure 4) which is the dynamic subset of the Global Maintenance System (GMS). To design this GMS, [10] has defined a prescriptive guide to help the maintenance agent to express his needs : The Maintenance Focus Area (MFA).

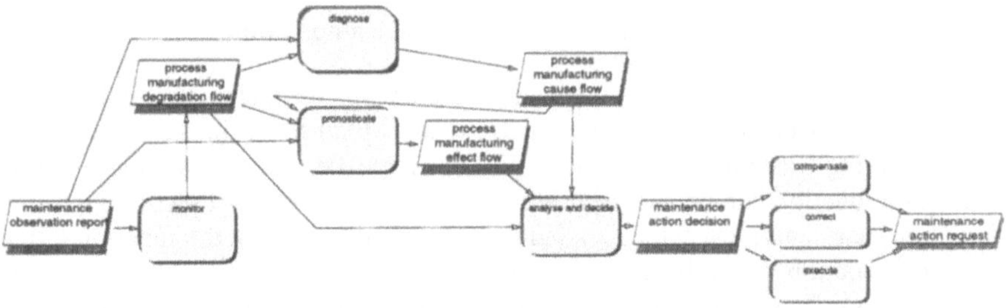

Figure 4: Activity model of predictive maintenance system.

In relation to the manufacturing system degradation, the PMS needs information of different natures :

- ♦ functioning variable (temperature, pressure, rotary speed,...)
- ♦ malfunctioning variable (noise, vibration, leak,...)
- ♦ macro variable including basic variables.

These variables can be produced from different devices, control measurement, maintenance measurement, specific processing.

On the basis of the degradation processed from these variables, the PMS can apply, to retrieve a normal state, an action through control actuator and maintenance agent.

As adding of new devices decreases the dependability of the system, it is better to use existing measurement, actuator, and specific monitoring processing to support first the variable production (past value, present value, operating time,...) and second the action implementation (compensation value, correction value, calibration value,...).

In this way, we propose to open the predictive maintenance system towards the other automation islands at the field level (Figure 5), in order to insure the information exchange for and from predictive maintenance.

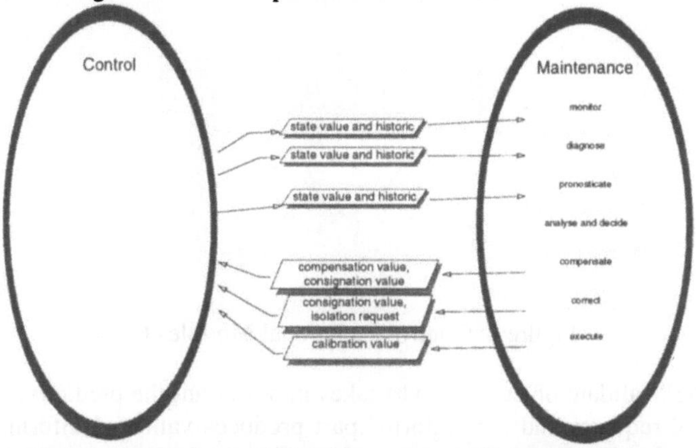

Figure 5: Openness of predictive maintenance towards control.

Indeed, the relevant information, for predictive maintenance, can be produced by the Supervisory Control which can also satisfy the maintenance action request. Supervisory Control integrates monitoring and objective validation functions on the control system.

4 Supervisory control for predictive maintenance

From a control point of view, the need of state and status information producing imposed the integration of monitoring processing in control part which is named "supervisory control". In this way, [11] proposed a generic behavioural module which adds, in control part, an on-line monitoring of the manufacturing process. The resulting reference models [3] and particularly the Control reference Model (MCC) structures the control module in three parts : to monitor, to react and to command.

The aim of the monitoring function is, among other things, to validate the right execution of a request, i.e. coherence between the real state of the manufacturing physical process and the state awaited after the request processing.

However, this control module is not adapted first to provide the right information in the right form to the predictive maintenance, and second to manage the possible conflict between the control and maintenance requests.

As solution, [12] [13] have proposed a structuring of the control modules, named Control Functional Modules CFM (Figure 6) in four parts which the roles related to the predictive maintenance are :

- "command" and "observe", which monitors the command processing and detects sensor failures,
- "validate objectives", which validates the control and maintenance requests and manage the conflict between these requests,
- "inform", which produces comprehensible information for predictive maintenance from data processed by the other parts,

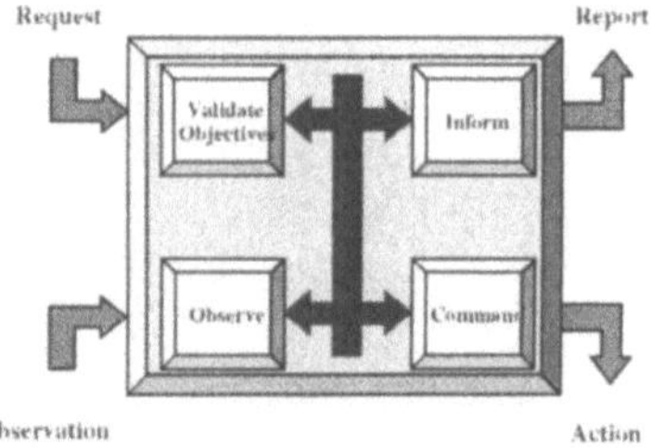

Figure 6: Control Functional Modules CFM.

Indeed, the "validate objectives" part takes into account the predictive maintenance and control requests and the "inform" part produces validated information for the predictive maintenance. Thus the informational flow exchange between the predictive maintenance and supervisory control is bi-directional.

5 Integrated Supervisory Control and Predictive Maintenance System

To ensure this bi-directional integration in an distributed and autonomous context, we have defined in [14] an open integrated intelligent module through an informational transaction which is associated to an information communication and storage.

The resulting integrated supervisory control and predictive maintenance system allows to satisfy the behavioural coherence and the total diffusion of information towards the different agents of the manufacturing system.

5.1 Information flow model

In this way, the supervisory control places, at disposal of predictive maintenance, necessary information for the four maintenance "observation" activities which are the maintenance management, the monitoring, the diagnosis and the prognosis. These activities need past and present value of process variable, to elaborate past and present value of degradation.

Moreover, the objective validation module of supervisory control produces degradation information related to functioning behaviour which is complementary to failure related to malfunctioning behaviour, and useful for predictive maintenance.

In complementary way from maintenance to supervisory control :

- the four observation activities of the predictive maintenance place, at disposal of supervisory control system, the failure information, future malfunctioning,...
- the three actuation activities of the predictive maintenance place, at disposal of supervisory control system, the compensation value, key-locking value, calibration value,...
- the decisional activity of the predictive maintenance informs the supervisory control system about availability of the manufacturing system.

The whole of previous information or action produced from predictive maintenance are consumed by the objective validation module to react on the functional behaviour of the manufacturing system (Figure 7).

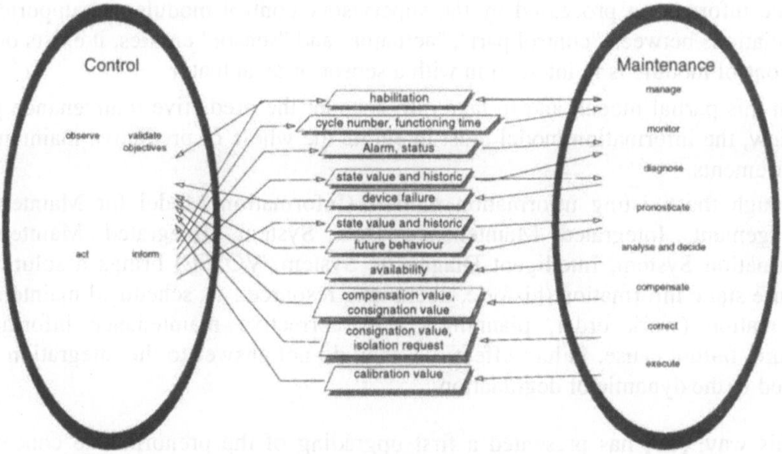

Figure 7:Information exchange between predictive maintenance and supervisory control.

5.2 Information Model

Therefore, the exchange of information flows between the supervisory control and the predictive maintenance implies to formalise a repository ensuring the coherence of the information encapsulated in these flows.

The finality of this repository is to store and process, on the one hand, the information exchanged between supervisory control and predictive maintenance, and on the other hand, the information processed by predictive maintenance. Indeed, in the conventional maintenance systems, information processing function of the predictive maintenance are broken up into several sub-functions supported by separated Information Technology systems without a real organisation and consistency in their implementation strategies. Each device of the global system can have a monitoring, a diagnosis and/or a prognosis systems [15] implying difficulties to have a consistent predictive maintenance system for the global manufacturing system.

To solve this double integration, we propose an information reference model which structures information processed by control and maintenance, and information exchanged between these two domains.

A first model, mainly from the control point of view, has been presented in [16] to be formalised and justified in [17]. It is structured by the feedback control loop of a manufacturing system and composed of :

- the "object" entity, which represents the physical object transformed by the manufacturing process part,
- the "actuator" and "sensor" entities, which represents the instrumentation manufacturing process,
- the "control part" entity, which represents the supervisory control modules.

As the information processed by the supervisory control modules is supported by the relations between "control part", "actuator" and "sensor" entities, it exists only if the control module is in interaction with a sensor or an actuator.

From this partial model, and to take into account the predictive maintenance point of view, the information model must integrate the whole of predictive maintenance requirements.

Although the existing information models (Information Model for Maintenance Management, Integrated Maintenance Data System, Integrated Maintenance Information System, Intelligent Diagnostic System, VOICE) brings a solution to capture static information (historic, document, resource,...), scheduled maintenance information (work order, planning,...) or corrective maintenance information (failure, failure cause, failure effect,...), they do not answer to the integration need related to the dynamic of degradation.

In this way, [10] has presented a first upgrading of the prenormative conceptual data schema for integrated Maintenance [9] developed on the GAMA ("Génie Automatique pour la Maintenance") and ESPRIT II-VOICE n°5510 projects, and the Information Model for Maintenance Management [7].

To take into account really the dynamic of the degradation, we propose a new improvement of this last model to be fully compliant with the global requirements of predictive maintenance system.

In a general way, to integrate the two control and maintenance models, it is necessary to implement explosion mechanism [18] identifying the bridge entities common to the control and maintenance informational models. The explosion of the "report on" relation leads to the adding of "process data from sensor" and "process failure from sensor" entities (Figure 8). In the same way, the explosion of the "control" relation allows to identify the "process action for actuator" entity.

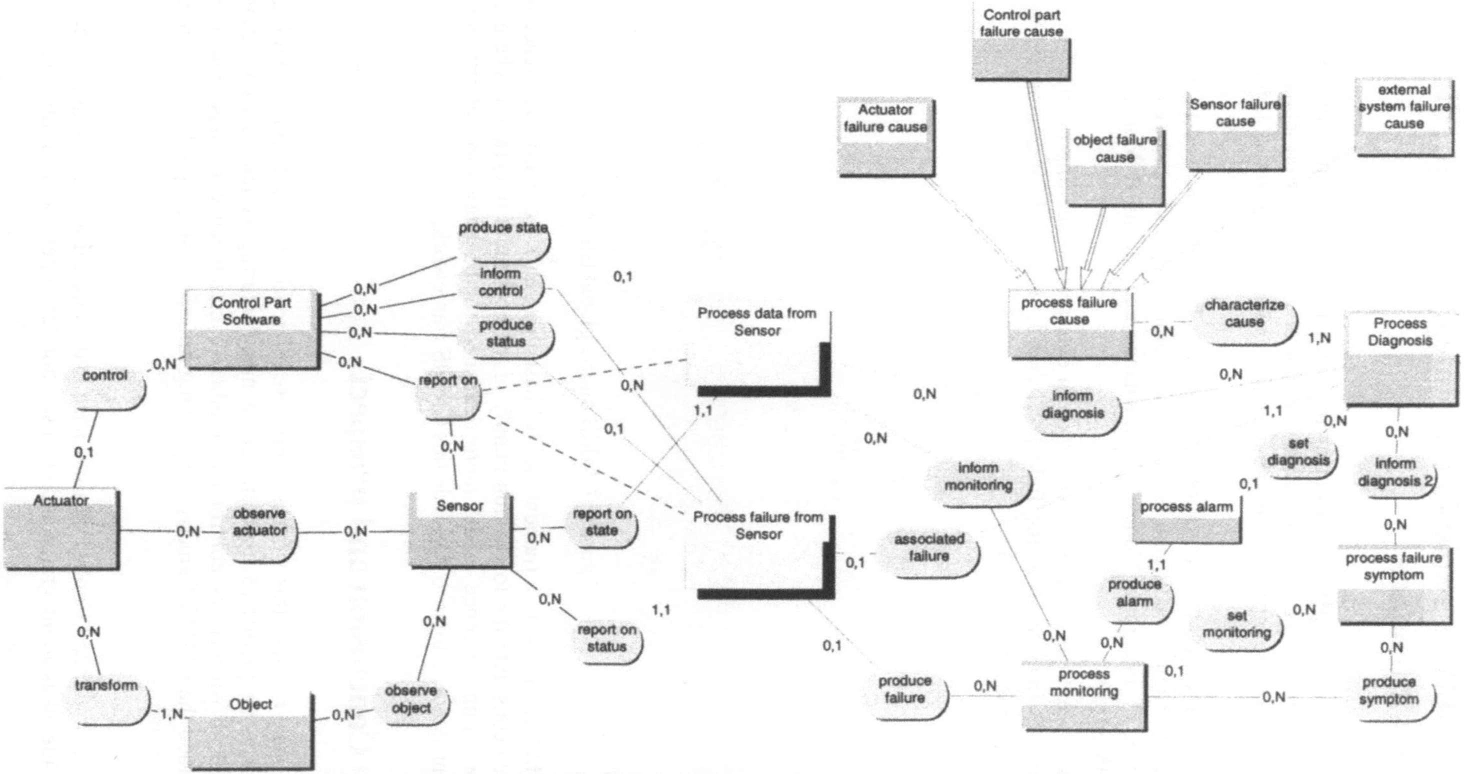

Figure 8: Partial information repository for integrated supervisory control and predictive maintenance.

6 CMMS : a solution for integration of predictive maintenance

This integrated Supervisory Control and Predictive Maintenance System is a partial solution on the one hand to the integration of the automation islands at the field level, and on the other hand to the openness towards the business level (linked with the GMS).

A proposal of a total integration and openness has been initiated in the ESPRIT II 2172 Distributed Intelligent Actuators and Sensors (DIAS) through the innovative concept of integrated Control, Maintenance and technical Management System (CMMS) [13].

This concept can be considered, from an enterprise modelling point of view, as a horizontal integration at the field level and as a vertical integration for business level openness (Figure 9).

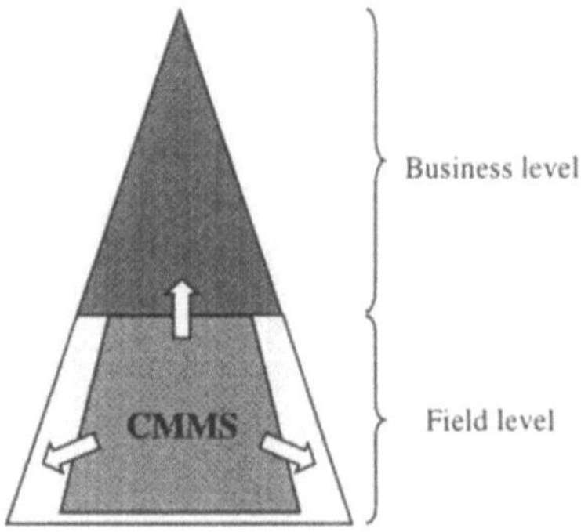

Figure 9: CMMS on the global Enterprise.

CMMS given rise to a framework [16] based on general or partial reference models which constitute the foundation mainly of the horizontal integration. Therefore our information reference model more dedicated to the integration of the predictive maintenance, allows to up-grade this CMMS framework.

7 Conclusion and prospect

The main result of this work is to propose reference models (Figure 7 and Figure 8) to integrate predictive maintenance in manufacturing system. This integration is based on supervisory control and predictive maintenance models, and allows to upgrade the CMMS framework by taking into account the predictive maintenance requirements.

So, the integrated Control, Maintenance and technical Management System can be considered as an integrated and open solution for Supervisory Production.

The main prospects are first to complete the information reference model to formalise behavioural reference models integrating both the normal and abnormal behaviours, and second to do to support the global framework by existing Enterprise software requiring an adaptability procedure or by developing a new one requiring an openness towards the existing Enterprise software.

References

1. Mobley K 1989 *Introduction to Predictive Maintenance.* Van Nostrand Reinhold, New York
2. Léger J-B, Iung B 1998 Methodological approach to modelling of degradation detection and failure diagnosis in complex production systems. In : Nayak P and Williams B, NASA Ames Research Center (eds) 1998 *Proceedings of the ninth International Workshop on Principles of Diagnosis*, DX98, Cap Cod, Massachusetts, USA, 24-27 May, pp 209-216
3. Lhoste P, Morel G 1996 From discrete event behavioural modelling to intelligent actuation and measurement modelling. In : Laboratory for Automation and Robotics of Patras (ed) 1996 *Proceedings of the Annual Conference of ICIMS-NOE*, ASI'96, Toulouse, France, 2-6 June, pp. 75-83
4. Van Schrick D 1997 Technical Fault and Quality Management – Terminology of Functions, Purposes and Means. In: Jongwon K, Seoul National University (ed) 1997 *Intelligent Manufacturing Systems*, Elsevier Science
5. Willsky, A S 1976 A survey of Design Methods for Failure Detection in Dynamic Systems. *Automatica* 12(6):601-611
6. Kramer M A, Fjellheim R 1995 Fault Diagnosis and Computer-Aided Diagnostic Advisors. In : Davis J, Stephanopoulos G, Venkatasubramanian V (eds) 1996 *International Conference on Intelligent Systems in Process Engineering*, Colorado, Vol 92. AIChE Symposium Series n°312, pp 12-24
7. Smit K , Slaterus W H 1992 *Information Model for Maintenance Management.* Cap Gemini Publishing, The Netherlands
8. Wylie, R, Orchard R, Halasz M, and Dubé F 1997 IDS: Improving aircraft fleet maintenance. In : Ford K (ed) 1997 *Proceedings of 14th National Conference on Artificial Intelligence and Innovative Applications of Artificial Intelligence,* July 27-31, Providence, Rhode Island. AAAI Press, pp 1078-1085
9. Koechlin J-B, Champaloux D 1993 A Conceptual Data Schema For integrated Maintenance : a map to design maintainable AMS. In : Croisier, Israel, Chavand (eds) 1993 *Proceedings of the 7th Annual European Computer Conference,* COMPEURO'93, Paris, France, 24-27 May. IEEE Computer Society Press, pp 112-118

10. Léger J-B, Iung B, Ferro Beca A, and Pinoteau J 1998 An innovative approach for new Distributed Maintenance System : application to Hydro Power plants of the REMAFEX project. To appear in *Computers In Industry*
11. Vogrig R, Baracos P, Lhoste P, Morel G, Salzemann B 1987 Flexible Manufacturing Shop Operation. *Manufacturing Systems* 16(1): 43-55
12. Belhimeur A 1989 Contribution à l'étude d'une méthode de conception des automatismes des systèmes de conduite des processus industriels, PhD thesis, Lille U.S.T.L. University (France)
13. Iung B, Lhoste P, Morel G, Roesch M 1991 Functional modelling of an Intelligent Actuator : Applicable to an ON/OFF or modulating electrical valve. In : *Proceedings of Workshop on CIM in the Process Industry*, Athens, Greece, 14-15 March
14. Neunreuther E, Iung B, Morel G, Leger J-B 1997 Engineering process modelling of an intelligent actuation and measurement system : from the users' needs definition to the implementation. In : Jongwon K, Seoul National University (ed) 1998 *Intelligent Manufacturing Systems*, Elsevier Science
15. Mobley K 1998 *Predictive Maintenance HandBook*. Plant Services on the Web
16. Mayer F, Morel G, Iung B, Leger J-B 1996 Integrated Manufacturing System meta-modelling at the shop-floor level. In : Laboratory For Automation and Robotics of Patras (ed) 1996 *Proceedings of the Annual Conference of ICIMS-NOE*, ASI'96, Toulouse, France, 2-6 June, pp 257-264
17. Neunreuther E 1998 Contribution à la modélisation des systèmes intégrés de production : application à la distribution du contrôle-commande et de la gestion technique sur les équipements de terrain, PhD thesis, Henri Poincaré Nancy 1 University (France)
18. Ducateau C F, Picavet M 1995 Progressive Adjusting Process for Data Modelling. In : *Proceedings of International Conference on Industrial Engineering and Production Management*, Marrakech, Morocco, 4-7 April, pp 235-244

14

A Geometric Calibration Method for Single-Head-Cone-Beam X-Ray Systems

K.J. Kyriakopoulos, P. Giannakos, V. Kallipolites and K. Domales

1 Introduction

1.1 Setup

A single-head cone-beam radiography system consists of: (i) the X-ray source, (ii) the object and its manipulator and (iii) the X-ray detector, as shown in figure 1. In this case, the detector consists of a fluoroscopic converter screen, a CCD imaging device of high resolution and sensitivity and an optical system that couples the CCD imaging device to the detector screen.

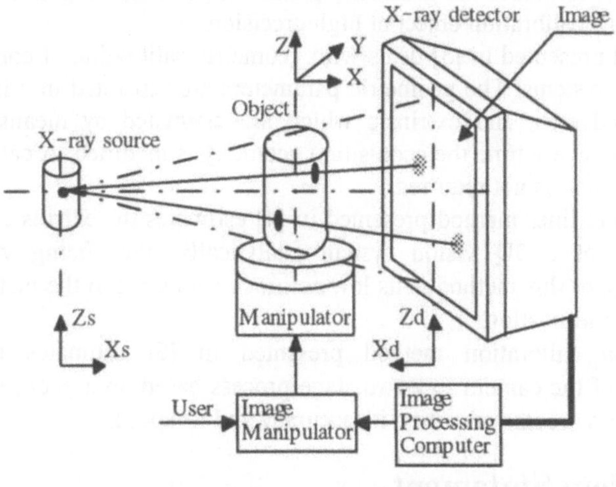

Figure 1. A single-head cone-beam radiography system.

The X-ray source and detector can be translated along the vertical axes Zs and Zd respectively and along the source to detector axes Xs and Xd respectively. The object is fastened to the manipulator, which can be translated along the parallel to the detector screen axis Y. The manipulator allows the rotation of the object around the vertical axis Z and the axis Y by the angles θ and β, respectively. θ is larger than -180° and less than or equal to 180° and β is limited to [-30°, 30°].

Angle φ describes the rotation of the object around the source to detector axis X and should be 0°, because the object cannot rotate around X.
For every pose, a digital radiogram is taken and the X-ray source position and orientation with respect to the object are estimated. The radiograms acquired are combined to provide a 3D map of the object's volume. Thus any defects of the object can be detected and defined in terms of size, shape and location, provided that their size is larger than the pixel size in the digital radiograms.

1.2 Existing Methods

There are two categories of methods presented here: methods used to calibrate tomography/radiography systems and methods used to calibrate the camera of vision systems. The geometric parameters of the camera or a tomography/radiography system can be seperated into **intrinsic** and **extrinsic** parameters. The intrinsic are pertinent to the aquisition device while the extrinsic indicate its position and orientation with respect to the examined object. The object used for geometric calibration purposes is called calibration object.
The alignment method presented in [1] is a geometric calibration method for microtomography systems. It provides estimates of the geometric parameters one by one, thus being cumbersome and time-consuming. It does not require a high precision calibration object.
The estimation of the projection matrix presented in [2] is a geometric calibration method for radiography systems that neither makes any assumptions on the X-ray source trajectory nor assumes that the distance of the X-ray source from the object is constant. It estimates the geometric parameters through nonlinear minimization and requires a calibration object of high precision.
The method presented in [3] deals with geometric calibration of cone-beam X-ray tomography systems. The geometric parameters are seperated into intrinsic which are measured once and extrinsic which are estimated by means of nonlinear minimization every time the acquisition geometry is modified. A calibration object of high precision is not required.
The vanishing lines method presented in [4] estimates the extrinsic parameters of the camera of a 3D vision system analytically, thus being very fast. The disadvantage of this method is its low accuracy compared to the methods based on nonlinear minimization.
The camera calibration method presented in [5] estimates the geometric parameters of the camera in a two-stage process based on a geometric constraint. Its advantages are its versatility, its accuracy and its speed.

2 Problem Statement

The purpose of the geometric calibration of an X-ray radiography system is to estimate its geometric parameters. We separate the geometric parameters into intrinsic and extrinsic. The intrinsic are measured only once because the geometry of the detection system does not change between different acquisitions. The extrinsic are estimated every time the position of the object changes w.r.t the detector or the X-ray source.

The geometry of the X-ray radiography system defined in figure 2 is general enough to describe all possible geometric configurations. (Od, ud, vd) is the detector coordinate system. Axis ud is parallel to the rows and axis vd is parallel to the columns of the detector screen. (F, u, v, w) is the X-ray source coordinate system. Point F, called the focal point, defines the position of the X-ray source. Axes u and v are parallel to ud and vd respectively. Axis w, called the focal or projection axis, is perpendicular to the detector screen. (O, x, y, z) is the object coordinate system and is tied to the object.

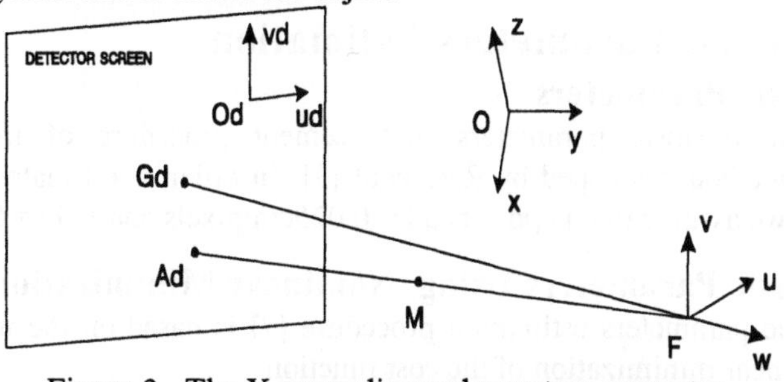

Figure 2. The X-ray radiography system geometry.

Let M be a point of the object space and Ad its projection onto the detector screen. The coordinates of M are in the object coordinate system (x_M , y_M , z_M) and in the X-ray source coordinate system (u, v, w). The coordinates of Ad in the detector coordinate system are (u_d, v_d). The coordinates of point M of the object coordinate system are expressed in the X-ray source coordinate system as:

$$(u, v, w)^T = R\cdot(x_M - x_F, y_M - y_F, z_M - z_F)^T \quad (1)$$

where R is the 3×3 rotation matrix defined by the Euler angles δ, ξ, ψ and (x_F , y_F , z_F) are the coordinates of F in the object coordinate system.

The three successive rotations of the object coordinate system with respect to the X-ray source coordinate system are described with the help of two intermediate coordinate systems (u′, y′, z) and (u′, v′, w) :

- ψ rotation counterclockwise about z gives (u′, y′, z).
- ξ rotation counterclockwise about u′ gives (u′,v′,w).
- δ rotation counterclockwise about w gives (u,v,w).

If the rotations are performed counterclockwise δ, ξ, ψ are positive. Each rotation is described by a matrix:

$$R(\delta)=\begin{pmatrix}\cos\delta & \sin\delta & 0\\ -\sin\delta & \cos\delta & 0\\ 0 & 0 & 1\end{pmatrix},\ R(\xi)=\begin{pmatrix}1 & 0 & 0\\ 0 & \cos\xi & \sin\xi\\ 0 & -\sin\xi & \cos\xi\end{pmatrix},\ R(\psi)=\begin{pmatrix}\cos\psi & \sin\psi & 0\\ -\sin\psi & \cos\psi & 0\\ 0 & 0 & 1\end{pmatrix}$$

R describing the 3D rotation of the object coordinate system is derived by :

$$R = R(\delta)\cdot R(\xi)\cdot R(\psi). \quad (2)$$

The coordinates (u_d, v_d) of the projection Ad of M with respect to the detector coordinate system are (in pixels)

$$u_d = pgd - (u\cdot FGd)/(w\cdot pepd) \quad (3a)$$

$$v_d = qgd - (v\cdot FGd)/(w\cdot pepd) \quad (3b)$$

where: (a) (pgd, qgd) are the coordinates of the perpendicular projection of the focal point onto the detector screen Gd wrt the detector coordinate system, (b) pepd and peqd are the sampling steps of the detector screen along ud and vd respectively and (c) FGd is the distance between focal point F and point Gd. The origin of the detector coordinate system is point Gd and pgd = 0 and qgd = 0 (in pixels).
The intrinsic parameters are pepd, peqd, pgd, qgd and FGd while the extrinsic are x_F , y_F , z_F and δ, ξ, ψ.

3 Geometric Parameters Estimation

3.1 Intrinsic Parameters

We use the intrinsic parameters measurement procedure of the geometric calibration method developed by Rizo et.al [3]. In order to estimate the extrinsic parameters we assume that pepd = peqd = 0.025cm/pixels and FGd = 200.0cm.

3.2 Extrinsic Parameters Using Nonlinear Minimization

The extrinsic parameters estimation procedure [7] is based on the method in [2] i.e. by nonlinear minimization of the cost function

$$\sum_{i=1}^{N}\left[\left(u_d - u_{center}\right)_i^2 + \left(v_d - v_{center}\right)_i^2\right]$$

where (u_d, v_d) are given by (3) and (u_{center}, v_{center}) are the measured projection coordinates of a calibration point. The calibration object is a regular grid of small spheres and the calibration points are the centers of these spheres (fig. 7). For minimization we use the Powell's zero order algorithm of conjugate directions [6].

3.3Extrinsic Parameters by means of Vanishing Lines

This method was developed for vision systems [5] where a camera with rotation angles limited to (-90°, 90°) moves as the calibration object remains still. The calibration object is a hexagon with parallel opposite sides. The perspective projection of each pair of parallel sides, that are not parallel to the image plane of camera, converges to a point in the image, called a **vanishing point**. The vanishing points that correspond to the three pairs of parallell sides lie on the **vanishing line** (fig. 3).
In order to apply the vanishing lines method we have to make a few modifications (see [8]) because one of the rotation angles is not limited in (-90°, 90°). The calibration object is a regular grid of spheres (fig. 4). The sphere centers define groups of parallel lines in the 3D object space, which are not parallel to the detector screen. Each line is defined by the centers of two or more spheres. In the image, the projections of the sphere centers define the perspective projections of the parallel lines belonging to each group. A vanishing point is determined by intersecting the lines, which are the perspective projections of the parallel lines belonging to the corresponding group. The coordinates of the vanishing point are the averages of the coordinates of these intersection points. The vanishing point that corresponds to a group of lines parallel to the y axis of the object coordinate

system (fig. 4) is called the depth vanishing point. The vanishing line equation is computed by least-square-error fitting from the vanishing points.

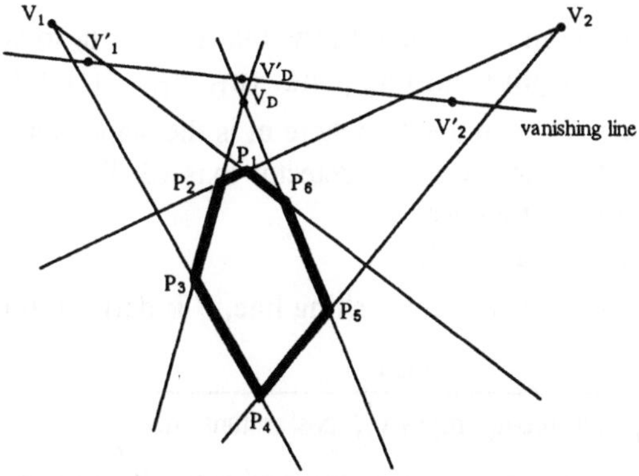

Figure 3. The vanishing points and the vanishing line produced by an image of the hexagonal calibration object.

The coordinates of the vanishing points and the vanishing line equation are computed wrt the detector coordinate system, whose origin is point Gd. Equation (1) is valid in the case of the vanishing lines method, but the object coordinate system (fig. 4) is different from that of our extrinsic parameters estimation procedure (fig. 7). As a result, rotation matrix R (eq. 2) is defined by the new Euler angles δ', ξ', ψ'. The coordinates of the focal point F expressed in the object

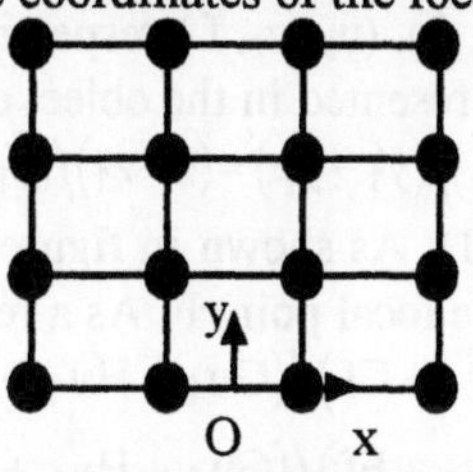

Figure 4. Calibration object & object coord. system of the vanishing lines method.

coordinate system of the vanishing lines method are (x'_F , y'_F, z'_F). For our convenience, matrix R is denoted as :

$$R = \begin{pmatrix} A & D & G \\ B & E & H \\ C & F & I \end{pmatrix}.$$

Estimation of the Euler angles δ', ξ', ψ'

Let $a_1x + b_1y = c_1$, $z = 0$ and $a_2x + b_2y = c_2$, $z = 0$ be the line equations of two different sets of parallel lines in the object coordinate system. Also let (u_1, v_1) (u_2, v_2) be the coordinates with respect to the detector coordinate system of the two vanishing points that correspond to these sets of parallel lines. Then (see [4], [8])

$$(u_i, v_i) = \left(\frac{f \cdot (-b_i \cdot A + a_i \cdot D)}{-b_i \cdot C + a_i \cdot F}, \frac{f \cdot (-b_i \cdot B + a_i \cdot E)}{-b_i \cdot C + a_i \cdot F} \right) \qquad (4)$$

where f = -FGd and i = 1, 2. Using (4) and resuming the terms of A through F from (2), we derive the vanishing line equation:

$$u \cdot \sin\delta' + v \cdot \cos\delta' = FGd/\tan\xi'. \tag{5}$$

By considering the slope value in (4) to be infinity, that is b_1 or b_2 to be zero, we obtain the coordinates of the depth vanishing point: $(u_D, v_D) = (D \cdot f/F,\ E \cdot f/F)$. The vanishing line equation is $v = m \cdot u + k$, where m is the slope and k is the vd value where the vanishing line intersects the detector axis vd. We obtain δ' from the slope of the vanishing line as follows:

$$\delta' = \arctan(-m). \tag{6}$$

Since the depth vanishing point is on the vanishing line, ξ' is derived from (5) as:

$$\xi' = \arctan\left(\frac{FGd}{u_D \cdot \sin(\arctan(-m)) + v_D \cdot \cos(\arctan(-m))}\right). \tag{7}$$

Resuming the terms of D and E from (2), we derive ψ' from the coordinates of the depth vanishing point:

$$u_D/v_D = D/E \Rightarrow \psi' = \arctan\left(\frac{\cos\xi' \cdot (u_D - v_D \cdot \tan\delta')}{u_D \cdot \tan\delta' + v_D}\right). \tag{8}$$

Estimation of the coordinates x'_F, y'_F, z'_F

Let P_1 (x_1, y_1, z_1), P_2 (x_2, y_2, z_2) be two points in the object coordinate system and P'_1, P'_2 their projections onto the detector screen. The coordinates of P'_1, P'_2 in the object coordinate system are (x'_1, y'_1, z'_1), (x'_2, y'_2, z'_2) respectively and in the X-ray source coordinate system are (u_1, v_1, f), (u_2, v_2, f) respectively.

Line (FP'_1) passes through P_1 and is represented in the object coordinate system as

$$(x - x_1)/(x'_1 - x'_F) = (y - y_1)/(y'_1 - y'_F) = (z - z_1)/(z'_1 - z'_F) \tag{9}$$

where (x'_1, y'_1, z'_1) are obtained from (1). As shown in figure 5 the intersection of line (FP'_1) (9) and plane $z = z_1 + h$ is the focal point F. As a result:

$$\begin{aligned} x'_F &= x_1 + h \cdot (Au_1 + Bv_1 + Cf)/(Gu_1 + Hv_1 + If) \\ y'_F &= y_1 + h \cdot (Du_1 + Ev_1 + Ff)/(Gu_1 + Hv_1 + If) \\ z'_F &= z_1 + h \end{aligned} \tag{10}$$

The angle between (FP'_1) and (FP'_2) is ω and its cosine is obtained by the inner product of the vectors $\overrightarrow{FP_1}$ and $\overrightarrow{FP_2}$

$$\cos\omega = \frac{(x_1 - x'_F)\cdot(x_2 - x'_F) + (y_1 - y'_F)\cdot(y_2 - y'_F) + (z_1 - z'_F)\cdot(z_2 - z'_F)}{\sqrt{\left[(x_1 - x'_F)^2 + (y_1 - y'_F)^2 + (z_1 - z'_F)^2\right]\left[(x_2 - x'_F)^2 + (y_2 - y'_F)^2 + (z_2 - z'_F)^2\right]}} \tag{11}$$

From (11), (10) we get a quadratic equation of h :

$$ph^2 + qh + r = 0 \tag{12}$$

where

$p = (1 - \cos^2\omega) \cdot (a^2 + b^2 + 1)^2$,

$q = -2 \cdot (1 - \cos^2\omega) \cdot (a^2 + b^2 + 1) \cdot [a(x_2 - x_1) + b(y_2 - y_1) + (z_2 - z_1)]$

$$r=[a(x_2-x_1)+b(y_2-y_1)+(z_2-z_1)]^2-\cos^2\omega(a^2+b^2+1)\cdot[(x_2-x_1)^2+(y_2-y_1)^2+(z_2-z_1)^2]$$

$a=\dfrac{Au_1+Bv_1+Cf}{Gu_1+Hv_1+If}$, $b=\dfrac{Du_1+Ev_1+Ff}{Gu_1+Hv_1+If}$ and $\cos\omega$ is computed by the inner product of $\overrightarrow{FP'_1}$ and $\overrightarrow{FP'_2}$.

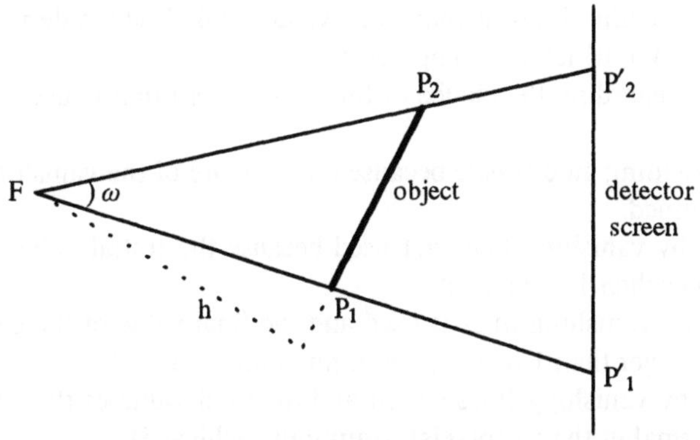

Figure 5. The spatial relation between the object points P_1 and P_2 and their projections P'_1 and P'_2.

If angle θ (section 1.1) takes a value in (-90°, 90°), the positive root of (12) is used to compute x'_F, y'_F, z'_F by (10), otherwise the negative root of (12) is used. There are as many estimates of x'_F, y'_F, z'_F as the number of all possible pairs of calibration points. These estimates are averaged to get the final values of the coordinates x'_F, y'_F, z'_F.

3.4 Extrinsic Parameters Estimation by Combination of Nonlinear Minimization of and Vanishing Lines

Powell's minimization algorithm requires as input an initial estimate of the extrinsic parameters. This is suggestedto be as close as possible to the real values of the extrinsic parameters. If the vanishing lines method is used to quickly provide good initial estimates for Powell's algorithm (fig. 6), this results to fewer iterations and faster convergence with the same or better accuracy.

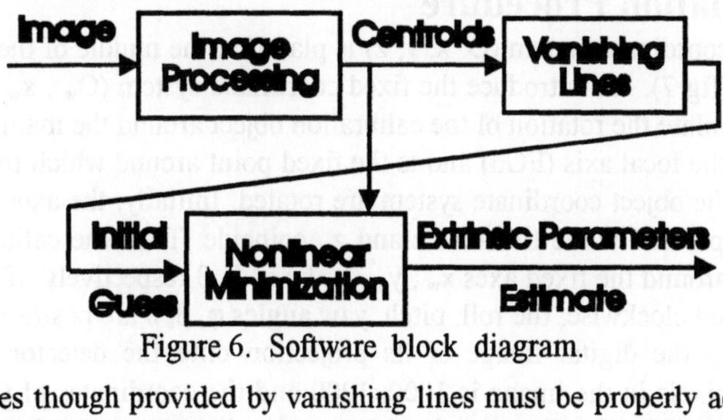

Figure 6. Software block diagram.

The estimates though provided by vanishing lines must be properly adjusted with respect to the object coordinate system of our procedure (fig. 7). This is done using

$$x_F = x'_F\ ,\ y_F = y'_F\ ,\ z_F = z'_F \tag{13}$$

$$R(\delta, \xi, \psi) = R(\delta', \xi', \psi') \cdot R(\delta_1, \xi_1, \psi_1) \tag{14}$$

where R is given by (2) and the Euler angles $\delta_1 = 0°$, $\xi_1 = 90°$, $\psi_1 = 90°$ describe the successive counterclockwise rotations applied to the object coordinate system of our estimation procedure so that its axes become parallel to those of the object coordinate system of the vanishing lines method. The values of δ, ξ, ψ are derived from the elements of R(δ, ξ, ψ), which are computed by (13).

There are 4 possible cases regarding the choice of the initial guess that is used by the Powell's algorithm:

1. Vanishing lines not providing an estimate because one or more of the vanishing points cannot be determined.
2. The estimate provided by vanishing lines not used because the initial value of the cost function is larger than 10^5 (pixels)^2.
3. The estimate provided by vanishing lines is used and the final value of the cost function is found to be larger than 1 (pixels)^2 (minimum not achieved).
4. The estimate provided by vanishing linesis used and the final value of the cost function is found to be smaller than 1 (pixels)^2 (minimum achieved).

Case 1 occurs when there are at least two parallel lines belonging to one or more groups of lines in the image of the calibration object. In this case it is possible that the values of δ′, ξ′, ψ′ derived from the vanishing line equation are incorrect and for this reason, the application of the vanishing lines method must stop. Limit values 10^5 (pixels)^2, 1 (pixels)^2 have been derived based on extensive simulations. In cases 1, 2 and 3 one of two fixed initial estimates is used. These correspond to two 3D object rotations described by the roll, pitch, yaw angles $\varphi = 0°$, $\beta = 0°$, $\theta = -2°$ or $2°$ (section 4). The initial guesses are $x_F = x_{1F} \cdot \cos\theta$, $y_F = x_{1F} \cdot \sin\theta$, $z_F = z_{Ow}$ = (grid height)/2, $\psi = 90° + \theta$, $\xi = 90°$, $\delta = 0°$, where x_{1F} is defined section 5.4. The values of φ, β, θ are called initial since they correspond to the initial estimate of the extrinsic parameters. The reason why we use two initial estimates instead of one is to ensure that the actual and the initial value of θ have the same sign in order to achieve minimization of the cost function. Powell's algorithm uses the preset initial estimate, for which the initial value of the cost function is the lowest.

4 Simulation Procedure

The object coordinate system (O, x, y, z) is placed at the middle of the bottom side of the grid (fig 7). We introduce the fixed coordinate system (O_w , x_w , y_w , z_w) in order to simulate the rotation of the calibration object around the manipulator axes. O_w lies on the focal axis (FGd) and is the fixed point around which the calibration object and the object coordinate system are rotated. Initially, the axes x and x_w , y and y_w are parallel while the axes z and z_w coincide. Then the calibration object are rotated around the fixed axes x_w , y_w , z_w by φ, β, θ respectively. If the rotations are performed clockwise, the roll, pitch, yaw angles φ, β, θ are positive.

We simulate the digital image of its projection onto the detector screen. The number of pixels in the image is 1000×1000 and the coordinates of Gd are pgd = qgd = 499 pixels with respect to the detector coordinate system when its origin is the sample interval at the lower left-hand corner (first row, first column) of the detector screen. The number of gray levels is 2^8 and hence any pixel in the image has an intensity level g , $0 \leq g \leq 2^8 - 1 = 255$. Note that the extrinsic parameters estimates obtained by Powell's algorithm become more accurate as the number of pixels increases above 1000×1000 and/or the pixel size decreases below pepd = peqd = 0.025cm/pixels. In order to simulate the detection error of the projections of the sphere centers onto the detector screen, we add gaussian noise to the initial image. Next, we apply thresholding to obtain a binary image to compute the projections of the sphere centers.

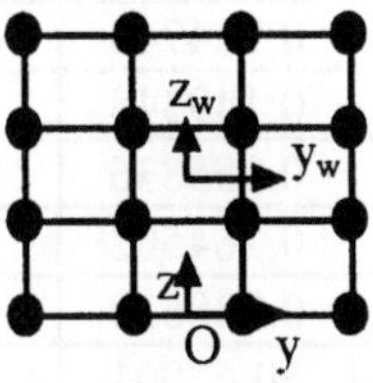

Figure 7. The grid and the fixed coordinate system.

5 Simulation Results

We performed simulations for the specification of the calibration object and to evaluate the extrinsic parameters estimates obtained by Powell's algorithm in terms of accuracy by the mean euclidean and the maximum absolute deviations between them and the real ones.

5.1 Calibration Object Specification

The optimal number of spheres and dimensions of the grid were determined in terms of accuracy and speed. The width-to-height ratio takes the values 1.0, 0.75, 0.5, the grid height is 16, 12, 8 cm and the number of spheres is 9, 16, 25. Each case of height, width-to-height ratio and number of spheres was examined according to a scenario of object rotations described by φ, β, θ (table 1).

φ	β	θ
2.0°	-1.0°	0.5°
2.0°	-1.0°	5.0°
2.0°	5.0°	5.0°
6.0°	5.0°	5.0°
5.0°	5.0°	10.0°
5.0°	10.0°	10.0°
10.0°	10.0°	10.0°
10.0°	10.0°	15.0°

Table 1: Scenario of calibration object rotations.

In the case of the first set of φ, β, θ the initial values corresponded to the initial position of the calibration object (fig. 7): $x_F = x_{1F}$, $y_F = 0.0$cm, $z_F =$ (grid height)/2, ψ = 90°, ξ = 90°, δ = 0°. At every other instance of φ, β, θ we used as initial extrinsic parameters estimate the extrinsic parameters estimate of the preceding instance of rotation. The calibration object was located at a fixed distance of $x_{1F} =$ 180.0cm from the X-ray source and the standard deviation of the gaussian noise was 20. The results (tables 2a, 2b, 2c) show that a grid of 16 spheres with dimensions 16cm × 16cm offers the optimum combination of accuracy and speed.

Grid of 9 spheres				
Grid height (cm)	Width to height ratio	Mean euclidean deviation	Mean maximum deviation	Average # of Powell iterations
16	0.5	0.321412	0.279383	31.625
16	0.75	0.263692	0.234748	27.125
16	1.0	0.171341	0.149348	27.125
12	0.5	0.409386	0.355836	32.750
12	0.75	0.314464	0.264502	29.750
12	1.0	0.292340	0.238650	28.875
8	0.5	62.03554	50.67001	35.000
8	0.75	0.720048	0.632073	31.875
8	1.0	0.526175	0.450633	26.500

Grid of 16 spheres				
Grid height (cm)	Width to height ratio	Mean euclidean deviation	Mean maximum deviation	Average # of Powell iterations
16	0.5	0.266882	0.224094	30.875
16	0.75	0.189462	0.146670	28.500
16	1.0	0.126421	0.105481	25.375
12	0.5	0.323150	0.292224	33.375
12	0.75	0.227124	0.188645	29.125
12	1.0	0.282707	0.234621	26.625
8	0.5	2.640790	2.389182	33.625
8	0.75	0.422631	0.378189	35.125
8	1.0	0.362476	0.324733	30.750

Tables 2a, 2b.

5.2 Magnification Ratio

We performed simulations in order to determine the effects of the magnification ratio FGd/x_{1F} when the calibration object is placed at a distance of x_{1F} = 180.0, 160.0, 140.0, 120.0, 100.0 and 80.0 cm from the X-ray source. For every value of x_{1F} we ran the scenario of calibration object rotations in table 1 and the standard deviation of the gaussian noise was 20. The results (tables 3a, 3b) show that accuracy improves as the magnification ratio increases, while speed is not affected.

5.3 Standard Deviation of the Gaussian Noise

We performed simulations in order to determine the effects of the standard deviation of the gaussian noise when it ranges from 10 to 36, following the scenario of table 1. The simulations showed that it has no effect on accuracy.

However, it increases the image processing time when it is larger than a value which depends on the thresholding value, because the number of clusters in the image becomes much larger than the number of spheres.

Grid of 25 spheres				
Grid height (cm)	Width to height ratio	Mean euclidean deviation	Mean maximum deviation	Average # of Powell iterations
16	0.5	0.183905	0.161773	32.000
16	0.75	0.154340	0.121784	27.500
16	1.0	0.109744	0.089505	27.500
12	0.5	0.266263	0.235622	33.000
12	0.75	0.189986	0.162818	26.250
12	1.0	0.215071	0.160173	29.500
8	0.5	0.598699	0.532401	34.625
8	0.75	0.755578	0.563565	23.625
8	1.0	0.377456	0.307242	30.250

Table 2c.

Magnif. ratio	Mean euclidean coord. dev. (cm)	Mean euclidean angle dev. (deg)
1.111111	0.343583	0.111827
1.250000	0.185672	0.068259
1.428571	0.217019	0.088499
1.666667	0.150418	0.072153
2.000000	0.126225	0.071665
2.500000	0.044172	0.031762

Table 3a: Mean euclidean deviations for various magnif. ratios.

Magnif. ratio	Mean max coordinate deviation (cm)	Mean max angle deviation (degrees)	Average # of Powell iterations
1.111111	0.324808	0.105835	28.875
1.250000	0.164929	0.062821	33.375
1.428571	0.197838	0.082202	30.375
1.666667	0.136090	0.065653	27.250
2.000000	0.118215	0.068884	26.875
2.500000	0.037957	0.027817	32.000

Table 3b: Mean maximum deviations and average number of Powell iterations for various magnification ratios

5.4 First Guess effects

We investigated the effects of the deviation between the first guess and the real values of the extrinsic parameters on the speed of our extrinsic parameters estimation procedure. The three cases of calibration object rotation (φ, β, θ) = (16°, 30°, 60°), (-16°, 30°,-60°), (-16°,-30°, 60°) were simulated. We used first guesses of increasing "distance" around those real rotations and found that it results in increasing the number of Powell iterations.

6 Suggestions for Further Research

The issue of updating the extrinsic parameters estimates as the object moves is currently researched. Estimation techniques, such as the Extended Second Order Kalman Filter, that can be continuously updated based on new information and can handle both the nonlinear and the stochastic nature of small variations in the object position and orientation are required. This technique can be applied to avoid the repeated use of Powell's algorithm, thus improving the computational efficiency.

REFERENCES

1. Sire P., Martin M. and Ph. Rizo, 1993 "X-ray cone beam CT system calibration", in *X-ray Detector Physics and Applications II Proceedings*, San Diego, CA.
2. Rougée A., Picard C., Ponchut C. and Trousset Y., 1992, "Geometrical calibration of X-ray imaging chains for three dimensional reconstruction", in *14th Annual International Conference of the IEEE EMBS Proceedings*, Rennes, France.
3. Rizo Ph., Grangeat P. and Guillemaud R., 1994, "Geometric calibration method for multiple-head cone-beam SPECT system", *IEEE Transactions on Nuclear Science*, Vol. 41, No. 6.
4. Wang L.-L., Tsai W.-H., 1991, "Camera calibration by vanishing lines for 3-D computer vision", *IEEE Transactions on Pattern Analysis and Machine Intelligence*, Vol. 13, No. 4.
5. Tsai R.Y., 1987, "A versatile camera calibration technique for high-accuracy 3D machine vision metrology using off-the-self TV cameras and lenses", *IEEE Journal of Robotics and Automation*, Vol. RA-3, No. 4, pp. 323-344.
6. Press W.H., Flannery B.P., Teukolsky S.A. and Vetterling W.T., 1988, *Numerical Recipes in C*, Cambridge: Cambridge University Press.
7. Giannakos P., Kallipolites V., 1996, "Geometric calibration method for a multi-radiograph system," Diploma Thesis, Control Systems Laboratory, Mechanical Eng. Dept., NTUA, (in Greek).
8. Domales K., 1996, "Extrinsic parameters estimation by vanishing lines for a multi-radiograph system," Diploma Thesis, Control Systems Laboratory, Mechanical Eng. Dept., NTUA, (in Greek).

15
A Tool for the Evaluation of Fieldbuses Reliability for Intelligence Distributed Automation System Design

B. Conrard, J.M. Thiriet and M. Robert

1 Introduction

Nowadays, the choice of a fieldbus is a complex task during the design of a Distributed Automation System. On the one hand, there is a lot of available kind of fieldbuses on the market. On the other hand, constraints and criteria are multiple and difficult to be evaluated. This chapter proposes a means to evaluate some dependability criteria. The security aspect is not taken into account in the present study.

2 Choosing a fieldbus

This section shows one conceivable way how designers choose a fieldbus. It is based on a set of criteria, which give a framework to evaluate each network structure. Some of these criteria are the cost, the reliability, the security, the available type of data transfer, the durability...

The first step is to fit the criteria to the needs. To succeed in it, the criteria are weighted according to the goals of the communication infrastructure [1]. For example, in some cases, security is more important than the system availability and the weight of security criterion is high. In some situations, some criteria can become constraints. For example, if physical data must be transmitted, all the fieldbuses that are limited to boolean data have to be eliminated. Globally, in this preliminary step, designers choose the weight of each criterion and identify the constraints according to the needs of the researched system.

3 Principle and Mission of a fieldbus

The main aim of a fieldbus is to obtain a reduction of the number of cables between the components in an automation system. Sharing the information on a same cable induces few additional advantages such as an easy data distribution between all components or control units [4]. To be used in an automation system, fieldbuses own mechanisms which allow them to respect the temporal constraints imposed by the physical process. Therefore, a new cause of error may appear with these communication systems.

A usual representation of this distributed real-time system is to consider the system as a distributed database. Then, each component can read a variable as if it is a local one. Conversely, by a unique writing on the interface of the component, a datum can be available for all other components. With such a representation, all problems concerning the data transport are let to the fieldbus.

Its mission is then, to put each variable at components disposal at the assumption that its content is not too old. Consequently, each variable owns a life time chosen by the designer according to the dynamic of the system. The problem induced by buses is that the messages which contain data, are transmitted sporadically and a loss of one of them may lead to a non-respect of temporal specifications [2]. If this case comes, the global system may stop if the error is detected and consequently, the system availability decreases. But if the error is not detected, the system may have an abnormal behaviour, and the security decreased. In the following, this chapter tries to help the designer to determine which fieldbus must be used and with which parameters.

4 Several models of transistory fault

In this part, only the transitory faults are considered. They are essentially electromagnetic perturbations or bad contacts. Their causes may be for example, an electromagnetic source, a connector submitted to vibrations, a bad condition of the communication medium...

4.1 Uniform failure rate

The first and the simplest way to treat the occurrence of faults, is that there is a uniform probability to occur at any moment. This hypothesis is based on the principle that a fault has the same probability to occur at any moment. According to the occurrence rate (λ) and the transmission duration (t) of a message, its probability to be disturbed is evaluated with the following formula (1):

$$P_{disturbed} = 1 - \exp(-\lambda t) \qquad (1)$$

If different fieldbuses are compared, other factors must be taken into account. More especially, according to the characteristics of each fieldbus, the occurrence rate must be adapted.

Other factors have an influence on the fault occurrence probability. One of them is the choice of the communication medium. The fault probability must be adjusted according to the immunity of the medium. Some tests have been made to evaluate this influence.

Non shielded pair	5.7	■■■■■■■■■■■
twisted pair	3.7	■■■■■■■
shielded pair	2.2	■■■■
shielded twisted pair	1.0	■■
optical fiber	≈0	

Values found in [3] which quotes [5].

Table 1: Relative Immunity of different medium against electromagnet perturbations.

Another factor is the transmission method. For instance, a data transmitting with a current signal rather than a tension signal is less immunised against electromagnetic perturbations. AS-Interface is for example one of these fieldbuses which control the current to diffuse their data.

However, this model is too simple to represent correctly the usual perturbations met in an industrial environment. A lot of perturbations actually continue during a delay long enough to influence several successive messages. The next sub-section discusses this case.

4.2 Fault coupled with a duration

Practically, this kind of fault is the combination of two phenomena. The first one is the probability of fault occurrence, the second one its duration. Such as previously, a fault has the same probability to appear at any moment. Consequently, a uniform rate of occurrence (λ_p) can be used. About the duration of each fault, the insufficiency of feedback induces that there is no specific law. In this chapter, the exponential and the log-normal laws are proposed. The advantage with these continuous laws, is that different transmit rates of different networks can be compared. With this principle of fault applied on a temporal period, some faults such as a hail of perturbations, can be stimulated.

This model of fault owns the following properties. On the one hand, the immunity of low speed network to some fast perturbations can be taken into account. On the other hand, the disturbance of several messages due to a prolonged fault can be treated simultaneously.

According to the transmit rate, if an occurred fault is short enough, no error appears. The duration limit of fault influence depends on two parameters the rate and the data encoding. For example, the Manchester code (used by FIP) uses 2 states to encode a byte. Then, in comparison, the limit of CAN which use one state per byte, should be 2 times longer. Arbitrary, the limit used by our method is fixed to the quarter of the duration of one electric state. In consequence, the previous occurrence rate (λ_p) could be adapted (λ) according to the immunity against perturbations.

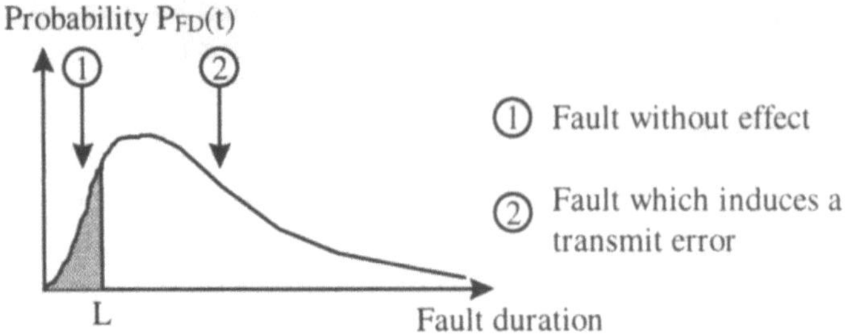

Figure 1: Probability of fault duration

$$\lambda = \lambda_P (1 - P) \quad \text{with } P = \int_0^L P_{FD}(t)\,dt \quad (2)$$

With this adjusted fault rate (λ) , the probability that a message is erroneous, is estimable. It is the sum of the probability that a fault occurs during the transmission and the one that a fault occurs before the beginning of the message transmission and affects it due to its duration. The first term only depends on the fault rate (λ) and the message transmission duration (T_{TD}) and is calculable with the formula (3). The second term only depends on the distribution of the duration fault, and is calculable with the formula (4).

$$P_1 = 1 - \exp(-\lambda T_{TD}) \qquad (3)$$

$$P_2 = \int_L^{\infty} \lambda P[T_{FD} > t]\,dt \qquad (4)$$

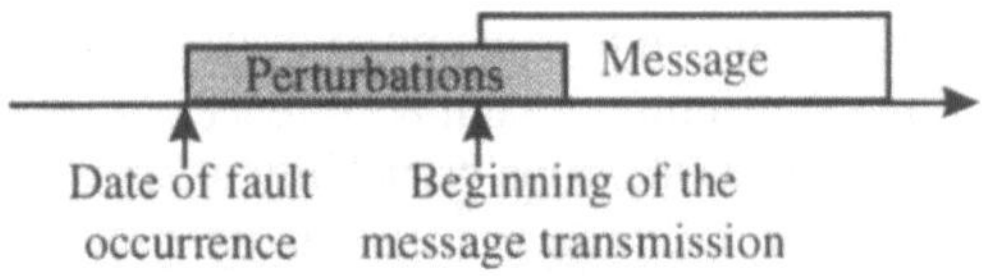

Figure 2: One long perturbation

$P[T_{FD} > t]$ is the probability that the duration of a fault (T_{FD}) is greater than a delay (t).

4.3 Hail of faults

In the previous model of fault, it is difficult to take into account the corrections due to the various precautions to avoid the fault according to the choice of the medium. Moreover, previously during a hail of faults no message must be diffused, however few messages may sometimes be correctly transferred. Consequently, an other way to model the fault, consists to consider periods such as previously, but with a high fault rate (λ_1). Their probability to occur and their duration can be established by the same previous way.

In this way, during the period of perturbation, an occurrence rate is applied such as in the section 4.1. Moreover, the sporadic faults can be integrated in this model with a second occurrence rate (λ_2) applied out of fault period. The following figure shows this principle.

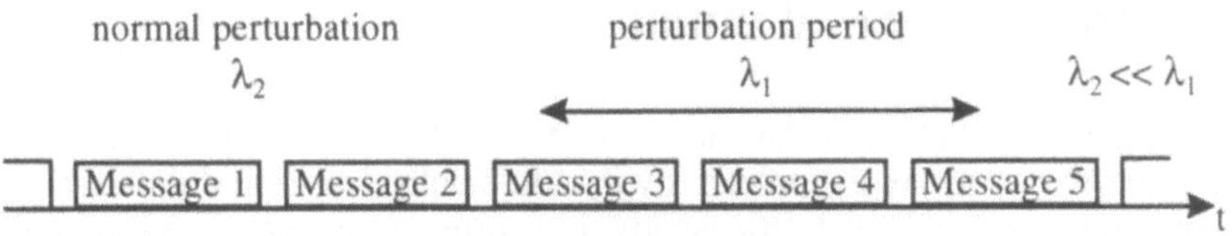

Figure 3: Perturbation period

With this model of fault, the previous corrective coefficients are applied to the occurrence rate (λ_1) and (λ_2) and consequently allow a comparison between different fieldbuses.

4.4 Application to reliability evaluation

Due to the complexity of the presented models of faults, to find a set of parameters is a good way to achieve the reliability evaluation. In the following section where this evaluation is performed, only some variables are needed. There are the probability to disturb the transmission of one message, and the ones concerning several successive messages. These last ones are necessary to take into account the relationship of dependence between several successive faults. These needed variables about fault probability are called P_0 for the probability to transmit correctly a message and P_{d1}, P_{d2}... P_{dn} for the ones of n disturbed successive messages due to the same perturbation. The other derived variables are $P_{d2..\infty}$, $P_{d3..\infty}$ ... $P_{dn..\infty}$ which correspond to the probability to disturb n or more messages, and P_d for the probability that one or more are disturbed. The following relationships combine these variables :

$$P_d = 1 - P_0 \quad (5)$$

$$P_d = P_{d1} + P_{d2..\infty} = \sum_{i=1}^{\infty} P_{di} \quad (6)$$

$$P_{dn..\infty} = P_{dn} + P_{d(n+1)..\infty} = \sum_{i=n}^{\infty} P_{di} \quad (7)$$

The evaluation of these parameters is relatively easy and requires one other variable concerning the delay (T_b) between two successive messages. The case of the evaluation of P_d was studied previously. Concerning the case of several messages, the evaluation of P_{dn} is presented below, with the fault model of the section 4.2.

Figure 4: Perturbation of several messages

$$P_{d2} = \int_{2L+T_b}^{\infty} \lambda P[T_{FD} > t]\,dt \quad (8)$$

5 Reliability evaluation

In our study, the reliability evaluation consists in a first approach to establish the failure rate, that is to say the probability that the fieldbus mission does not respect the temporal specifications. Consequently, if a correct or incorrect state is applied at each reading according to the respect of the temporal constraints, it is the probability of the transition from correct to incorrect that is researched for an estimation. A representation of this principle is shown by a fault tree on the following figure. It's important to take into account the dependence between successive readings because a fault occurrence can have consequences for several reading cycles. And, one aim of this study is to determine the mean time between failure (MTBF) and not the failure rate of each reading.

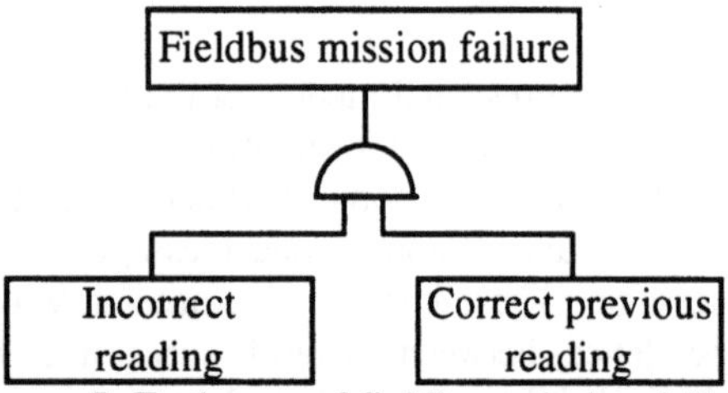

Figure 5: Fault tree of fieldbus mission failure

In the following, this fault tree is applied to the usual fieldbuses in order to compare them.

5.1 Periodical fieldbus

With this kind of fieldbuses, the data transmission is realised periodically and independently of the needs of the systems. These fieldbuses are characterised by predictable behaviour. A unique station often monitors the data transmission. It's the case of FIP, AS-Interface, Interbus-S... But several stations can sometimes achieve this activity of bus controller. In this case, they monitor the exchanges one after the other during a predefined length of time. For instance, the periodic transmission between a master and a slave, in the case of Profibus, follows this rule.

The fundamental principle is that the operations of data reading or writing are independent of the bus activity.

In the case that data reading and the data writing are periodical (but not necessary at the same frequency), the following figure shows the timing diagram of network activities.

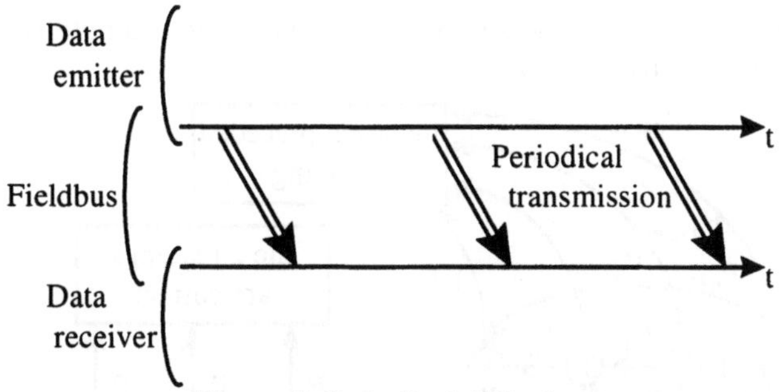

Figure 6: Periodical diffusion

Such transmission is characterised by the following parameters: period of writing for the emitter (P_w), period of diffusion realised by the fieldbus (P_{dif}), period of reading for the receiver (P_r) and transmit duration (T_d). the specification gives the time life of the datum to diffuse: T_L. Some values are chosen by the designer, but some of them are determined by the components. This last constrain results in dissimilar temporal parameters that can induce a complex evaluation. The following method takes into account this possibility that there can be a great difference between these temporal parameters. An example of this principle, a sensor with a predefined period of 100 ms gives its measure to a controller which runs at a period of 500 ms thanks to a fieldbus whose period is 250 ms. The problem is to determine if this period is sufficient to respect the time life of 500 ms despite the possibility that one error occur.

To resolve it, the proposed method requires the establishment of a set of triple (n_1,n_2,n_3) where n_1 is the number of successive disturbed messages after n_2 messages preceded by n_3 messages of whose one or more are correctly transmitted. On the following figure, the triple (3,1,4) means that in this temporal configuration, if 3 messages is disturbed after 1 message (without importance) and after one of 4 messages correctly transmitted, the data reading becomes "fail".

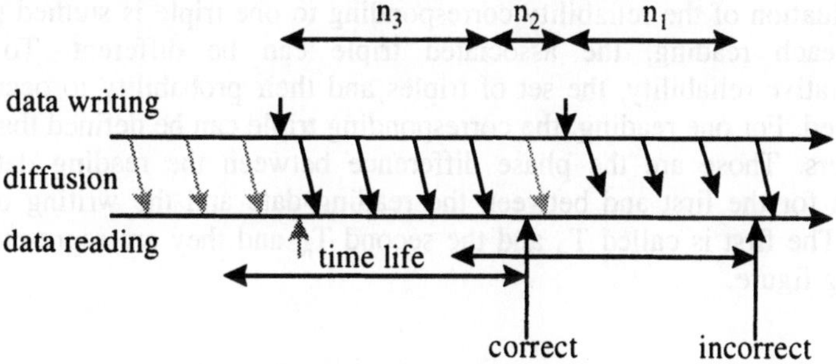

Figure 7: The triple (3,1,4)

With the use of one of the previous models of fault, the probability that the fieldbus do not respect the temporal specification, is feasible. To achieve this evaluation, the probabilities of failure of 1 or several successive messages and the probability of a non-disturbed message are needed. These values are called P_{d1}, P_{d2}, P_{d3}... P_{dn} where n is the number of disturbed messages and P_0 the probability to not be disturbed. Finally, with the help of a graph whose an example is shown below, the evaluation is relatively easy.

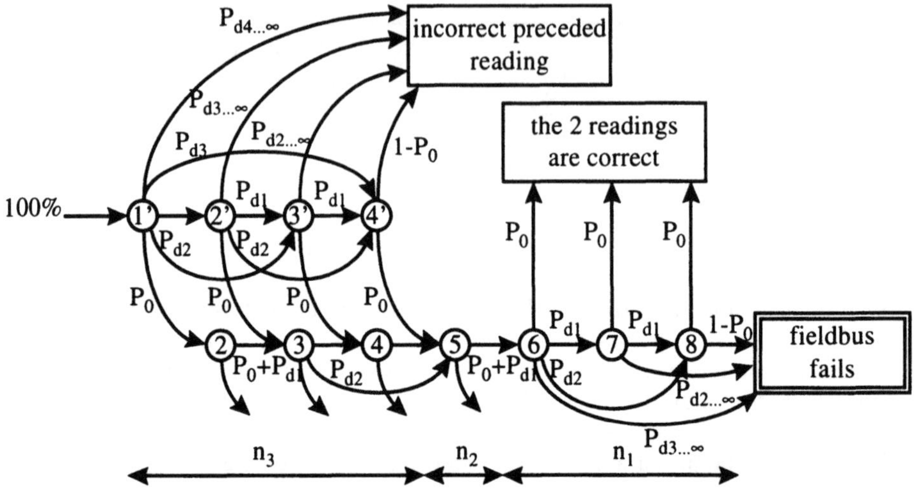

Figure 8: Graph to evaluate the probability of failure rate in the case of the triple (3,1,4)

The above graph represents the ways to reach at the state "fieldbus fails". The states 1', 2', 3', 4' are the states corresponding to when 0, 1, 2, 3 messages are not correctly diffused for the preceded reading. The state 5 is representative of a correct previous reading. And the states 6, 7, 8 express the incorrect transmission of the studied reading. The final state "disturbed preceded reading" and "the 2 reading are correct" is represented in order to explain the graph. With such a graph, the calculation is made easier by the fact that each state of the graph could be estimated successively with the order 1', 2', ...n_3' then 1, 2, ... (n_1+n_2+n_3).

The evaluation of the reliability corresponding to one triple is studied previously, but at each reading, the associated triple can be different. To obtain a representative reliability, the set of triples and their probability to occur must be established. For one reading, the corresponding triple can be defined thanks to two parameters. Those are the phase difference between the reading date and the diffusion for the first and between the reading date and the writing one for the second. The first is called T_{rd} and the second T_{rw} and they are represented on the following figure.

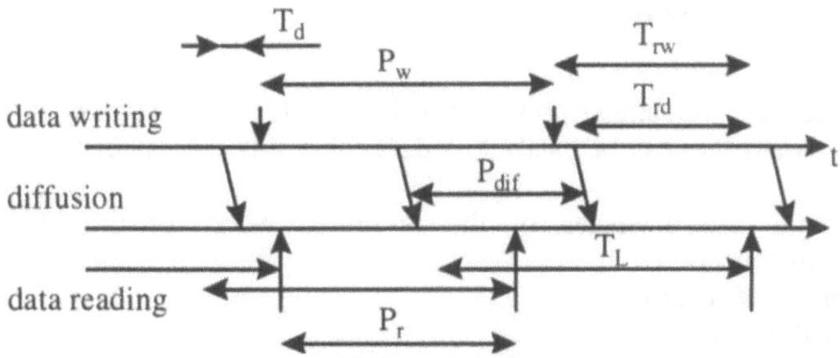

Figure 9: Temporal diagram

The date of the oldest usable writing (T_{OW}) can be established with the next formula. That date is relative to the date of reading.

$$T_{OW} = T_{rw} + P_w[1 + \mathrm{Int}(\frac{T_L - T_{rw}}{P_w})] \quad (9)$$ where Int() is the integer part.

With it, the number of messages (n_1) can be evaluated according to the delay T_{rd}:

$$n_1 = 1 + \mathrm{Int}(\frac{T_{ow} - T_{rd}}{P_{dif}}) \quad (10)$$

Consequently, the number of messages depend on two variables at each reading, T_{rw} and T_{rd} and their probabilities of value are uniform on their interval. The others are constant at each reading. The different values of n and their probabilities can be found thanks to the table below.

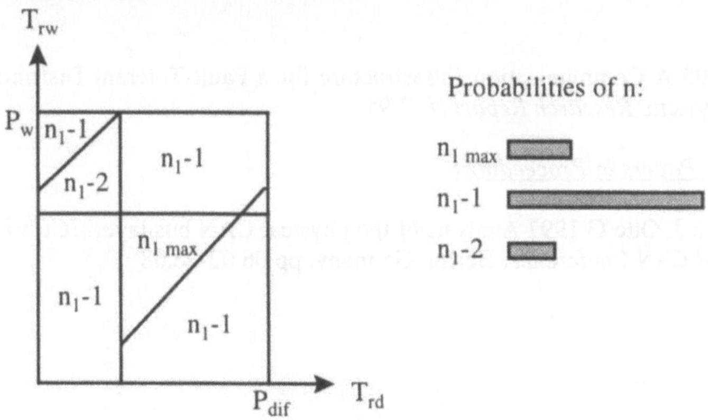

Figure 10: Table of the number of messages (n)

The evaluation of n_2 and n_3 uses the same principle. In the previous table, each zone is then associated with a triple (n_1,n_2,n_3).

5.2 Simple Example

The proposed example is the same one than in the beginning of the section 5.1. A sensor with a predefined period of 100 ms gives its measure to a controller which

6 Conclusion

The method in development uses a spreadsheet which is very easy to use. Various values of parameters can be easily tested. By this way the calculation is completely automatic.

The main insufficiencies of the presented work, which are presently in development concern:
- the problems of security
- the aperiodic cases
- the gigue phenomena
- the lack of feedback

References

Journal Articles

1. Bayart M, Robert M et al. 1998 Dans la jungle des réseaux de terrain vers un guide de choix dédié Automatisation d'application. *REE* 3: 44-48

2. Cauffriez L, Defrenne J 1995 Viabilité de l'information et réseau à diffusion. *Diagnostic et sûreté de focntionnement, Vol 5,* 2: 219-247

3. Navet N , Song Y Q , Thomesse J P 1998 Le réseau CAN et les erreurs de transmission. *Terrain* 16 : 15-18

Books

4. Kopets H 1995 A Communication Infrastructure for a Fault-Tolerant Distributed Real-Time System. *Research Report Nr 7/95*

Book Chapters or Papers in Proceedings

5. Barrenscheen J, Otte G 1997 Analysis of the physical CAN bus layer. *ICC'97, 4th International CAN Conference*, Berlin, Germany, pp 06.02-06.08

Annex: Example of datasheet for the evaluation of fieldbus failure rate.

Failure Analyse

failure rate with the fieldbus WorldFIP

Environment description:

Reading period for the consummer:	**1**	s
Life time of the datum:	**1**	s
Period of diffusion realised by the fieldbus:	**0,5**	s
Transmission protocol, length of 1st frame:	**64**	bits
Transmission protocol, lenght of 2nd frame:	**128**	bits
Resumption method :	**FAUX**	True/False
Probability of non-activation :	**0%**	probabilité
Resumption protocol, length of 1st frame:	**0**	bits
Resumption protocol, lenght of 2nd frame:	**0**	bits
Number of bus:	**2**	1 ou 2
Perturbation of one bit on one only bus:	**0,0010%**	probability/bits
Perturbation of one bit on all buses:	**0,0005%**	probability/bits

Failure rate:

Mean time between failure :	1081494	s
equivalent (max: 9:06:07)	---------	h:mm:ss
equivalent:	12,52	days
Faimure rate of transmission:	0,000%	probability
Number of diffusion at each cycle:	2	
in:	100%	probability
Number of diffusion at each cycle:	3	
in:	0%	probability
Transmission et resumption failed:	0,10%	probability
Resumption failed:	100,00%	probability
Transmission failed:	0,10%	probability
Frame 1 erroneous:	0,03%	probability
Frame 2 erroneous:	0,06%	probability
Frame 1 disturbed on all buses:	0,03%	probability
Frame 2 disturbed on all buses:	0,06%	probability
Frame 1 disturbed on bus 1:	0,06%	probability
Frame 2 disturbed on bus 1:	0,13%	probability
Frame 1 disturbed on bus 2:	0,06%	probability
Frame 2 disturbed on bus 2:	0,13%	probability
Frame 1 erroneous during resumption :	0,00%	probability
Frame 2 erroneous during resumption :	0,00%	probability

PART III

SYSTEM FORECASTING, OPTIMIZATION AND CONTROL

16

Industrial Forecasting Using Knowledge-Based Techniques and Artificial Neural Networks

S.G. Tzafestas and N. Mekras

1 Introduction

Most techniques used for forecasting can be classified in under three main categories. The first treats the demand as a time series and predicts it using different time series analysis techniques [1], [2]. In the time series forecasting approach several methods (like, for example, the exponentially smoothing technique) have been developed to give different weight to more recent data, to monitor the forecasting error and to adjust the smoothing factor to the evolution of the error using Trigg's warning signal [2]. General problems with the time series approach include the inaccuracy of prediction and numerical instability due to lack of information about external factors that influence the model and which are not included in the time series historical data [3], [4]. The second approach is based on regression techniques which accept that demand is heavily dependent upon external factors. Using linear or non-linear regression analysis a functional relationship is created between the demand and selected external variables. However, this relationship usually depends also upon temporal past data making the relationship non-stationary time dependent [3]. The third approach which has appeared more recently is based on Artificial Neural Networks [3], [5]. It is a very interesting approach that gives sufficient results with small forecasting errors. The ANNs can generate the forecasting result after they have been trained and have learned from past historical data, that might concern temporal or any other functional relationship between the model's inputs and the real demand of previous historical periods. During the training phase ANNs require heavy computations until they minimize the forecasting error, arriving at a state that corresponds to a global minimum of the error function [7]. After been trained, the ANN is used to forecast the demand for future time periods. The Multi-Layer Perceptron (MLP) combined with the Backpropagation (BP) training algorithm is one of the most common Artificial Neural Networks used for forecasting [3], [6]. In most cases the ANNs are trained with a large amount of data from the past and no-emphasis is given to the selection of data which are going to be used for training . Also the large amount of training data makes the continuous retraining generally difficult and almost impossible in short-term forecasting where the time period is rather short (for example an hour) [4].

This chapter presents a forecasting method which employs Knowledge Based (KB) techniques and Artificial Neural Networks (ANNs). The Knowledge Based techniques are used to select an appropriate and efficient data set for training an ANN, given a limit of the forecasting error, and a pattern of the ANN's inputs at the forecasting periods. Also, the KB system monitors and records the error and when necessary it selects a new data set for retraining the Neural Network. In this way adequate data sets are being created for several input patterns, which are stored, used and adjusted in any changes of the model's behavior. Further more a minimization of the size of the training data set is achieved making the method time-efficient especially in cases of short-term forecasting.

Regarding the implementation of the proposed forecasting system, the PROLOG language and the MATLAB software package have been used for the development of the Knowledge Base and the Neural Network, respectively.

2 Description of the method

This section presents a description of the proposed method including the architecture and contents of the Knowledge Base (KB), and the structure of the Neural Network.

Rules and frames [8], [9], [10] are the main Knowledge Representation Structures (KRS) used for creating the KB. The knowledge used is classified in two main categories :

A) Knowledge used for data selection.
B) Knowledge used for data classification.

For the data selection process, the rules which are used mainly concern the following :

- The size of the training data set.
- The periodicity of data and the existence of seasonal patterns.
- The limits and the trends (increasing or decreasing) of the training data in relation with the limits of the forecasting input data.

- The evolution of the forecasting error in relation to the current training data set.
- The dimension of the input array (number of ANN inputs) in relation to the forecasting error.
- The size of the ANN (number of hidden layers and neurons) in relation to the size of training data set and the needed training epochs.

For the data classification process the following KRS are used :

- Rules for creating classes of data sets (like limits of the class. increasing or decreasing trends. size of input array).
- Frames and sub-frames for representing classes and sub-classes of data sets with generic and more specific characteristics.
- Rules for extracting the data set classes out of the frame structures.

Using the previous knowledge the KB system finds a training data set, out of the whole data set available. which gives a minimum mean absolute forecasting error for specific input patterns that correspond to a number of continuous forecasting periods. In this way a training data set with small size is selected that gives the best results for the specific forecasting ANN's inputs. This data set, the trained ANN and their corresponding forecasting inputs are stored and used again when similar input patterns appear. During forecasting, the KB system monitors continuously the forecasting error and, using specific rules. decides if it is necessary to retrain the ANN with a new updated data set. In this case the KB system, updates the data set and retrains the ANN until a minimum error is achieved again, giving the possibility to the ANN to be adjusted to any changes of the model's behavior.

One important aspect that should be considered, for the process to succeed. is that at each training cycle it must be assured that using the backpropagation algorithm the error function will reach the global minimum [7], otherwise the overall method will give wrong results for the error and the size of the data set.

3 Implementation of the method

Several types of ANNs have been used for forecasting purposes the most common of which is the Multi-Layer Perceptron (MLP) which is trained using the backpropagation algorithm. Two very interesting applications of MLP Neural Networks for electrical load forecasting are described in [4], [6]. In [4] the structure of the ANN is rather simple and for the forecasting of the total daily load and the hourly load includes one hidden layer with 5 and 10 neurons respectively. Other MLPs with more hidden neurons do not provide better results for the same type of forecasting [4]. Also, the large amount of training data (more than 2000 input-output data arrays), makes the whole training process time consuming even in a SUN SPARK station [4].

An ANN with almost a similar structure has been used for a prototype implementation of the method proposed here, i.e. an MLP ANN with the backpropagation training algorithm. Note that there is not any restriction for the application of the method to other types of ANNs. The structure our ANN is a 2-5-1 network and it uses as transfer functions, for the two first layers, the hyperbolic tangent sigmoid function and for the third (output) layer the linear function. The ANN's architecture is given in the Figure 1.

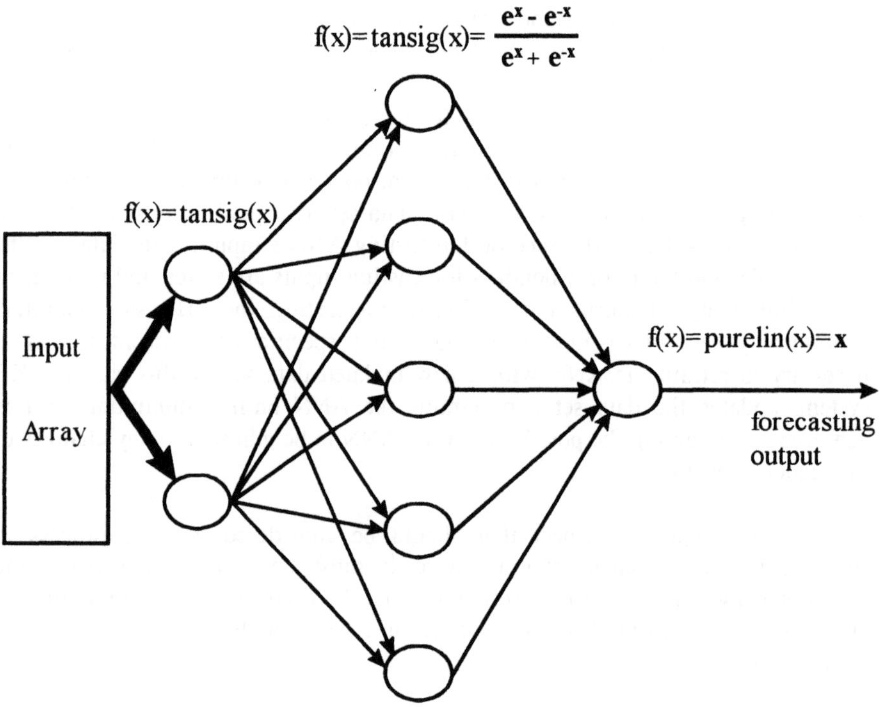

Figure 1. Neural Network architecture

For testing the method a *Gas consumption database* (drawn from [11]) has been used. This Gas consumption database contains data on the average monthly gas consumption (in gallons per capita) and the average minimum temperatures (in degrees Celcius) in the city of Hamilton, New Zealand, for the period January 1988 to July 1992. The knowledge based (KB) system provided the ANN with data sets selected from the four-year data (Jan 1988 until Dec 1991) and testing data for a six month period (Jan 1992 until Jun 1992).

The input data array for each period includes :

1. Month at period t
2. Monthly average of minimum temperatures
3. Monthly Gas consumption at period t
4. Monthly Gas consumption at period t-1
5. Monthly Gas consumption at period t-2

The overall architecture of the system is given in the Figure 2.

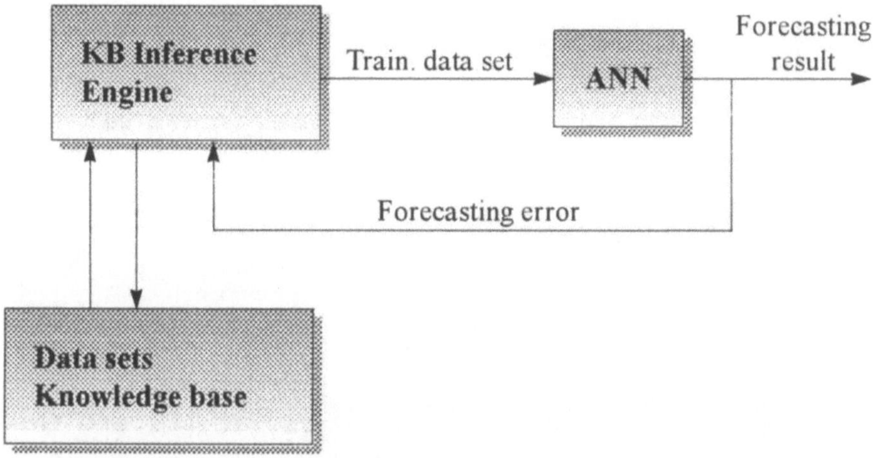

Figure 2. Knowledge Based Forecasting ANN system architecture

4 Performance of the method

For the given ANN the result obtained - mean absolute error - was the best for a relatively small data set and with data being excluded by the rules. Using data sets with more and extra data, or the whole data set available, without employing the excluding rules, worst results were obtained, as it can be seen in the next figures. Applying the method for several forecasting periods, various input patterns of the forecasting period can be stored with their related training data set. When forecasting and retraining of the ANN is necessary, this training data set forms the input to the KB system that tries to add or exclude data in an effort to minimize the error which is continuously monitored by the KB system. Figures 3a and 3b show in detail how the method reached the minimum mean absolute error, that corresponds to the best training data set for the given 6-month forecasting period.

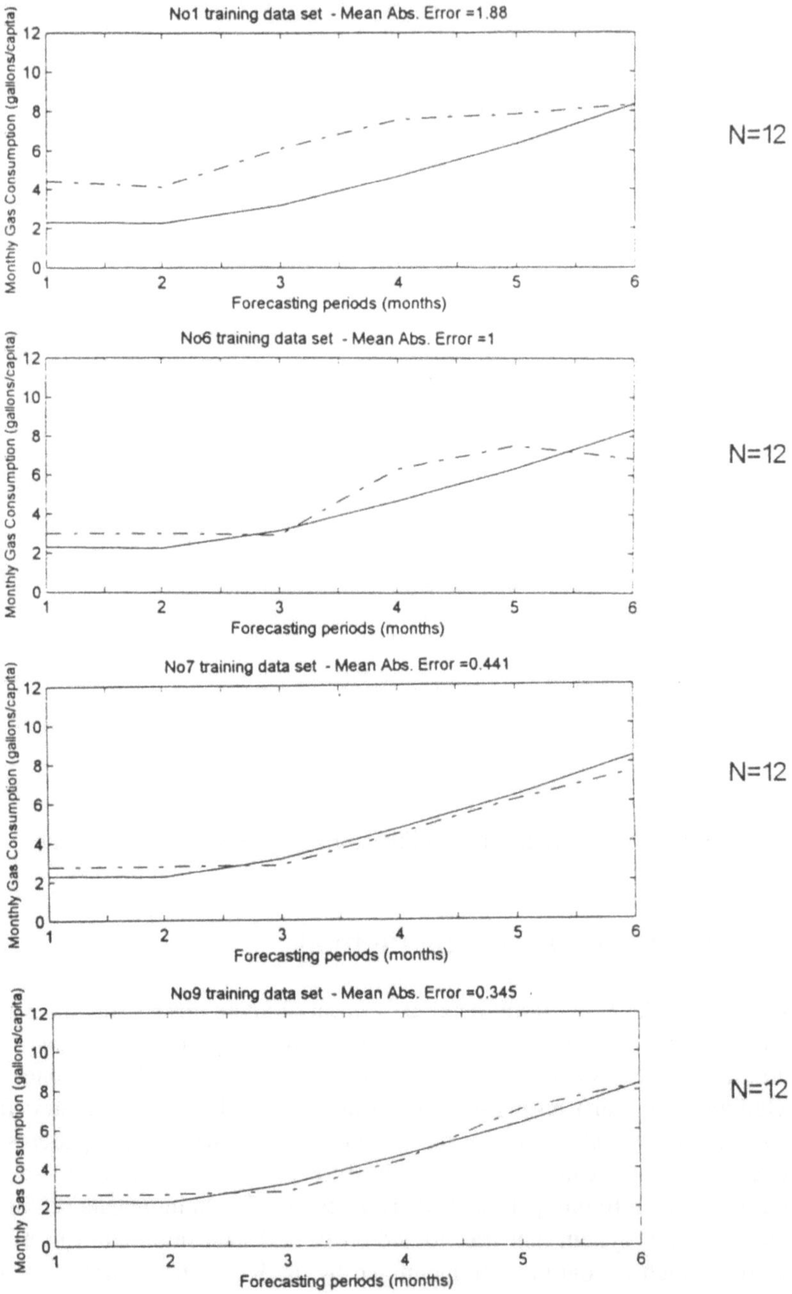

Figure 3a. Forecasting results (N: size of training data set)

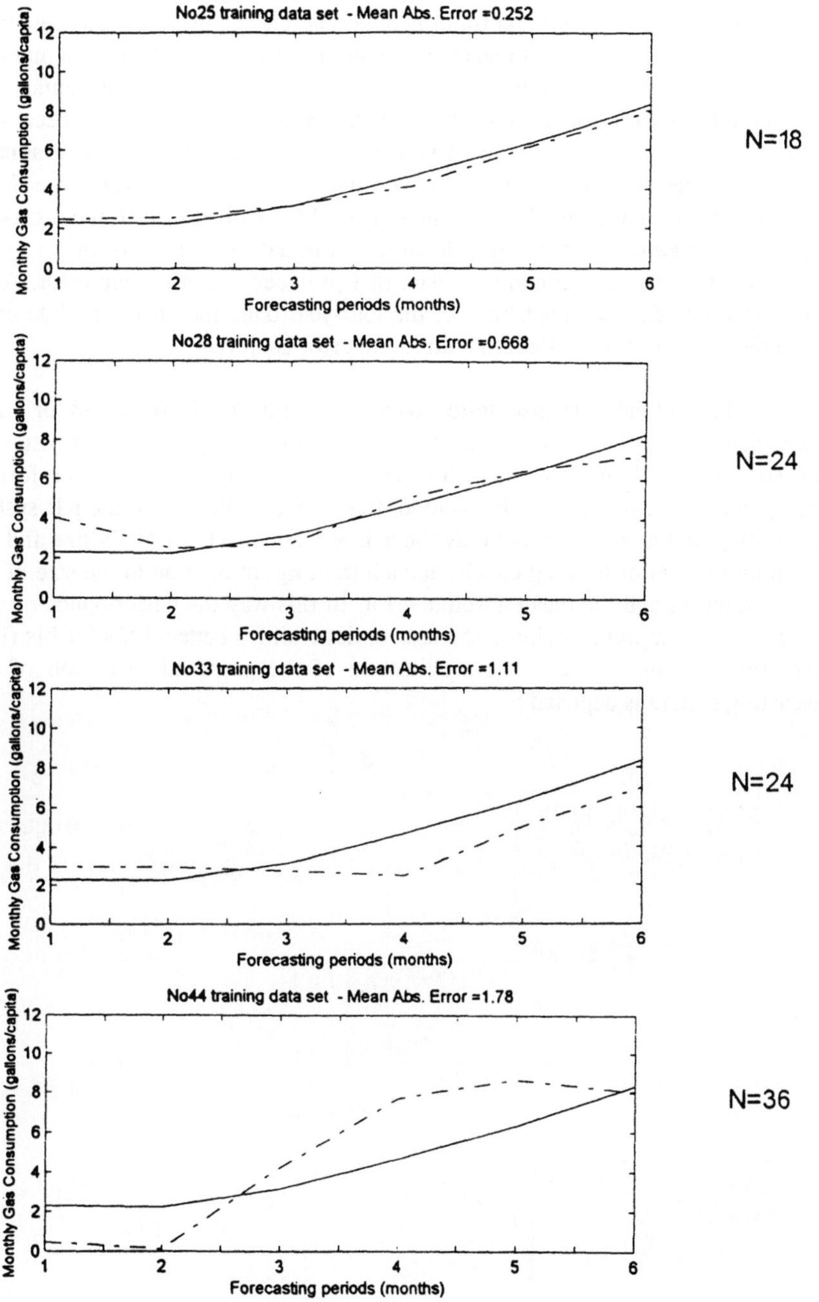

Figure 3b. Forecasting results (N: size of training data set)

The diagrams on the two previous figures show successive forecasting results that correspond to several training data sets selected by the Knowledge Base system and the way that the forecasting results have matched the real demand at each month of the forecasting period. A data set with size N=18 is found, that gives the minimum mean absolute error for the testing period, as it can been seen in the first diagram of Figure 3b. Using the KB system, the method gradually optimizes the result starting from a relative small data set and increasing the size it finds the data set that gives the minimum error. As it can be seen in the last three diagrams of Figure 3b, increasing the size of the data set leads to bigger error, even if the new data sets contain the data that provided the minimum error. Also, using the whole data set available, i.e. the four-year data, the mean absolute error was higher and even worst than the minimum error achieved.

The method was also tested using different ANNs with less or more hidden neurons and in all cases one data set was found that gave the best result for each specific ANN. A very interesting aspect should be considered for further development and extension of the method, that concerns the use of the KB system for selecting and monitoring, not only the data set, but also the ANN's size and the maximum number of training epochs at each training, in relation to the size of the training data set and the mean absolute error. In this way the performance of the method will be improved helping the user to determine a better ANN for his (her) forecasting. In the Figure 4 the architecture of the proposed extension of the forecasting system is depicted :

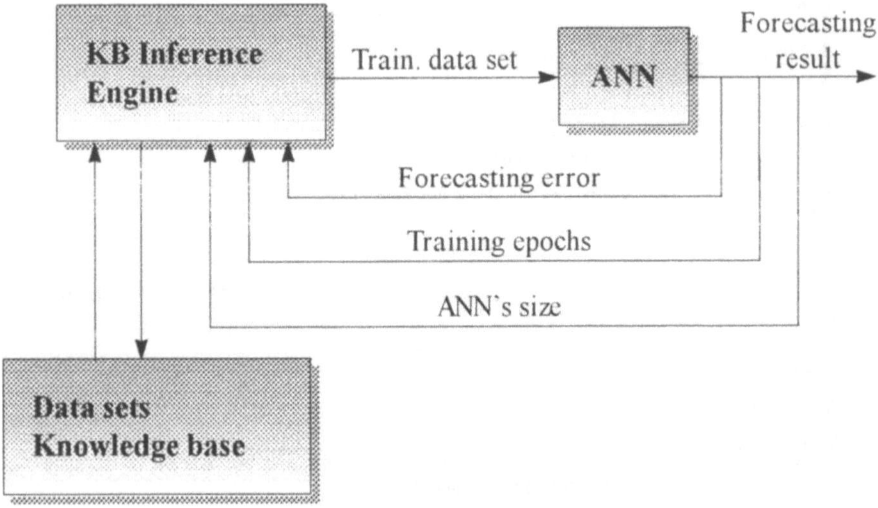

Figure 4. Knowledge Based Forecasting system extended architecture

5 Conclusions

In this chapter a forecasting method that merges Knowledge Based techniques and Artificial Neural Networks has been presented. The Knowledge Based system provides the ANN with training data selected out of the whole historical data set available. This selection is based on rules that concern data periodicity, seasonal patterns, data limits, decreasing or increasing trends, error evolution and other relevant factors that express the relations which exist between historical data and input data for the forecasting period. In this way subsets of data are created that correspond to similar input patterns and are used for training the ANN in several cases.

The method was implemented and a prototype system was developed, using PROLOG and MATLAB, that gave sufficient results. The minimum mean absolute error for the forecasting testing period, was achieved when the ANN was trained with a relatively small data set which was selected by the Knowledge Base. Training the ANN with other data sets with more data or with the whole data set available gave worst results. The ANN used, was a Multi-Layer Perceptron (MLP) trained with the BP algorithm, but the method can also be applied to other types of ANNs.

In the proposed method, emphasis was given on the selection and creation of appropriate training data sets for several input and output forecasting patterns. In this way, the size of the training data set and the size of the ANN are reduced, making easier the implementation of ANNs on smaller computers (for example PCs) with acceptable running times.

References

1. Bowerman B L, O'Conell R T 1993 *Forecasting and Time Series : An Applied Approach*. Duxbury Press, Belmont California.

2. Lewis C D 1981 *Forecasting, in Operations Management in Practice*. Philip Allan Publishers Ltd., Oxford

3. Song Y H, Johns A, Aggarwal R 1995 *Computational Intelligence Applications to Power Systems*. Science Press, Kluwer Academic Publishers, New York

4. Park D C, El-Sharkawi M A, Marks R J et al 1991 Electric load forecasting using an Artificial Neural Network. *J IEEE Transactions on Power Systems*, Vol. 6, No. 2 : 442-449

5. Swingler K 1996 *Applying Neural Networks*. Academic Press Ltd., London

6. Bakirtzis A, Petridis B, Kiartzis S, Maissis A 1995 A neural network short term load forecasting for the Greek power system. In *CIGRE '95 Conference Proceedings*, Athens

7. Hagan M T, Demuth H B, Beale M 1995 *Neural Network Design*. PWS Publishing Co, Boston

8. Tzafestas S 1996 *Introduction to Artificial Intelligence and Expert Systems*, National Technical University of Athens, Athens

9. Frost R A 1987 *Introduction to Knowledge Base Systems*. William Collins & Sons Ltd., London

10. Malpas J 1987 *PROLOG: A Relational Language and its Applications*. Prentice-Hall, Englewood Cliffs, New Jersey

11. Gonzales L 1992 Report on Temperature Effect on Gas Volumes, Department of Finance and Quantitative Analysis, University of Otago, New Zealand

17
DAPHNE: a neural network based short-term load forecasting program. Application to an autonomous power system.

S.J. Kiartzis, S.E. Papadakis, J.B. Theocharis, A.G. Bakirtzis and V. Petridis

1 Introduction

Accurate short term load forecasting (STLF) is a necessary part of resource management for a power generation company. The more precise the hourly load forecast, the more closely the power generation assets of the company can be configured to minimize the cost. Significant forecasting errors can lead to either overly conservative or overly risky scheduling, which can in turn induce heavy economic penalties [1]. Deregulation and consequent increase in competition makes a company's ability to accurate forecasts an important contributor to its future success [2]. Automating the load forecasting process is a profitable goal and neural networks provide an excellent means of doing the automation [3].

A variety of STLF models have been reported in the literature, including statistical and regression based methods [4-5] and expert systems applications [6], while neural networks [7-10] and fuzzy systems [11-13] have also been employed. A significant number of power companies world-wide has embarked on neural network based STLF models, which offer significant benefits over the traditional (statistical or regression) techniques, such as reduced modeling effort and high accuracy [1,8,9,10,14].

This chapter presents DAPHNE a neural network based short term load forecasting software package. Forecast results from the off-line application of DAPHNE to the autonomous power system of the island of Crete are also reported.

2. The DAPHNE Program

The Demand Advanced Prediction Hellenic Neural Environment (DAPHNE) program is designed to provide hourly electric load forecasts for the Greek Public

Power Corporation (PPC) power system (Figure 1), for a period of one to seven days ahead. DAPHNE is a Windows based program that depends on neural networks technology and utilizes load and weather data to provide load forecasts. The program's main functions are:

- creation of training patterns
- training of the neural network (based on the predefined structure)
- load forecasting
- statistical evaluation of load forecasts

2.1 Neural network structure

The proposed short term load forecasting model is based on a fully connected three layer feedforward neural network. The neural network consists of 64 input neurons, 48 hidden neurons and 24 output neurons representing next day's 24 hourly forecasted loads. Temperature is a crucial parameter that affects load consumption behaviour therefore, it plays a key role as an input to the neural network. Both historical temperature and load data and temperature forecasts are employed in the network's input set. The neural network inputs and outputs are described in Table I.

The first 48 inputs represent historical hourly load data for today and yesterday. Inputs 49-52 are maximum and minimum daily temperatures for today and temperature forecasts for the forecast day. Inputs 53-54 represent the square value of the deviation of daily maximum temperature and daily maximum temperature forecast from the region of the cooling and heating threshold temperatures ($T_{c,min}$ = 17°C, $T_{c,max}$ = 26°C) [9], that is :

$$CT = \begin{cases} \left(T - T_{c,min}\right)^2 & \text{if } T < T_{c,min} \\ 0 & \text{if } T_{c,min} \le T \le T_{c,max} \\ \left(T - T_{c,max}\right)^2 & \text{if } T_{c,max} < T \end{cases}$$

Input 55 is the forecast of daily maximum temperature change. The following two inputs (56-57) represent the day of the year (referring to the forecast day) and are given in the form: $\sin(2\pi d/365)$ and $\cos(2\pi d/365)$, where d=1, ..., 365 is the day of the year. The last seven inputs, 58-64, represent the day of the week, bit encoded. Other input variables were also tested but they did not improve the performance of the proposed model.

2.2 Input data pre-processing

Short-term load forecasting in an autonomous power system proved to be a rather difficult task due to frequent irregularities in the available historical load data. It was found that almost 2.5% of load data were missing or disturbed and finally replaced by the introduced pre-processing procedure. Load data are missing or are false due to unrecorded samplings or to bad measurements [15].

Table I. Definition of the ANN inputs and outputs.

Inputs	Description
1-24	L (d-1, h), h = 1, 24
25-48	L (d-2, h), h = 1, 24
49-50	T_{max} (d-1), T_{min} (d-1)
51-52	$\hat{T}_{max}$ (d), $\hat{T}_{min}$ (d)
53	CT (d-1)
54	$C\hat{T}(d)$
55	$\hat{T}_{max}(d) - T_{max}(d-1)$
56-57	sin(2πd/365), cos(2πd/365)
58-64	Day of the Week
Outputs	**Description**
1-24	$\{\hat{L}(d, h),\ h = 1,24\}$

d = day index, h = hour of day index

L = load

$\hat{L}$ = load forecast

T = temperature

T_{min} (T_{max})= minimum (maximum) temperature

$\hat{T}_{min}$ ($\hat{T}_{max}$)= min. (max.) temperature forecast

CT = square value of the deviation of daily maximum temperature from the comfort temperature region

$C\hat{T}$ = square value of the deviation of daily maximum temperature forecast from the comfort temperature region.

In the case of the autonomous power system we examined (island of Crete), irregularities in load data are also due to the frequent load interruptions, especially during the summer months, attributed to delays in the construction of new generating plants.

Two kinds of problems are observed in neural network based STLF, because of irregularities in load data :

a) During the neural network training, irregular load data result in large training errors (i.e. the neural network cannot be trained to a desired small tolerance).
b) During the consulting phase, irregular load data, when introduced as inputs to neural network (even to a correctly trained network) result in large forecast errors.

In order to overcome these difficulties "raw" historical load data are pre-pro-cessed before they are introduced to the neural network either as training patterns or as test inputs. Data pre-processing is one of DAPHNE's subsidiary functions (Figure 2).

2.3. Current day and next week forecasts

The basic neural network load forecasting model described in previous section assumes that load and weather data are available until 12 midnight of the previous day and produces load forecasts for the current day. As new load data become available, during the current day, load forecasts can be refined so as to take into account the most recently available data. To this purpose, in addition to the basic forecasting model defined in Table I, 23 auxiliary neural network models are defined for refined current day's forecasts. The m-th auxiliary model assumes that load data are available until hour m of the current day and produces the forecast of the remaining current day's loads. The structure of the auxiliary neural network models is similar to the one of Table I except for the fact that the number of neural network outputs is reduced to 24-m, whereas the 48 first inputs represent the 48 most recently available loads.

With the combined use of the basic neural network model and the m-th auxiliary model next day's load forecasts can be obtained at any hour, m, of the current day taking into account the most recently available load data. First the auxiliary model is used in order to provide the refined forecasts for the remaining loads of the current day (hours m+1, ..., 24) which are then used in order to complete the input data set of the basic neural network model that gives next day's load forecasts. This procedure complies with standard operating practices : system dispatchers need next day's load forecasts some time between 8 am and 12 noon of the current day when next day's generation scheduling is performed.

A number of Energy Management System operations like unit commitment and hydrothermal co-ordination require load forecasts in a time horizon of a week. Next week's load forecasts are obtained with the recursive use of the basic neural network forecasting model by replacing unavailable neural network load inputs with already obtained forecasts and using one week ahead weather forecasts.

3. Test Results

The proposed neural network based model for STLF was applied to the power system of the island of Crete. It is an autonomous power system with a summer peak load of about 300 ÌW during 1997 and is supplied by the Public Power Corporation (PPC).

Table II. Comparison of forecast errors with and without data pre-processing (testing period : the first 100 days of 1998).

type of load data		training error (%)	forecast error (MW)	forecast error (%)	maximum error (MW)
training	consulting				
pre-processed data	"raw" data	2.00	7.09	2.71	249.2
pre-processed data	pre-processed data	2.00	6.18	2.42	45.2

Table III. Statistical analysis of load forecast errors for the first 100 days of 1998.

DAYS AHEAD	AVERAGE (MW)	ERROR (%)	MAX.ERROR (MW)	HOURS ERROR GREATER THAN 5MW	10MW	15MW	20MW	25MW
1	6.18	2.42	45.2	485	142	37	4	0
2	8.38	3.26	49.9	724	251	90	17	0
3	9.86	3.84	56.9	858	342	117	37	4
4	11.02	4.29	60.5	912	381	176	85	18
5	12.29	4.76	68.6	1005	464	231	96	29
6	13.24	5.14	68.5	1089	508	262	101	36
7	13.55	5.28	80.5	1175	516	225	80	38

The neural network was trained using "raw" load data from the period of 1991 to 1997, as they were provided from PPC Since load forecasts were characterized by big errors the data pre-processing procedure described in section 2.2 was applied. and the resulting forecast errors were significantly lower. The neural network was trained to the predefined tolerance of 2.00%, leading to an average forecast error (for the first 100 days of 1998) of 6.18 MW or an equivalent average percent forecast error of 2.42%, expressed with respect to daily peak (Table II). Figure 3, gives a sample daily load curve forecast. In Table III, statistical analysis of load forecast errors for the first 100 days of 1998, for next week STLF is presented. In Figure 4 the user interface for the DAPHNE's forecast accuracy statistics, is presented.

4. Conclusions

"DAPHNE" is a windows-based neural network short term load forecasting software package. DAPHNE can provide load forecasts with lead times up to a week while an auxiliary model is employed to haddle incomplete current day load information. Input data pre-processing is introduced in order to improve forecast accuracy.

Acknowledgement

This research was funded by the Secretary General of Research and Technology of the Greek Ministry of Industry, Energy and Technology and the Public Power Corporation (PPC) of Greece. The authors would like to thank Mr Katsigiannakis. Mr Stefanakis and Ms Gigantidou of PPC for supporting this research and providing the necessary data.

References

1 IEEE/Power Engineering Society 96TP112-0 1996 A tutorial course on Artificial Neural Networks with Applications to Power Systems. (eds) M. El-Sharkawi and D. Niebur *IEEE*

2 D. Niebur et al. 1995 Artificial neural networks for Power Systems. CIGRE TF38.06.06 Report *Electra* 159:77-101

3 W.J. Gerber 1997 CoSS, STRICOM, US Army, research paper (http://www-mskeg.stricom.army.mil/papers/Gerber/).

4 G. Gross and F.D. Galiana 1987 Short term load forecasting. *Proc. IEEE* 75:1558-1573

5 A.D. Papalexopoulos and T.C. Hesterberg 1990 A Regression-Based Approach to Short Term System Load Forecasting. *IEEE Trans. on Power Systems* 5:1535-1547

6 S. Rahman and R. Bhatnagar 1988 An expert system based algorithm for short term load forecast. *IEEE Trans. on Power Systems* 3:392-399

7 D.C. Park, M.A. El-Sharkawi, R.J. Marks, L.E. Atlas and M.J. Damborg 1991 Electric Load Forecasting Using an Artificial Naural Network. *IEEE Trans. on Power Systems* 6: 442-449

8 O. Mohammed, D. Park, R. Merchant, T. Dinh, C. Tong, A. Azeem, J. Farah and C. Drake 1995 Practical Experiences with an Adaptive Neural Network Short Term Load Forecasting System. *IEEE Trans. on Power Systems* 10: 254-265

9 A.D. Papalexopoulos, S. How and T.M. Peng 1994 An Implementation of a Neural Network based Load Forecasting Model for the EMS. *IEEE Trans. on Power Systems* 9: 1956-1962

10 A.G. Bakirtzis, V. Petridis, S.J. Kiartzis, M.C. Alexiadis and A.H. Maissis 1996 A Neural Network Short Term Load Forecasting Model for the Greek Power System. *IEEE Trans. on Power Systems* 11:858-863

11 A.G. Bakirtzis, J.B. Theocharis, S.J. Kiartzis and K.J. Satsios 1995 Short Term Load Forecasting Using Fuzzy Neural Networks. *IEEE Trans. on Power Systems* 10:1518-1524

12 D. Srinivasan, C.S. Chang and A.C. Liew 1995 Demand Forecasting Using Fuzzy Neural Computation, with special emphasis on weekend and public holiday forecasting. *IEEE Trans. on Power Systems* 10:1897-1903

13 S. Papadakis. J. Theocharis, S. Kiartzis and A. Bakirtzis 1998 A Novel Approach to Short-term Load Forecasting using Fuzzy Neural Networks. *IEEE Trans. on Power Systems* 13:480-492

14 A. Khotanzad, R.C. Hwang, A. Abaye and D.J. Maratukulam 1995 An Adaptive Modular Artificial Neural Network Hourly Load Forecaster and its Implementation at Electric Utilities. *IEEE Trans. on Power Systems* 10:1716-1722

15 A. Piras, A. Germond, B. Buchenel, K. Imhof and Y. Jaccard 1996 Heterogeneous Artificial Neural Network for Short Term Electrical Load Forecasting. *IEEE Trans. on Power Systems* 11:397-402

Figure 1. The DAPHNE Program.

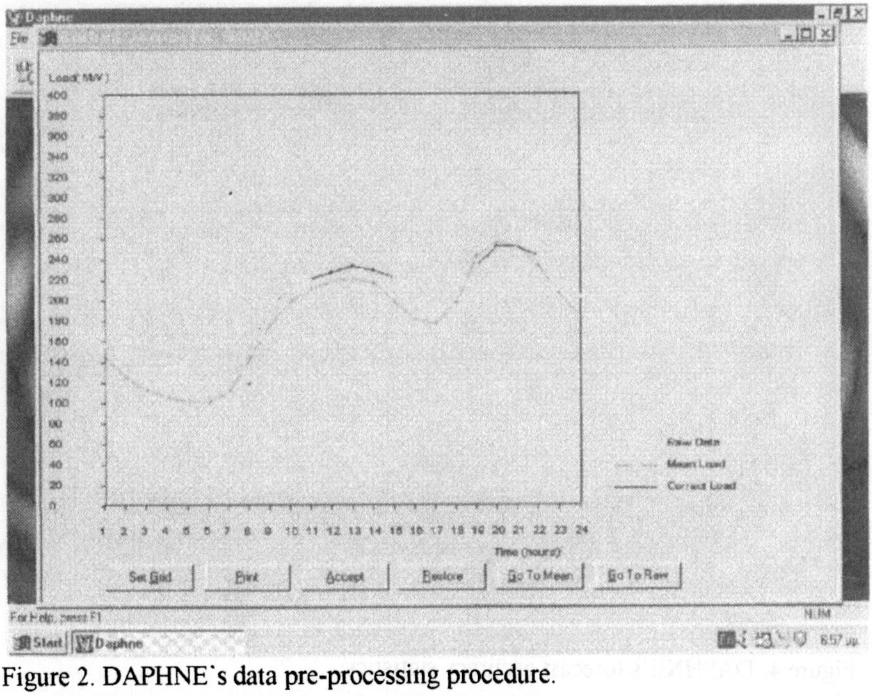

Figure 2. DAPHNE`s data pre-processing procedure.

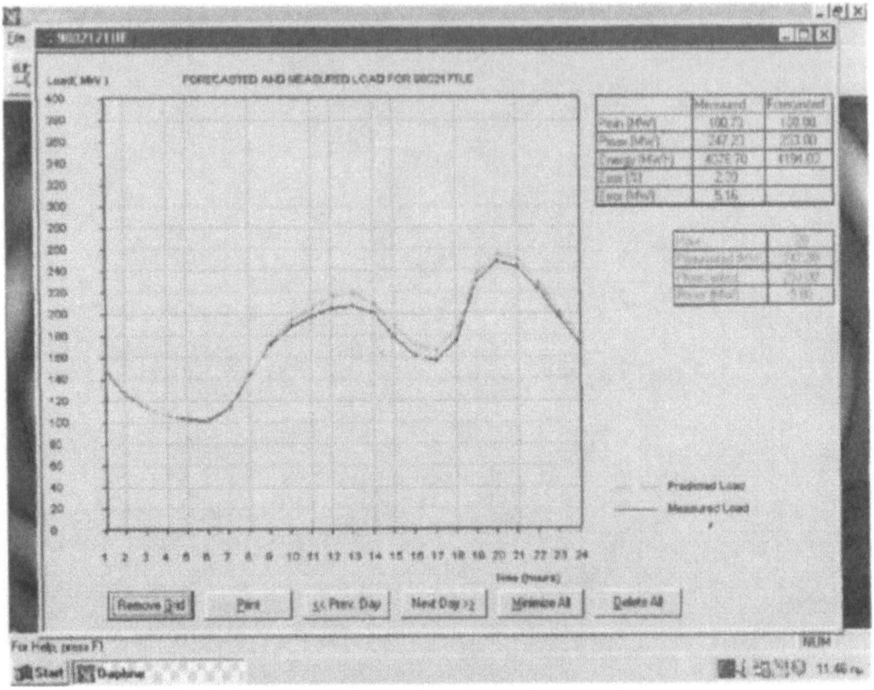

Figure 3. DAPHNE's sample daily load curve forecast.

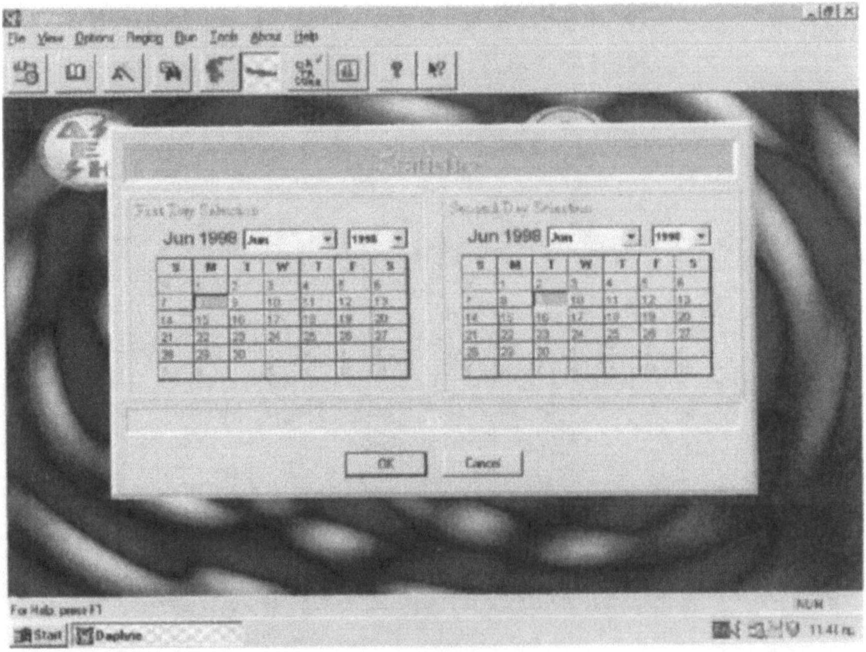

Figure 4. DAPHNE's forecast accuracy statistics.

18
Process Optimization System Used in Metallurgical Plants

F. Hartescu and C. Danilov

1 Description of Technological Processes

Agglomeration Factory
The raw materials used in the technological process are iron ore, coke, limestone and dolomite. The production result at the agglomeration factory is exclusively destined for feeding blast furnaces. The agglomeration factory is composed of the following technological fluxes: measuring station, agglomeration machine, and sorting and dispatching device

Iron ore, coke, limestone, the returning agglomerate from sorting, and iron offal, are piled in containers of measuring station. The raw materials are extracted from containers with running belts conducted by the programmable logic controllers (PLC) that receives signals from the electronic measuring device. Then, they are homogenized and sent to the agglomeration machine.

The agglomeration machine does the pyro-metallurgical process of sintering, the resulting agglomerate being riddled at high temperature, then cooled, sorted and dispatched to the blast furnaces. The agglomerate is sifted and sorted by size and granulation.

Cowper Stove
The role of the cowper stoves (pre-heaters) is to heat the air up to 1100 ^{0}C and to supply this to the blast furnace. In almost all the cases there are four pre-heaters. They have a focus where gaseous combustible burns. The body of the pre-heater has a refractor cellular structure that keeps the warmth made by burning gases.

The pre-heater has two fazes of working:

- The heating period, when the refractor structure is heated at a high temperature by burning the gaseous combustible into the focus
- The blowing period, in which the cold air circulating on the contrary of the burned gases in the recent period takes the heat accumulated by the refractor in the heating period

2 The Informational Structure of the System

The informational system is designed first for data and technical parameter acquisition and then for operative supplying of information required for the process control, for exploitation and maintenance of the equipment. The entry information is received as digital or analog signals from transducers through the process interface. As a result of data processing, the computing system elaborates the output information, analog or numerical, that can be used for controlling, adjusting or for displaying on video terminals.

The system has a database for data and programs on the mass memory (one or more magnetic hard drives).

The automation system receives the following information:
From electrical powering it receives on/off information regarding the following events: starting / stopping the devices; device status (working / not working); alarms; damages and the nature of damages; existence of power and controlling tensions

From weighing machines in containers, on the running belt, or on the measuring devices the system receives analog or digital signals regarding: status; damage; kind of material; debit of transported or measured material; quantity transported or measured; place of stocking or transporting; proportion of different components of material.

From measuring, controlling and adjusting devices the system receives analog or digital signals regarding: temperatures taken in different zones of thermal processing; debits of water and gas; dynamical analysis of different gases; speed of equipment

From the superior hierarchical level, the system receives data regarding: production planning; programs of raw material supplying; material, fuels and energy consumption; maintenance programs

3 Description of the techniques

Typically, classical control techniques use analog PID (PI, filtered PID) controllers, or on/off controllers for slower processes. Controllers are used in structures like:

- Simple adjusting with imposed reference
- Cascade adjusting
- Combined adjusting by perturbation and reference

These types of controllers have the advantage that they can be easily implemented, but very often their parameters are fixed. That's why classical control

structures can be used only for those subsystems of technological installation that have very well defined models, or have modest control requirements.
From all the diverse modern control techniques we have chosen to implement some algorithms by computer programs. This implementation has a great flexibility: it is very easy to modify the dates of the program that contains the adjusting parameters.

4 Hardware and Software Architecture

The hardware platform used as a support of the control programs is a personal computer with a Pentium 100 MHz processor, 32 Mb RAM, 64 bits video graphic accelerator, SCSI 1Gb HDD, data acquisition card, one multiplex for 16 serial ports and a network of 16 PLC-s. On this machine we have used Windows NT operating system. This system offers multitasking facilities for parallel managing aspects regarding data acquisition, data transmission, adjusting and data displaying. Using multitasking we could manage more control loops on the same computer even though some of the processes are controlled only by PLC-s.

The development platform used was Visual C++ 4.0 because it can compile programs for Windows NT operating system and it has the following facilities:

- Permits creation of separate threads with different adjusting algorithms for each of the processes.
- Offers communication methods inter-threads for transferring data between data acquisition processes and controlling processes. Communications are made using message boxes and critical sections.
- Permits realization of communication modules in a TCP/IP network. That makes possible implementation of the hierarchical architecture.
- Offers the possibility to create a user-friendly interface for the product, in Windows environment.

The only disadvantage of this platform is that it hasn't any dedicated functions for complicated mathematical calculus needed in adjusting algorithms, so that we had to write them ourselves.

The process computer is placed in the control room of technological installation because it must be protected from the vibrations and from the environment of the technological process.

5 Process Optimization

The optimization problem is "given a dynamic system (S) that evolves on a finite time interval $[T_0...T_f]$ and the performance index J (u) it is required a command that minimizes J". In this standard formulation there are no restrictions, but it follows a movement in the states space for minimum energy consumption. Using the computer we have chosen dynamic optimization, that relies on the idea of finding a procedure for generating a "relaxing" array defined by the following condition:

$X1, X2, X3, \ldots Xn$, so that $f(X1) \geq f(X2) \geq f(X3) \geq \ldots$ with the property that if $Xk \rightarrow X$, then $f(Xk) \rightarrow f(X)$, where X is the extreme value of the objective function in a specific domain.

The relaxing arrays generating procedure has two components:

- Choosing the descending direction of the function with the modified Newton method. This is a gradient method that means successive approximations of the function.
- Determining the step on the descending direction.

6 Control System Functions at the Agglomeration Factory

The automation system works in a collecting-processing way, interfering with and conducting the controlling processes. The system does the following functions:

- Supervision of technological equipment and devices
- Displaying and printing any damage that may occur, showing the moment of appearance, the kind and the place of the damage
- Displaying and printing the alarms at the moment of the occurrence
- Showing the entries and outputs from the programmable logic controllers for debugging any anomaly or damage
- Controlling the measuring equipment and supervising the process of combining of raw materials
- On demand, displaying the parameters of the technological process
- Calculating the consumption of the raw materials by sorts, indicating the norms of consuming for combustibles and electrical energy
- On demand, transmitting data regarding the level of production, the structure of consumption, etc., to the superior hierarchical level

Some of the subsystems of the technological installation are variable in time due to action of stochastic perturbations. That determines large variations of the parameters of the attached mathematical model, but fortunately the structure of the model remains the same.

The control takes place at the level of the central computer, and also at the level of the PLC-s. We have used a hybrid adjusting method, i.e. some adjusting loops are implemented with classical algorithms (PI, PID), and the others are adaptive.

The system is composed by:

- Classical algorithms running on the PLC-s that also have implemented data acquisition modules, and analog and digital command modules. Programs are written in the PLC's language (PL 7-2).
- Controlling algorithms implemented with the process computer. Some of them are typical numerical algorithms, and the others are adaptive control algorithms.

7 Control System Functions at the Cowper Stove

The primary objective of numerical controlling of the pre-heaters is optimization of the technological process. The purpose of this optimization is obtaining a high efficiency and combustible saving. The computer controls and effectively optimizes the technological process. The intelligent control of the process takes place at the level of the process computer. It realizes the following functions:

- Computing the quantity of warmth required for heating the pre-heaters. The optimization of the technological process imposes an exact correlation between the quantity of warmth required for heating and the quantity of warmth taken by the air blown into the furnace By this correlation it can be determined a minimum combustible consumption for heating. The algorithm of optimization is:

$$W = \int_{t=0}^{t=t_s} Q_w(t)dt$$

Where Q_w is a thermodynamic equation depending on the quantity of warmth taken by the air blown in a blowing period t_s.

- Computing the caloric power of the combustible gaseous mixture. Pre-heaters are supplied with a mixture of gases in different proportions. It is required to calculate the minimum debit of mixture, function of caloric power of components.

  The relation is: $F_b = \dfrac{Q_W}{H}$, where:

  F_b is the gaseous debit
  H is the caloric power of the mixture of gases
  Q_W is the quantity of warmth
- Computing the debit of air required. For a complete burning, the proportion of gaseous combustible mixture and air in the focus must be optimal. This value is calculated starting from the quantity of air required by any of the components of the combustible mixture.

  The relation of calculus is:

$$F_a = \frac{k_1F_1 + k_2F_2}{F_1 + F_2} F_b, \text{ where:}$$

  F_a is the debit of air
  F_b is the debit of combustible mixture
  k_1 , k_2 are variable coefficients
  F_1 is the debit of the first component of the combustible mixture
  F_2 is the debit of the second component of the combustible mixture

Depending on the values obtained, the computer commands and adjusts the debits of gaseous combustible and air, and the time between the two working fazes of the pre-heaters.

8 Conclusion

Implementation of this system in Romanian metallurgical plants has the following advantages: obtaining a high efficiency and combustible saving, limited efforts for developing a new application in a short period of time, and high performance of the system in solving the demands of applications.

References

1. Hartescu F.: *RT-ARCH – A New Approach in Real Time Application Design*, IASTED Mini and Microcomputers and Their Applications, Lugano, Switzerland, June 1990
2. Astrom K., J. Wittenmark: *Adaptive Control*, Addison Wesley Company, 1989
3. Dumitrache I.: *Intelligent Techniques for Control Applications*, The 9th CSCS Conference, May 1993
4. Selic B., P.T. Ward: *The Challenges of Real Time Software Design*, Embedded Systems Programming, Miller Freemans Inc., Oct. 1996
5. Selic B., G. Gulliksons, P.T. Ward: *Real Time Object Oriented Modelling*, John Wiley & Sons, 1994
6. Cichocki A., R. Unbehauen, *Neural Netwarks for Optimization and Signal Processing*, John Wiley & Sons, Ltd. 1993

19
Knowledge-Based Control Synthesis of Discrete Event Dynamic Systems

F. Capkovič

1 Introduction

An approach to dealing with the problem of the control synthesis for discrete event dynamic systems (DEDS) is presented in this chapter. It simultaneously utilizes both the analytical model of the system to be controlled and the knowledge representation expressing the control task specifications (like criteria, constraints, etc.) by means of a knowledge base (KB) in analytical terms. Both the model and the KB are created by means of an analogy with corresponding kinds of Petri nets (PN), oriented graphs (OG) and the combination of them. As to the occurrence there are distinguished two principal kinds of the discrete events here: 1) the spontaneous discrete events (their occurrence cannot be influenced from the outside - e.g. some specific singularities or some failures); 2) the events which can be influenced from the outside - i.e. controllable events. The problem of the DEDS control synthesis is that of finding the most suitable sequence of the controllable discrete events which will be able to transfer the system in question from an initial state into a desirable terminal one or into a desirable terminal set. The control task specifications (like criteria, constraints, etc.) are many times given verbally or in another nonanalytical form and frequently can be fuzzy. To handle them in the control synthesis procedure a suitable form of the representation of knowledge about them must be found. The PN represent the apparatus of discrete mathematics able to express in analytical terms not only the model of the system to be controlled but also the KB (in spite of the fact that knowledge to be represented is given in nonanalytical terms). The PN-based approach to creating the system model yields the linear discrete k-invariant system (where k is the discrete step of the system dynamics development) suitable especially for expressing the controllable discrete events. The OG-based approach to DEDS modellling presented here is suitable especially for expressing the spontaneous discrete events and yields the linear discrete k-variant system. When PN are understood to be OG with two kinds of nodes (positions and transitions) and two kinds of edges (oriented arcs emerging from the positions and entering the transitions on one hand and oriented arcs emerging from the transitions and entering the positions on the other hand), the hybrid model of the system is derived here by means of combinig both of the previous approaches. In such a model the spontaneous discrete events are expressed by means of the system parameters. The same or very analogical approaches are used also for knowledge representation. The analytical models of the KB are expressed in the form of the PN-based model (the linear K-invariant logical or/and fuzzy system with K being the discrete step of the KB truth propagation), OG-based model (the linear K-variant logical or/and fuzzy system, especially suitable e.g. for expressing the rules in the form of both-sided implications), and hybrid model combining the previous two ones. The approach will be illustrated on the example of a transport subsystem of flexible manufacturing systems (FMS).

2 Modelling DEDS in Analytical Terms

There are many different approaches to the problem of the DEDS modelling. They depends on the properties of the modelled object that should be especially underlined. An overview can be found e.g. in [1]. The PN-based approach is one of them. It represents a set of techniques based on different kinds of PN. Especially the ordinary PN (OPN) are very suitable because they yield very simple (linear) analytical model for the wide class of DEDS and make possible to use different methods of linear algebra for the system analysis and for the control synthesis.

In this section both the OG-based model and the hybrid one will be introduced. However, the PN-based approach used in [2]-[5] should be concisely mentioned before.

2.1 The PN-Based Model

When an analogy between the DEDS subprocesses or operations and the PN positions on one hand as well as an analogy between the discrete events occurring in DEDS and the PN transitions on the other hand the simple PN-based model of the DEDS dynamics can be derived

$$\mathbf{x}_{k+1} = \mathbf{x}_k + \mathbf{B}.\mathbf{u}_k \quad , \quad k = 0, N \tag{1}$$

$$\mathbf{B} = \mathbf{G}^T - \mathbf{F} \tag{2}$$

$$\mathbf{F}.\mathbf{u}_k \leq \mathbf{x}_k \tag{3}$$

where

$\mathbf{x}_k = (\sigma_{p_1}^k, ..., \sigma_{p_n}^k)^T$; $k = 0, N$ is the state vector of the system in the step k. $\sigma_{p_i}^k$, $i = 1, n$ is the state of the elementary subprocess or operation p_i in the step k. $(.)^T$ symbolizes the vector or matrix transposition and N is an integer representing formally the number of different state vectors occurring during the DEDS dynamics development.

$\mathbf{u}_k = (\gamma_{t_1}^k, ..., \gamma_{t_m}^k)^T$; $k = 0, N$ is the "control" vector of the system (expressing the occurrence of the controllable discrete events) in the step k. $\gamma_{t_j}^k$, $j = 1, m$ is the state of the occurrence of the elementary event t_j in the step k.

$\mathbf{F}$ is the $(n \times m)$-dimensional structural matrix of constant elements (integers) expressing causal relations between the subprocesses and the discrete events

$\mathbf{G}$ is the $(m \times n)$-dimensional structural matrix of constant elements expressing the mutual causal relations between the discrete events and the subprocesses.

Such a model is suitable especially for expressing the controllable discrete events. Namely, in general

$$\mathbf{x}_k = \underbrace{\underbrace{\underbrace{\mathbf{x}_0 + \mathbf{B}.\mathbf{u}_0}_{\mathbf{x}_1} + \mathbf{B}.\mathbf{u}_1}_{\mathbf{x}_2} + \ldots + \mathbf{B}.\mathbf{u}_{k-2}}_{\mathbf{x}_{k-1}} + \mathbf{B}.\mathbf{u}_{k-1} \tag{4}$$

Thus, inserting the spontaneous discrete events among the components of the control vectors is not appropriate.

2.2 The OG-Based Model

The ordinary OG having only one type of nodes and one type of edges can also serve as an analogon for modelling DEDS. Suppose that the nodes correspond with the PN positions and the edges include the PN transitions (see Fig. 1). Hence,

$$\langle P, \Delta \rangle \quad ; \quad P \cap \Delta = \emptyset \tag{5}$$

where

$P = \{p_1, ..., p_n\}$ is a finite set of the PN positions with p_i , $i = 1, n$, being the elementary positions.

$\Delta \subseteq P \times P$ is a set of the oriented arcs among the positions. It can be expressed by the arcs incidence matrix $\boldsymbol{\Delta} = \{\delta_{ij}\}$, $i = 1, n$; $j = 1, n$ where the element δ_{ij} represents formally the occurrence of the arc oriented from the

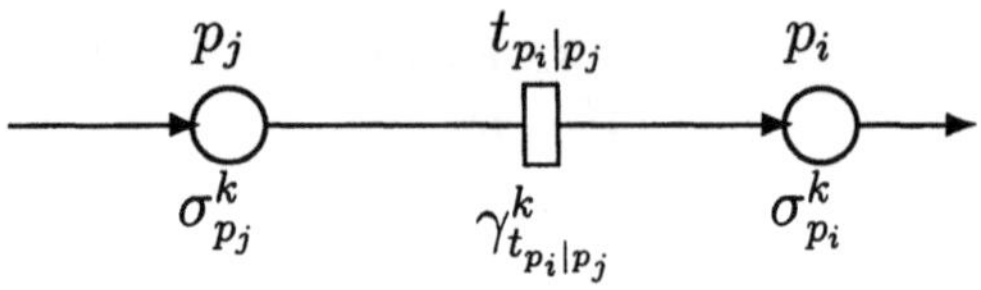

Figure 1. The causality interconnection among two positions

position p_i to the position p_j. The dynamics of such a system can be formally described in analytical terms as follows

$$\mathbf{x}_{k+1} = \mathbf{x}_k \ \underline{or} \ \boldsymbol{\Delta}_k \ \underline{and} \ \mathbf{x}_k \tag{6}$$

Such a model is suitable especially for expressing the spontaneous discrete events in the form of the system parameters. Namely, in general

$$\mathbf{x}_k = \boldsymbol{\Delta}_{k-1}.\boldsymbol{\Delta}_{k-2}.\ldots.\underbrace{\boldsymbol{\Delta}_1.\underbrace{\boldsymbol{\Delta}_0.\mathbf{x}_0}_{\mathbf{x}_1}}_{\mathbf{x}_2}\Bigg\}_{\mathbf{x}_{k-1}} \tag{7}$$

Thus, inserting the controllable discrete events among the system parameters is not appropriate.

2.3 The Combined Model

To expess simultaneously both the controllable discrete events and the spontaneous ones the following hybrid model of DEDS combining both of the above mentioned models can be created

$$\mathbf{x}_{k+1} = \mathbf{A}.\mathbf{x}_k + \mathbf{B}.\mathbf{u}_k \quad , \quad k = 0, N \tag{8}$$

$$\mathbf{A}_k = \mathbf{I}_n + \boldsymbol{\Delta}_k \tag{9}$$

$$\mathbf{B} = \mathbf{G}^T - \mathbf{F} \tag{10}$$

$$\mathbf{F}.\mathbf{u}_k \leq \mathbf{x}_k \tag{11}$$

where

k is the discrete step of the DEDS dynamics development.

$\mathbf{x}_k = (\sigma^k_{p_1}, \ldots, \sigma^k_{p_n})^T$; $k = 0, N$ is the n-dimensional state vector of the DEDS in the step k; $\sigma^k_{p_i}$, $i = 1, n$ is the state of the elementary subprocess p_i in the step k (in the PN analogy it is the state of the elementary position) - its activity is expressed by 1 and its passivity by 0.

$\mathbf{u}_k = (\gamma^k_{t_1}, \ldots, \gamma^k_{t_m})^T$; $k = 0, N$ is the m-dimensional control vector of the DEDS in the step k; $\gamma^k_{t_j}$, $j = 1, m$ is the state of occurring the controllable discrete event t_j in the step k. In the PN analogy it is the state of the elementary transition (1 - enabled, 0 - disabled).

$\mathbf{A}_k$ is the $(n \times n)$-dimensional system matrix consisting of the $(n \times n)$-dimensional identity matrix $\mathbf{I}_n$ and the matrix $\boldsymbol{\Delta}_k$ being the $(n \times n)$-dimensional functional matrix expressing the causal relations between subprocesses that are influenced by spontaneous (i.e. uncontrollable) discrete events occurring in the system.

$\mathbf{B}$ is the $(n \times m)$-dimensional structural matrix of constant elements expressing the causal relations between subprocesses and controllable discrete events. It is given by means of the $(n \times m)$-dimensional matrix $\mathbf{F}$ of constant elements expressing causal relations between the subprocesses and the controllable discrete events and $(m \times n)$-dimensional matrix $\mathbf{G}$ of constant element s expressing the mutual causal relations between the controllable discrete events and the subprocesses.

T symbolizes the matrix or vector transposition.

3 The Problem of Control Synthesis

To find a suitable control vector $\mathbf{u}_k$ in the step k the analysis of the control possibilities should be performed and the most suitable one (with respect to the control task specifications) should be chosen. The complete set of the potentional control possibilities can be obtained as follows

$$\overline{\mathbf{x}}_k = \underline{neg}\, \mathbf{x}_k = \mathbf{1}_n - \mathbf{x}_k \quad (12)$$

$$\mathbf{v}_k = \mathbf{F}^T \underline{and}\, \overline{\mathbf{x}}_k \quad (13)$$

$$\mathbf{w}_k = \underline{neg}\, \mathbf{v}_k = \mathbf{1}_m - \mathbf{v}_k \quad (14)$$

where

$\underline{neg}$ is the operator of logical negation.

$\underline{and}$ is the operator of logical multiplying.

$\mathbf{1}_n$ is n-dimensional constant vector with all its elements equaled to 1.

$\mathbf{v}_k, \mathbf{w}_k$ are, respectively, m-dimensional auxiliary vector and m-dimensional vector of the base for the control vector choice (it expresses the possible candidates for the control vector $\mathbf{u}_k$ in the step k.) When only one of its components is different from zero, it can be used (when (3) or (11) is met) to be the control vector, i.e. $\mathbf{u}_k = \mathbf{w}_k$. When there are several components of the $\mathbf{w}_k$ different from zero, the control vector $\mathbf{u}_k$ has to be chosen on the basis of additional information about the actual control task. The choice of the control vector can be made either manually by an operator during the process of the control synthesis or automatically by means of the rule-based KB (predefined by an expert in the domain) expressing the specifications of the control task in question. The latter approach is expected rather than the former one.

4 The Knowledge Representation

When OPN are replaced by means of logical PN (LPN) or/and fuzzy PN (FPN) defined in [6] the PN-based knowledge representation can be obtained in the form of the uniform linear logical or/and fuzzylogical system

$$\boldsymbol{\Phi}_{K+1} = \boldsymbol{\Phi}_K \ \underline{or} \ \boldsymbol{\Theta} \ \underline{and} \ \boldsymbol{\Omega}_K \ , \quad K = 0, N_1 \tag{15}$$

$$\boldsymbol{\Theta} = \boldsymbol{\Gamma}^T \ \underline{or} \ \boldsymbol{\Psi} \tag{16}$$

$$\boldsymbol{\Psi} \ \underline{and} \ \boldsymbol{\Omega}_K \leq \boldsymbol{\Phi}_K \tag{17}$$

Combining such a model with the OG-based one the analytical hybrid model is obtained as follows

$$\boldsymbol{\Phi}_{K+1} = \boldsymbol{\Lambda} \ \underline{and} \ \boldsymbol{\Phi}_K \ \underline{or} \ \boldsymbol{\Theta} \ \underline{and} \ \boldsymbol{\Omega}_K \ , \quad K = 0, N_1 \tag{18}$$

$$\boldsymbol{\Theta} = \boldsymbol{\Gamma}^T \ \underline{or} \ \boldsymbol{\Psi} \tag{19}$$

$$\boldsymbol{\Lambda} = \mathbf{I}_{n_1} \ \underline{or} \ {}^1\boldsymbol{\Delta} \tag{20}$$

$$\boldsymbol{\Psi} \ \underline{and} \ \boldsymbol{\Omega}_K \leq \boldsymbol{\Phi}_K \tag{21}$$

In these models

$\underline{or}$, $\underline{and}$, $\underline{neg}$ are, respectively, the operator of logical additioning, multiplying and negation.

${}^1\boldsymbol{\Delta} = \{{}^1\delta_{ij}\}$, where ${}^1\delta_{ij} \in \{0,1\}\,, i = 1, n_1\,; j = 1, n_1$ in the analogy with the LPN or ${}^1\delta_{ij} \in < 0, 1 >$ in the analogy with the FPN. The element ${}^1\delta_{ij}$ represents the absence (when 0), presence (when 1) or a fuzzy measure of existence of the mutual causal relation (being the both-sided implication) between the statement S_i and the statement S_j.

$\boldsymbol{\Psi} = \{\psi_{ij}\}$, where $\psi_{ij} \in \{0,1\}\,, i = 1, n_1\,; j = 1, m_1$ in the analogy with the LPN or $\psi_{ij} \in < 0, 1 >$ in the analogy with the FPN. The element ψ_{ij} represents the absence (when 0), presence (when 1) or a fuzzy measure of existence of the causal relation between the input statement S_i and the rule R_j.

$\boldsymbol{\Gamma} = \{\gamma_{ij}\}\,, \gamma_{ij} \in \{0,1\}\,,$ or $\gamma_{ij} \in < 0, 1 >, i = 1, m_1\,; j = 1, n_1$ analogically expresses the occurrence of the causal relation between the rule R_i and its output statement S_j.

$\boldsymbol{\Phi}_K = (\Phi^K_{S_1}, ..., \Phi^K_{S_{n_1}})^T\,; K = 0, N_1$ being the state vector of the KB (the state of the statements truth propagation) in the step K; $\Phi^K_{S_i}\,, i = 1, n_1$ is the state of the truth of the elementary statement S_i in the step K. The false statement is expressed by the state 0 and the true statement by 1. When the statement is true with a fuzzy measure the state acquires its value from the real interval $< 0, 1 >$. T symbolizes the vector or matrix transposition and N_1 is an integer representing the number of different situations during the KB dynamics development.

$\boldsymbol{\Omega}_K = (\omega^K_{R_1}, ..., \omega^K_{R_m})^T\,; K = 0, N_1$ being the control vector of the KB (i.e. the state of the rules evaluability) in the step K, where $\omega^K_{R_j}\,, j = 1, m_1$ is the state of the elementary rule R_j evaluability in the step K. When the rule is not able to be evaluated the state is 0, when the rule is able to be evaluated the state is 1. When the rule is able to be evaluated with a fuzzy measure the state acquires its value from the interval $< 0, 1 >$.

The knowledge inference is uniform for both of the models:

$$\overline{\boldsymbol{\Phi}}_K = \underline{neg}\,\boldsymbol{\Phi}_K = \mathbf{1}_n - \boldsymbol{\Phi}_K \tag{22}$$

$$\boldsymbol{\Upsilon}_K = \boldsymbol{\Psi}^T \underline{and}\, \overline{\boldsymbol{\Phi}}_K \tag{23}$$

$$\boldsymbol{\Omega}_K = \underline{neg}\,\boldsymbol{\Upsilon}_K = \mathbf{1}_m - \boldsymbol{\Upsilon}_K = \underline{neg}(\boldsymbol{\Psi}^T\, \underline{and}\, (\underline{neg}\,\boldsymbol{\Phi}_K)) \tag{24}$$

where

$\boldsymbol{\Upsilon}_K$ is a m-dimensional auxiliary vector pointing out (by its nonzero elements) the rules that cannot be evaluated, because there is at least one false (of course in the LPN analogy) statement among its input statements.

$\boldsymbol{\Omega}_K$ is a m-dimensional "control" vector pointing out the rules that have all their input statements true and, consequently, they can be evaluated in the step K of the KB dynamics development. This vector is a base of the inference, because it contains information about the rules that can contribute to obtaining the new knowledge. These rules correspond to the nonzero elements of the vector $\boldsymbol{\Omega}_K$.

In case of the FPN analogy the sense of the rules evaluability is a little different, because all of the KB rules are always evaluable with a fuzzy measure given by the vector $\boldsymbol{\Omega}_K$ components.

5 The System Model and Knowledge Base

In any step of the control synthesis procedure the cooperation of the KB and the system model is performed. The KB decision in any step k of the DEDS dynamics development depends on both the actual situation inside of the system itself (i.e. on the present state of the system $\mathbf{x}_k$, on the previous control vector $\mathbf{u}_{k-1}$) and the actual state of knowledge inference (i.e. on the actual state of statements truth propagation or/and knowledge inference $\boldsymbol{\Phi}_K$, etc.). In any step k of the sytem dynamics development several steps $K = 0, N_1$ of the KB dynamics development can occur. In general the KB decision directly influences the choice of the control vector $\mathbf{u}_k$ as follows

$$\mathbf{u}_k = \mathcal{F}(\mathbf{x}_k,\ \mathbf{u}_{k-1},\ \boldsymbol{\Phi}_K) \tag{25}$$

or more exactly

$$\mathbf{u}_k = \mathcal{F}(\boldsymbol{\Phi}_K(\boldsymbol{\Omega}_K(\boldsymbol{\Phi}_{K-1},\ \mathbf{u}_{k-1},\ \mathbf{x}_k))) \tag{26}$$

where

$\mathcal{F}(.)$ is a symbolical operator. It cannot be expressed in analytical terms. It depends on the actual case of the DEDS, on actual case of the control task specifications, etc.

6 The Illustrative Example

To demonstrate utilization of the hybrid model consider two autonomous mobile robots $\Re_1$, $\Re_2$ or two automatically guided vehicles of a flexible manufacturing system placed into a complicated working space given on Fig. 2 consisting of five rooms denoted by numbers 0, 1,..., 4 connecting by the door-

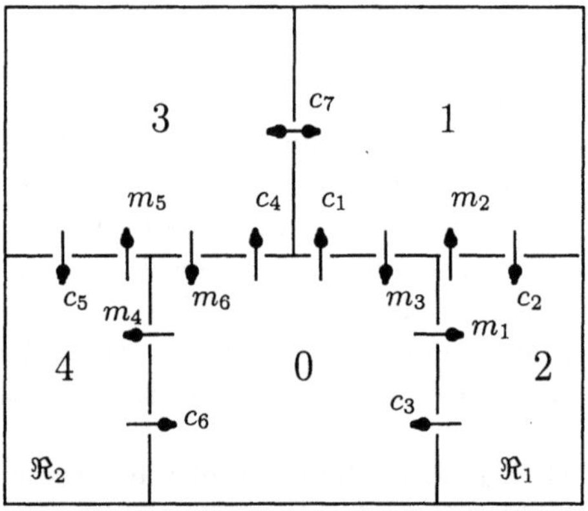

Figure 2. The maze structure.

ways exclusively for the $\Re_1$ denoted by $c_i, i = 1,7$ and the doorways exclusively for the $\Re_2$ denoted by $m_j, j = 1,6$. $\Re_1$ is initially in the room 2 and $\Re_2$ in room 4. Each doorway can be traversed only in the direction indicated. Each door (with the exception of the door c_7) can be opened or closed by means of control actions. The door c_7 is uncontrollable (or better, it is continuously open in both directions). Only discrete events generated by sensors in the doors can be observed. They indicate that a device is just running through. The control task specifications are (see e.g. [7]):

1. The devices never occupy the same room simultaneously.
2. It is always possible for both of them to return to their initial positions (i.e. $\Re_1$ to the room 2 and $\Re_2$ to the room 4).
3. The controller should enable the devices to behave as freely as possible with respect to the constraints imposed.

The rooms 0 - 4 of the maze will be represented by the PN positions p_1 - p_5 and the doorways by the PN transitions. The permanently open door c_7 should be replaced by means of two PN transitions or two OG edges. The representation of the working space (maze) is given on Fig. 3. The initial state vectors of $\Re_1$ and $\Re_2$ are

$$ {}^c\mathbf{x}_0 = (0\,0\,1\,0\,0)^T \, , \; {}^m\mathbf{x}_0 = (0\,0\,0\,0\,1)^T \qquad (27) $$

6.1 The Combined Model of the System

The parameters of the combined model for $\Re_1$ and $\Re_2$ are: $n = 5$, $m_c = 6$, $m_m = 6$

$$\mathbf{F}_c = \begin{pmatrix} 1&0&0&1&0&0 \\ 0&1&0&0&0&0 \\ 0&0&1&0&0&0 \\ 0&0&0&0&1&0 \\ 0&0&0&0&0&1 \end{pmatrix} \quad \mathbf{G}_c = \begin{pmatrix} 0&1&0&0&0 \\ 0&0&1&0&0 \\ 1&0&0&0&0 \\ 0&0&0&1&0 \\ 0&0&0&0&1 \\ 1&0&0&0&0 \end{pmatrix}$$

$${}^c\boldsymbol{\Delta}_k = \begin{pmatrix} 0&0&0&0&0 \\ 0&0&0&{}^c\delta_{24}^k&0 \\ 0&0&0&0&0 \\ 0&{}^c\delta_{42}^k&0&0&0 \\ 0&0&0&0&0 \end{pmatrix} \quad {}^c\mathbf{A}_k = \begin{pmatrix} 1&0&0&0&0 \\ 0&1&0&{}^c\delta_{24}^k&0 \\ 0&0&1&0&0 \\ 0&{}^c\delta_{42}^k&0&1&0 \\ 0&0&0&0&1 \end{pmatrix}$$

$$\mathbf{F}_m = \begin{pmatrix} 1&0&0&1&0&0 \\ 0&0&1&0&0&0 \\ 0&1&0&0&0&0 \\ 0&0&0&0&0&1 \\ 0&0&0&0&1&0 \end{pmatrix} \quad \mathbf{G}_m^T = \begin{pmatrix} 0&0&1&0&0&1 \\ 0&1&0&0&0&0 \\ 1&0&0&0&0&0 \\ 0&0&0&0&1&0 \\ 0&0&0&1&0&0 \end{pmatrix}$$

$${}^m\boldsymbol{\Delta}_k = \mathbf{\emptyset}_5 \; ; \; {}^m\mathbf{A}_k = \mathbf{I}_5$$

The structure of the $\Re_1$ and $\Re_2$ control vectors is the following

$$\begin{aligned} {}^c\mathbf{u}_k &= (c_1^k, c_2^k, c_3^k, c_4^k, c_5^k, c_6^k)^T \\ {}^m\mathbf{u}_k &= (m_1^k, m_2^k, m_3^k, m_4^k, m_5^k, m_6^k)^T \end{aligned}$$

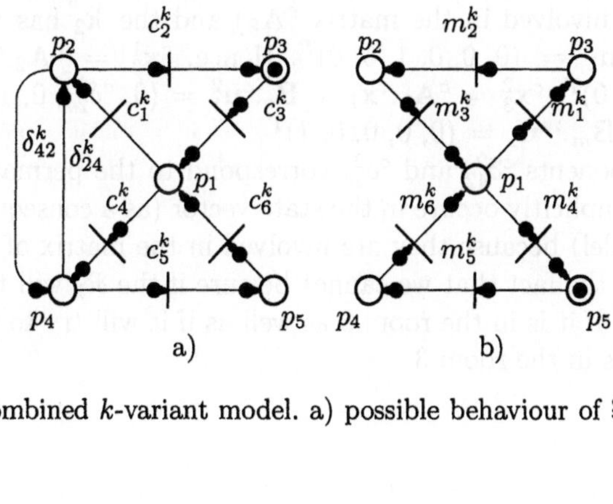

Figure 3. The combined k-variant model. a) possible behaviour of $\Re_1$; b) possible behaviour of $\Re_2$

6.2 The Combined Model in the Control Synthesis

Let us start from the initial states ${}^c\mathbf{x}_0$ and ${}^m\mathbf{x}_0$ given by (27). It can be found that the doors c_3^0 and m_5^0 can be opened in the step $k = 0$. Namely, the control base vectors ${}^c\mathbf{w}_0 = (0, 0, c_3^0, 0, 0, 0)^T$ and ${}^m\mathbf{w}_0 = (0, 0, 0, 0, m_5^0, 0)^T$ were obtained. It means that both the $\Re_1$ and the $\Re_2$ have only single control possibility ${}^c\mathbf{u}_0 = (0, 0, 1, 0, 0, 0)^T$ and ${}^m\mathbf{u}_0 = (0, 0, 0, 0, 1, 0)^T$ respectively. Because both of them satisfy the condition (11), as well as they do not contradict the control task specification represented by means of the KB, they can be realized in the step $k = 0$ - i.e. $c_3^0 = 1$ and $m_5^0 = 1$. Consequently, ${}^c\mathbf{x}_1 = {}^c\mathbf{A}_0.{}^c\mathbf{x}_0 + \mathbf{B}_c.{}^c\mathbf{u}_0 = (1, 0, 0, 0, 0)^T$ and ${}^m\mathbf{x}_1 = {}^m\mathbf{A}_0.{}^m\mathbf{x}_0 + \mathbf{B}_m.{}^m\mathbf{u}_0 = (0, 0, 0, 1, 0)^T$. By means of these state vectors the control synthesis procedure gives us the control base vectors ${}^c\mathbf{w}_1 = (c_1^1, 0, 0, c_4^1, 0, 0)^T$ and ${}^m\mathbf{w}_1 = (0, 0, 0, 0, 0, m_6^1)^T$ respectively. Consequently, (for the step $k = 2$) in the step $k = 1$ the control interferencies c_1^1, c_4^1 and m_6^1 should be simultaneously realized. It means that the $\Re_1$ has two alternative control possibilities ${}^c\mathbf{u}_1^1 = (1, 0, 0, 0, 0, 0)^T$, ${}^c\mathbf{u}_1^2 = (0, 0, 0, 1, 0, 0)^T$ and the $\Re_2$ has only one control possibility ${}^m\mathbf{u}_1 = (0, 0, 0, 0, 0, 1)^T$. All of them satisfy the (11) as well as they are not in any contradiction with the mentioned knowledge base. Hence, ${}^c\mathbf{x}_2^1 = {}^c\mathbf{A}_1.{}^c\mathbf{x}_1 + \mathbf{B}_c.{}^c\mathbf{u}_1^1 = (0, 1, 0, 0, 0)^T$, ${}^c\mathbf{x}_2^2 = {}^c\mathbf{A}_1.{}^c\mathbf{x}_1 + \mathbf{B}_c.{}^c\mathbf{u}_1^2 = (0, 0, 0, 1, 0)^T$ and ${}^m\mathbf{x}_2 = {}^m\mathbf{A}_1.{}^m\mathbf{x}_1 + \mathbf{B}_m.{}^m\mathbf{u}_1 = (1, 0, 0, 0, 0)^T$.

By means of these state vectors the control synthesis procedure yields the following control base vectors for the step $k = 3$ ${}^c\mathbf{w}_2^1 = (0, c_2^2, 0, 0, 0, 0)^T$, ${}^c\mathbf{w}_2^2 = (0, 0, 0, 0, c_5^2, 0)^T$ and ${}^m\mathbf{w}_2 = (m_1^2, 0, 0, m_4^2, 0, 0)^T$ respectively. Hence, in the step $k = 2$, the control interferencies c_2^2, c_5^2, m_1^2, m_4^2 should be simultaneously realized. In both the $\Re_1$ model and the $\Re_2$ one are two alternative control possibilities satisfying the condition (11). The knowledge base chooses the simultaneous realization of c_2^2 and m_4^2. Namely, owing to the control task specifications c_2^2 has priority to m_1^2 and m_4^2 has priority to c_5^2. Consequently, the $\Re_1$ has two alternative control possibilities ${}^c\mathbf{u}_2^1 = (0, 1, 0, 0, 0, 0)^T$, ${}^c\mathbf{u}_2^2 = (0, 0, 0, 0, 0, 0)^T$ (this one is necessary with respect to the spontaneous events involved in the matrix ${}^c\mathbf{A}_k$) and the $\Re_2$ has the single control possibility ${}^m\mathbf{u}_2 = (0, 0, 0, 1, 0, 0)^T$. Hence, ${}^c\mathbf{x}_3^1 = {}^c\mathbf{A}_2.{}^c\mathbf{x}_2^1 + \mathbf{B}_c.{}^c\mathbf{u}_2^1 = (0, 0, 1, {}^c\delta_{42}^2, 0)^T$, ${}^c\mathbf{x}_3^2 = {}^c\mathbf{A}_1.{}^c\mathbf{x}_1 + \mathbf{B}_c.{}^c\mathbf{u}_1^2 = (0, {}^c\delta_{24}^2, 0, 1, 0)^T$ and ${}^m\mathbf{x}_2 = {}^m\mathbf{A}_1.{}^m\mathbf{x}_1 + \mathbf{B}_m.{}^m\mathbf{u}_1 = (0, 0, 0, 0, 1)^T$.

The components ${}^c\delta_{42}^2$ and ${}^c\delta_{24}^2$ correspond to the permanently open door c_7^k and they implicitly occure in the state vector (as a consequence of using the k-variant model) because they are involved in the matrix of parameters ${}^c\boldsymbol{\Delta}_k$. They express the fact that we cannot be sure if the $\Re_1$ will try to steal in the room 3 (${}^c\delta_{42}^2$) if it is in the room 3 as well as if it will try to steal in the room 1 (${}^c\delta_{24}^2$) if it is in the room 3.

6.3 Representing the Control Task Specifications

The control task specifications were given verbally. It is very difficult to work with them in the control synthesis process. Consequently, in the first place a rule-based knowledge about them should be formed and secondly the analytical model of the KB should be built.

The staments of the KB (the pieces of knowledge) can be the following:

$S_1 = ({}^c\mathbf{x}_{k+1} = {}^m\mathbf{x}_{k+1})$; $S_2 = ({}^c\mathbf{x}_{k+1} = {}^c\mathbf{x}_0)$; $S_3 = ({}^m\mathbf{x}_{k+1} = {}^m\mathbf{x}_0)$
$S_4 = ({}^c\mathbf{x}_{k+1} = {}^m\mathbf{x}_0)$; $S_5 = ({}^m\mathbf{x}_{k+1} = {}^c\mathbf{x}_0)$; $S_6 = ({}^c\mathbf{x}_{k+1} \neq {}^m\mathbf{x}_{k+1})$
$S_7 = ({}^c\mathbf{x}_{k+1} \neq {}^c\mathbf{x}_0)$; $S_8 = ({}^m\mathbf{x}_{k+1} \neq {}^m\mathbf{x}_0)$; $S_9 = ({}^c\mathbf{x}_{k+1} \neq {}^m\mathbf{x}_0)$
$S_{10} = ({}^m\mathbf{x}_{k+1} \neq {}^c\mathbf{x}_0)$; S_{11} = there is only one vector ${}^c\mathbf{u}_k$
S_{12} = there is only one vector ${}^m\mathbf{u}_k$; S_{13} = there are several vectors ${}^c\mathbf{u}_k$
S_{14} = there are several vectors ${}^m\mathbf{u}_k$; S_{15} = accept ${}^c\mathbf{u}_k$; S_{16} = accept ${}^m\mathbf{u}_k$
S_{17} = eliminate ${}^c\mathbf{u}_k$; S_{18} = eliminate ${}^m\mathbf{u}_k$; S_{19} = solution does not exist
S_{20} = take another ${}^c\mathbf{u}_k$; S_{21} = take another ${}^m\mathbf{u}_k$
S_{22} = (accept ${}^c\mathbf{u}_k$ and eliminate ${}^m\mathbf{u}_k$) or (accept ${}^m\mathbf{u}_k$ and eliminate ${}^c\mathbf{u}_k$)
S_{23} = I do not know

The $IF - THEN$ rules of the KB ($\circ$ symbolizes **and**) can be set in the following form:

R_1: *IF* $(S_1 \circ S_7 \circ S_{12} \circ S_{13})$ *THEN* $(S_{16} \circ S_{17} \circ S_{20})$
R_2: *IF* $(S_1 \circ S_8 \circ S_{11} \circ S_{14})$ *THEN* $(S_{15} \circ S_{18} \circ S_{21})$
R_3: *IF* S_6 *THEN* $(S_{15} \circ S_{16})$
R_4: *IF* $(S_1 \circ S_2 \circ S_{14})$ *THEN* $(S_{15} \circ S_{18} \circ S_{21})$
R_5: *IF* $(S_1 \circ S_3 \circ S_{13})$ *THEN* $(S_{16} \circ S_{17} \circ S_{20})$
R_6: *IF* $(S_1 \circ S_5 \circ S_{14})$ *THEN* $(S_{15} \circ S_{18} \circ S_{21})$
R_7: *IF* $(S_1 \circ S_4 \circ S_{13})$ *THEN* $(S_{16} \circ S_{17} \circ S_{20})$
R_8: *IF* $(S_4 \circ S_6 \circ S_{12} \circ S_{13})$ *THEN* $(S_{16} \circ S_{17} \circ S_{20})$
R_9: *IF* $(S_5 \circ S_6 \circ S_{11} \circ S_{14})$ *THEN*$(S_{15} \circ S_{18} \circ S_{21})$
R_{10}: *IF* $(S_2 \circ S_6)$ *THEN* $(S_{15} \circ S_{16})$; R_{11}: *IF* $(S_3 \circ S_6)$ *THEN* $(S_{15} \circ S_{16})$
R_{12}: *IF* $(S_1 \circ S_2 \circ S_{12})$ *THEN* S_{19}; R_{13}: *IF* $(S_1 \circ S_3 \circ S_{11})$ *THEN* S_{19}
R_{14}: *IF* $(S_1 \circ S_2 \circ S_5 \circ S_{11} \circ S_{12})$ *THEN* S_{19}
R_{15}: *IF* $(S_1 \circ S_3 \circ S_4 \circ S_{11} \circ S_{12})$ *THEN* S_{19}
R_{16}: *IF* $(S_1 \circ S_{11} \circ S_{12})$ *THEN* S_{19}
R_{17}: *IF* $(S_1 \circ S_7 \circ S_8 \circ S_{13} \circ S_{14})$ *THEN* S_{22}
R_{18}: *IF* $(S_4 \circ S_6 \circ S_{13})$ *THEN* S_{20}; R_{19}: *IF* $(S_5 \circ S_6 \circ S_{14})$ *THEN* S_{21}

As to the analytical model of the KB the structural matrix of the KB - i.e. the matrix $\boldsymbol{\Theta}$ or the matrices $\boldsymbol{\Gamma}$ and $\boldsymbol{\Psi}$ can be writen. The dimensionalities of them are $n_1 = 23$ and $m_1 = 19$. However, the matrices are too large in order to introduced them here in the limited space of this chapter.

7 Conclusions

A knowledge-based control synthesis of DEDS was presented in this chapter. The k-variant OG-based model and the hybrid one (combining the k-invariant PN-based model and the OG-based one) were built in order to model DEDS to be controlled. The KB was described very analogically by means of the analytical model based on the LPN or/and the FPN as well as by means of the hybrid model. The models as well as the analytical expression of knowledge inference are uniform for both logical and fuzzy knowledge. Simultaneous utilizing of the DEDS model and the model of the KB in the control synthesis procedure was outlined. To illustrate the proposed approach the example of the knowledge-based control synthesis for the transport subsystem of FMS was introduced. It utilizes both the hybrid model of the controlled object and the PN-based expression of the control task specifications that were given verbally.

The presented approach seems to be very suitable for knowledge-based control synthesis of the wide class of DEDS simultaneously containing both the controllable discrete events and the spontaneous ones.

References

1. Ho Y Ch (ed) 1989 *Dynamics of Discrete Event Systems.* IEEE Press, Piscataway (Proceedings of the IEEE, Vol. 77, No. 1)

2. Čapkovič F 1994 Petri nets-based approach to the maze problem solving. In: Balemi S, Kozák P, Smedinga R (eds) 1994 *Discrete Event Systems: Modelling and Control.* Birkhäuser Verlag, Basel-Boston-Berlin (Progress in Systems and Control Theory, Vol. 13), pp 173-179

3. Čapkovič F 1994 Computer-aided design of intelligent control systems for discrete event dynamic systems. In: Mattson S E, Gray J O and Cellier F (eds) 1994 *IEEE/IFAC Joint Symposium on Computer-Aided Control System Design.* IEEE Press, Piscataway, pp 55-60

4. Čapkovič F 1995 Petri net-based approach to intelligent control synthesis of FMS. In: 1995 *INRIA/IEEE Symposium on Emerging Technologies and Factory Automation, Vol. 1.* IEEE Computer Society Press, Los Alamitos, USA, pp 323-331

5. Čapkovič F 1996 Knowledge-based control of DEDS. In: 1996 *IFAC World Congress 1996, Vol. J.* Compact Disc, Elsevier Sciences Ltd., Pergamon, London, pp 347-352

6. Looney C G 1988 Fuzzy Petri nets for rule-based decisionmaking. *IEEE Trans Syst Man Cybern* 18:178-183

7. Wonham W M, Ramadge P J 1987 On the supremal controllable sublanguage of a given language. *SIAM J Cont Optim* 25:637-659

20

Static and Dynamic Control of Discrete Event Dynamic Systems

M.N. Varvatsoulakis and P.N. Paraskevopoulos

1 Introduction

A discrete event system (DES) is a dynamic system representing processes for which the evolution of the state is triggered by the occurrence of physical events. Typical instances include flexible industrial systems, communication protocols, traffic control systems e.t.c. at some level of abstraction. The synthesis of a supervisor for a discrete event process requires 1) A model that provides a complete description of the behavior of the process, and 2) A model of the desired or legal behavior. This work investigates such structures from the viewpoint of control theory [6].

The principal features of discrete event processes in this work is that they are discrete, asynchronous and deterministic. The supervisory control theory introduced by Ramadge-Wonham [1],[3] provides a formal framework for the study of such systems.

In the general case a DES (plant to be controlled) is modeled as an automaton that generates a formal language over a finite alphabet U whose elements are the automaton's transitions or events. Events labeled by elements in a fixed subset U_c of U are declared to be controllable, what means they can be disabled by an external controller called supervisor. Disablement is made to depend on the past history (string) of generated events, in such a way that a design specification of the controlled system behavior is satisfied. Under suitable conditions the control law can be optimized, in the sense of minimally restricting plant behavior.

It is natural to enhance the abstract automaton model of Ramadge-Wonham by exploiting algebraic regularity of internal system structure when it exists. An obvious instance of such structure has already been presented in [5] by exploiting arithmetic additivity over the integers. The state of a manufacturing shell might be the current contents of various buffers and the numbers of machines in various modes of operation. When the unit finishes operation the status vector machines and buffer contents would be suitably incremented. Similar examples are furnished by various kinds of computer systems, traffic systems e.t.c.

It is possible to go further by considering not only vector addition systems but generalized transition functions. In this way any component of the next state vector after one event's occurrence is given as a combination of one or more current components' values.

To design control policies for such discrete event systems the desirable controlled system behavior is given in terms of predicates over the state set. In this way after one controllable event's occurrence the supervisor processes information related to the current state in order to decide the enablement or disablement of event.

The supervisors considered in this work are static if the control action is determined by the present system state only else they are dynamic i.e. based on the previous history of the system before it reaches the state. For the dynamic supervisors a memory mechanism must be provided in order to track necessary information and based on that control action is determined.

The modeling and terminology framework for discrete event systems is defined in section 2. In sections 3 and 4 we refer to the general concepts related to static and dynamic supervisory control respectively and present the control problem in the general case. In section 5 we present topics of a general case for solving problems of control with an application example of a manufacturing cell. Finally in section 6 the conclusions are provided as with as topics for further investigation.

2 Modeling and Terminology

The class of systems we consider is an extension of deterministic finite-state machine generators following the framework of Ramadge-Wonham [1]. A finite-state machine generator represents a discrete event dynamic system that responds to generated spontaneous events producing internal state transitions and output symbols. Events are considered to occur spontaneously (no auxiliary forcing mechanism is postulated), asynchronously and instantaneously.

The internal state of the system is described by state variables ranging over a finite set SX. Therefore the finite state space is defined as the set of n-dimensional vectors:

$X = [x_1, x_2, \dots, x_n]$ $x_i \in SX$, $i = 1, 2, \dots, n$ $n \in N$

Definition 2.1. $G = (X,U,Y,f,h,X_0,X_f)$ is a finite-state machine generator where

X is the state space

U is the events alphabet

Y is the output alphabet

$f: X \times U \rightarrow X$ is the state transition function

$h: X \times U \rightarrow Y$ is the output function

$X_0 \in X$ is the initial state

$X_f \subseteq X$ is the set of final states representing completed tasks

The behavior of G is stated as:
$X(k+1) = f(X(k), U(k))$
$Y(k+1) = h(X(k+1), U(k))$
Here $X(k+1) \in X$ is the state after the k event, $U(k) \in U$ is the k event and $Y(k+1)$ is the produced output symbols.

Definition 2.2. An automaton G is called deterministic if
$\forall\, x_1, x_2, x_3 \in X, u \in U: x_1 = f(u,x_3), x_2 = f(u,x_3) \Rightarrow x_1 = x_2$

Definition 2.3. A sequence of states X(k)X(k-1)...X(0) is called a path of G if a sequence of event symbols U(k)U(k-1)...U(0) exists such that X(k)=f(X(k-1), U(k-1)) for k=1,...,n. When X(k)=X(0) this path is called a cycle. If a G has no cycles is called acyclic.

3 Static Supervisory Control

We denote by U^* the set of all finite traces or strings of elements of U, including the empty trace e. $U_c \subseteq U$ will denote the set of controllable events while $U_u \subseteq U$ will denote the set of uncontrollable events and $U_c \cup U_u = U$, $U_c \cap U_u = \emptyset$. The behavior of G is described in terms of the formal language $L(G) \subseteq U^*$ i.e. the set of all finite traces generated by G.

Let $g = \{0,1\}^{Uc}$ be the set of all binary assignments to the elements of U_c. Each assignment $t \in T$ i.e. each function $g: U_c \rightarrow \{0,1\}$ is a control pattern. An event u is said to be enabled by g if g(u) = 1, or disabled by g if g(u) = 0.

The control objective is to design a controller that switches control patterns in such a way that a given generator coupled with it behaves in obedience to various constraints. Such a controller will be called a supervisor.

Formally a supervisor is a pair CG=(G',r) where $G'=(Z,U,Y,f',Z_0,Z_f)$ is a deterministic automaton with state space Z, initial state Z_0, final states Z_f, event alphabet U and transition function $f': ZxU \rightarrow Z$. We define $r: Z \rightarrow g$ as a total function that maps supervisor states into control patterns.

We interpret CG as a device that executes a sequence of state transition functions (according to f') in response to an appropriate string $u \in U^*$. Thus we may couple G to CG in a feedback loop by allowing the state transitions of G' to be forced by G and requiring G to be constrained by the successive control patterns determined by the states of G'.

The desirable controlled behavior of the complete coupled system is given in terms of a formal language $K=L(G/CG) \subseteq L(G)$. There exists a supervisor CG such that for the closed-loop system G/CG have L(G/CG)=K if K is closed and controllable i.e. $K=\overline{K}$ and $\overline{K}S_U \cap L(G) \subset \overline{K}$. That means each prefix of a string in K belongs also in K and no uncontrollable events are disabled along a string path.

The control specifications of a discrete event system can be stated in terms of its state trajectories by using predicate calculus. A predicate on a state space X is a function $P:X\rightarrow\{0, 1\}$. Let P(X) be the family of all predicates on X. The operators "$\neg$" (negation), "$\wedge$" (conjunction), "$\vee$" (disjunction) are defined as follows:

$(\neg P)(x) = 1 \Leftrightarrow P(x) = 0$

$(P_1 \wedge P_2)(x) = 1 \Leftrightarrow P_1(x) = 1$ and $P_2(x) = 1$

$P_1 \vee P_2 = \neg((\neg P_1) \wedge (\neg P_2))$

Then $(P(X), \neg, \wedge, \vee)$ is a Boolean algebra.

By the foregoing relations a partial order is induced on P(X) according to

$P_1 \leq P_2 \Leftrightarrow P_1 \wedge P_2 = P_1$

For each $u\in U$ the transformation $wlp_u:P(X)\rightarrow P(X)$ is defined for a predicate $P\in P(X)$ by:

$$wlp_u(P)(X) := \begin{cases} 1 \text{ if } f(u,x) \text{ is defined and } P(f(u,x))=1 \\ \quad \text{or } f(u,x) \text{ not defined} \\ 0 \text{ otherwise} \end{cases}$$

For a DES G and a predicate P, P is controllable with respect to G if:

$\forall u\in U_u : P \leq wlp_u(P)(X)$

For a DES and a predicate P on X assume that $X_0 \in P$. A predicate R(G,P) is defined which characterizes the set of states reachable from Q_0 by way of states satisfying P:

1. $Q_0 \in R(G,P)$
2. If $Q\in R(G,P)$, $u\in U$, f(u,x) is defined and $f(u,x)\in P \Rightarrow f(u,x)\in R(G,P)$
3. Every state in R(G,P) is obtained as previous.

Theorem 3.1:Let P be a predicate on X with $X_0\in P$. There exists a state-feedback controller C for G such that R(G/CG)=P if and only if P is controllable i.e. $(\forall u \in U_u)\, P \leq R(G,P) \wedge wlp_u(P)$.

Theorem 3.2:The supremal element of the set $sup(P):=\{P'|P'\leq P$ and P is controllable$\}$ always exists.

Theorem 3.3: Let [P] be the set of states X satisfying $(\forall u \in U_u^*) f(u,x) \text{exists} \Rightarrow f(u,x) \in P$. Assume that $X_0\in P$. Then supSR(P) = R(G,[P]).

To construct the supremal controllable predicate of P sup(P) i.e. the set of states for which no uncontrollable transition is permitted to reach states not satisfying P we present the following algorithm :

Algorithm 3.4.

1. Let $R_0=\{ x\in X: P(x)=0 \}$ i.e. is the set of states not satisfying P.
2. Let $S_0=\{ x\in X: P(x)=0 \}$ i.e. is the set of states that will be examined for uncontrollable transitions the first time.
3. Let $R_{i+1}=R_i\cup\{x\in X: \exists u\in U_u, \exists x'\in S_i, f(u,x)=x'\}$ i.e. is the instantaneous set of states that uncontrollable transitions lead to states not satisfying P.
4. Let $S_{i+1}=R_{i+1}- R_i$ i.e. is the set of new states that will be examined.

5. Terminate if $S_{i+1}=\varnothing$ or $x_0 \in S_{i+1}$ else goto 3.
6. If $S_{i+1}=\varnothing$ then sup(P)=X- S_{i+1} else sup(P)= $\varnothing$.

Given a feedback controller r we define, for each $u \in U$, the u-component of f to be the predicate r_u on X with $r_u(X):=1$ if $u \in r(X)$ and 0 otherwise.

Theorem 3.5: Assume that the legal specification for a DES is given by a predicate P of the form $P = \bigwedge_{i=1}^{k} P_i$ where P_i are called subspecifications. For each subspecification we synthesize a subcontroller $r_i: X \rightarrow g$ which satisfies $R(C_i/G)=supSR(P_i)$. A modular controller is then formed as $r_u := \bigwedge_{i=1}^{k} r_{i,u} (u \in U)$. The modular controller satisfies $R(\bigwedge_{i=1}^{k} r_i/G)=supSR(P)$.

4 Dynamic Supervisory Control

The state feedback controllers discussed in the previous section are static in the sense that their control actions are fully determined by the current state of the plant and are independent of how this state is reached. Such «memoryless» controllers are sufficient in forcing the plant to behave in accordance with a legal-behavior specification given in the form of a legal state range i.e. a predicate on state space X. However, in many cases, legal-behavior specifications also refer to the history of the plant, thus cannot be formalized merely as predicates on X. Controllers accessing only the current plant state are not powerful enough to enforce such «dynamic» specifications-auxiliary information about the plant history must be made available to controllers. Auxiliary devices, called memories, can be constructed to record the required history information. This leads to dynamic control structure.

Definition 4.1. SCG = (X,U,f,g,h,X_0,X_f) is a static controlled generator where
X is the finite state space
U is the events alphabet
Y is the output alphabet
$f: X \times U \rightarrow X$ is the state transition function
$h: X \times U \rightarrow Y$ is the output function
$g: X \rightarrow \{0, 1\}$ is the control function
$X_0 \in X$ is the initial state
$X_f \subseteq X$ is the set of final states representing completed tasks

The behavior of the SCG is stated as:

$X(k+1) = f(X(k), U(k))$ if $g(X(k))=1$
$X(k+1) = X(k)$ if $g(X(k))=0$

$Y(k+1) = h(X(k+1), U(k))$ if $g(X(k))=1$

$Y(k+1) = Y(k)$ if $g(X(k))=0$

Control of a SCG consists of switching the control function $g(X(k))$ that generally depends on current state observation according to defined conditions. If $g(X(k))=0$ the system is blocked and no state transition occurs unless $g(X(k))=1$. Therefore every new event symbol $U(k)$ is rejected i.e. is not recorded in the event string $U(k)U(k-1)\ldots U(0)$ if $g(X(k))=0$. This controlled process is static.

Let $X^*(k)=X(k)X(k-1)\ldots X(0)$ the k sequence of states and $U^*(k)=U(k)U(k-1)\ldots U(0)$ the k sequence of event symbols.

Definition 4.2. DCG = (X,U,f,g,h,X_0, X_f) is a dynamic controlled generator where

X is the finite state space

U is the events alphabet

Y is the output alphabet

$f : X \times U \rightarrow X$ is the state transition function

$h : X \times U \rightarrow Y$ is the output function

$g : X^* \times U^* \rightarrow \{ 0, 1 \}$ is the control function

$X_0 \in X$ is the initial state

$X_f \subseteq X$ is the set of final states representing completed tasks

The behavior of the DCG is stated as:

$X(k+1) = f(X(k), U(k))$ if $g(X^*(k), U^*(k))=1$

$X(k+1) = X(k)$ if $g(X^*(k), U^*(k))=0$

$Y(k+1) = h(X(k+1), U(k))$ if $g(X^*(k), U^*(k))=1$

$Y(k+1) = Y(k)$ if $g(X^*(k), U^*(k))=0$

The control function $g(X(k)X(k-1)\ldots X(0),U(k)U(k-1)\ldots U(0))$ depends on full state and event string observation. This controlled process is dynamic.

5 Application Examples

5.1. Production Network

The modeling and control of a typical production network is investigated here. The network operates as follows:

Machines in groups 1 and 2 receive parts from a common buffer and after processing deposit the finished part to collecting points 3 and 4 respectively. Machines in group 5 receive parts from points 3 and 4 for assembling. The final part is deposited to point 6 for further processing in machines group 7. The final product is transferred to the inspection unit 8 which can either output the part as a finished product or reject it for reworking by group 4 again.

A modular approach will be used to modeling the production network. We first model modules of the network individually and then compose them to to form the model of the complete system.

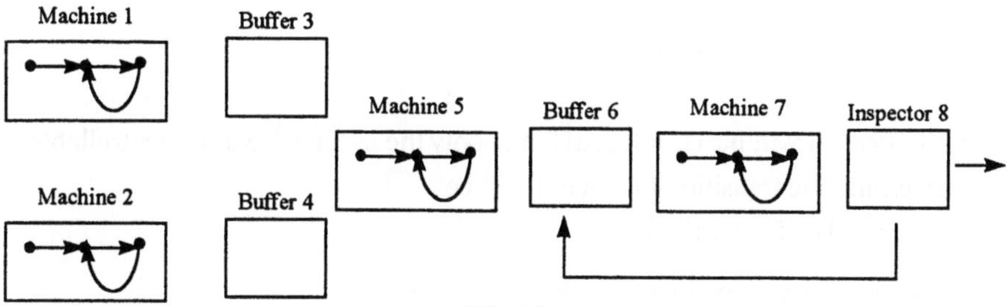

[Fig.1]

5.1.1. Machine Groups

The state of each machine can be described by four state variables and the state space of a machine is:

$$X_i = \begin{pmatrix} x_i^1 \\ x_i^2 \\ x_i^3 \\ x_i^4 \end{pmatrix} x_i^j \in R \; (i = 1,2,5,7)$$

The event set will be $U_i = \{u_i^1, u_i^2, u_i^3, u_i^4, u_i^5\}$ with u_i^1=start working, u_i^2=finish processing, u_i^3=discharges a processed part, u_i^4=break down, u_i^5=repair and only the events u_i^1, u_i^2 will be controllable. The transition function will be as follows:

$$f_{u_i^1} = \begin{pmatrix} x_i^1 - 1 \\ x_i^2 + 1 \\ x_i^3 \\ x_i^4 \end{pmatrix} \quad f_{u_i^2} = \begin{pmatrix} x_i^1 \\ x_i^2 - 1 \\ x_i^3 + 1 \\ x_i^4 \end{pmatrix} \quad f_{u_i^3} = \begin{pmatrix} x_i^1 + 1 \\ x_i^2 \\ x_i^3 - 1 \\ x_i^4 \end{pmatrix} \quad f_{u_i^4} = \begin{pmatrix} x_i^1 \\ x_i^2 - 1 \\ x_i^3 \\ x_i^4 + 1 \end{pmatrix} \quad f_{u_i^5} = \begin{pmatrix} x_i^1 \\ x_i^2 + 1 \\ x_i^3 \\ x_i^4 - 1 \end{pmatrix}$$

For the state transition to be possible it must be $\begin{pmatrix} x_i^1 >= 0 \\ x_i^2 >= 0 \\ x_i^3 >= 0 \\ x_i^4 >= 0 \end{pmatrix}$.

The discrete event model for machine i will be the generator $G_i = \{U_i, X_i, f_i, X_{0i}\}$.

5.1.2. Buffers

The state of each buffer can be modeled as a scalar system and the state space will be:

$$X_i = \left(x_i^1\right) x_i^1 \in R\,(i = 3,4,6)$$

For buffer 3 it will be $U_i = \{u_1^3, u_5^1\}$ and only the event u_1^3 will be controllable.

Concerning the transition function it will be :

$f_{u_i^3} = (x_i^1 + 1) \quad f_{u_5^1} = (x_i^1 - 1)$

For the state transition to be possible let $(x_i^1 \geq 0)$.

Additionally for buffer 6 it will be $f_{u_8^2} = (x_i^1 + 1)$.

The discrete event model for each buffer will be the generator $G_i = \{U_i, X_i, f_i, X_{0i}\}$.

5.1.3. Inspection Unit

The state of inspection unit is described by 2 variables and the state space will be:

$$X_8 = \begin{pmatrix} x_8^1 \\ x_8^2 \end{pmatrix} x_i^j \in R\,(i = 8), (j = 1,2)$$

It will be $U_8 = \{u_8^1, u_8^2, u_8^3\}$ and only the event u_8^1 will be controllable.

The transition function will be:

$$f_{u_7^3} = \begin{pmatrix} x_i^1 + 1 \\ x_i^2 \end{pmatrix} \quad f_{u_8^1} = \begin{pmatrix} x_8^1 - 1 \\ x_i^2 + 1 \end{pmatrix} \quad f_{u_8^2} = \begin{pmatrix} x_i^1 \\ x_i^2 - 1 \end{pmatrix} \quad f_{u_8^3} = \begin{pmatrix} x_i^1 \\ x_i^2 - 1 \end{pmatrix}$$

For the state transition to be possible it must be $\begin{pmatrix} x_8^1 >= 0 \\ x_8^2 >= 0 \end{pmatrix}$.

The discrete event model for the inspection unit will be $G_8 = \{U_8, X_8, f_8, X_{08}\}$.

5.1.4. Control policies

The specification is that no buffer overflow must happen and that at most one part be inspected at a given time. Consequently the predicate conditions will be :

$P_1 = (x_3^1 \leq k_1)$

$P_2 = (x_4^1 \leq k_2)$

$P_3 = (x_6^1 \leq k_3)$

$P_4 = (x_8^1 \leq 1)$

The supervisor is static and is designed to satisfy the overall performance specification. It will be $P = \bigwedge_{i=1}^{4} P_i$ and we'll have:

$f_{1,u}(X) = 0 \Leftrightarrow u = u_1^3 \text{ and } x_3^1 \geq k_1$

$f_{2,u}(X) = 0 \Leftrightarrow u = u_2^3 \text{ and } x_4^1 \geq k_2$

$f_{3,u}(X) = 0 \Leftrightarrow (u = u_5^3 \text{ or } u = u_8^1) \text{ and } x_6^1 + x_8^2 \geq k_3$

$f_{4,u}(X) = 0 \Leftrightarrow u = u_7^3 \text{ and } x_8^1 \geq 1$

5.2. Control in the presence of uncontrollable transitions

The computations in this section are intended to illustrate the application of Algorithm 3.4. An automaton $G = \{U, X, f, X_0\}$ is displayed below with $U_c = \{a,b,c\}$, $U_u = \{d,e\}$. In the sequel we compute the supremal controllable predicate of P where $X_{12}, X_{13} \in P$.

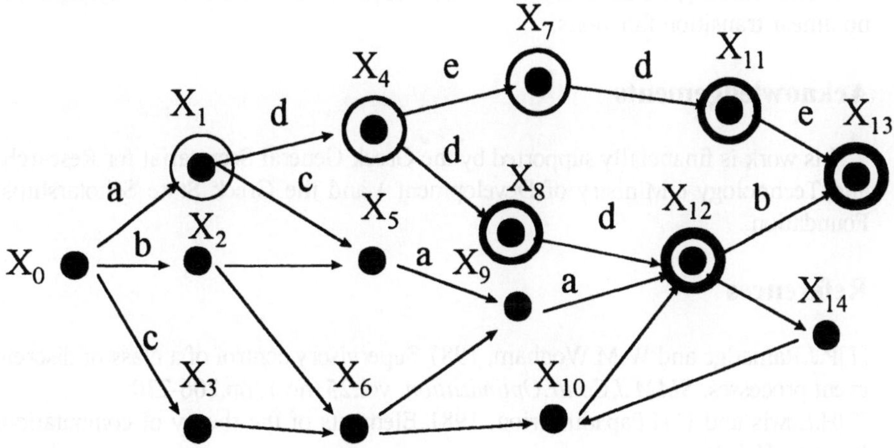

[Fig.2]

-$R_0 = \{ X_{12}, X_{13} \}$

-$S_0 = \{ X_{12}, X_{13} \}$

-$R_1 = \{ X_{12}, X_{13} \} \cup \{ X_8, X_{11} \}$

-$S_1 = \{ X_8, X_{11} \}$

.

.

-$R_3 = \{ X_{12}, X_{13}, X_8, X_{11}, X_7, X_4 \} \cup \{ X_1, X_4 \}$

-$S_3 = \{ X_1 \}$

-$R_4 = R_3$

-The Algorithm is terminated and $\sup(P) = \{ X_0, X_2, X_5, X_3, X_6, X_{10}, X_{14} \}$

6 Conclusions

A discrete event system (DES) with a vector state space is a structured model, which is economical in modeling systems with inherent internal structure - especially, manufacturing, computer, communication and traffic systems.

In this work [6] we gave the general concepts and developed basic results for a state feedback theory of DES. The state-feedback theory logic in Ramadge-Wonham theory was extended to include such concepts as controllability with static and dynamic control. The control problem was given as a set of predicates over the state set to be true. Predicates can concisely represent an infinite state space, hence many of the discrete event systems including timing issues, queues with unbounded buffers e.t.c. can be modeled in this framework. The weakest liberal precondition has been presented as a fundamental concept for describing the state space evolution of a DES. The notion of controllability of a required predicate describing the set of legal states has been defined, and it has been shown that controllability is a necessary and sufficient condition for the existence of a supervisor that guarantees the invariance of the required predicate under system evolution.

Subjects for further investigation include design methods for dynamic controllers and generalization of vector represented discrete event systems to nonlinear transition functions.

Acknowledgements

This work is financially supported by the Greek General Secretariat for Research and Technology (Ministry of Development) and the Greek State Scholarships Foundation.

References

[1]P.J.Ramadge and W.M.Wonham, 1987 Supervisory control of a class of discrete event processes. *SIAM J.Contr.Optimization,* vol.25, no.1, pp.206-230.

[2]H.Lewis and C.H.Papadimitriou, 1981 Elements of the theory of computation, Prentice-Hall Inc.

[3]W.M.Wonham and P.J.Ramadge, 1988 Modular supervisor control of discrete event systems, *Mathematics of Control, Signals and Systems,* vol.1, no.1, pp.13-30.

[4]R.Kumar,V.Gong and S.Markus, 1993 Predicates and Predicate Transformers for Supervisory Control of Discrete Event Systems, *IEEE Trans.Automatic Control,* vol.38, no.2, pp.232-247.

[5]Y.Li and W.M.Wonham, 1993 Control of Vector Discrete-Event Systems I-The Base Model, *IEEE Trans.Automatic Control,* vol.38, no.8, pp.1214-1227.

[6]M.N.Varvatsoulakis 1998 Learning and Control of Discrete Event Processes, Ph.D. Thesis, National Technical University of Athens, in preparation.

21

Min-Max Control with Application to Discrete Event Dynamic Systems

P. Maragos and S.G.Tzafestas

1 Introduction

Large classes of dynamic phenomena such as material flow in manufacturing systems, traffic flow in transportation or communication networks, and related scheduling problems can be viewed as *discrete event dynamical systems (DEDS)*; see the papers in [7] for surveys. An efficient approach [3,9] to model large classes of DEDS has been based on the minimax algebra [4] and describes the time dynamics of such DEDS by using nonlinear state space equations which algebraically resemble the linear (sum-product) equations of linear control but in which the addition and multiplication have been replaced by maximum and addition, respectively.

In our work we propose the following general algebraic model for such nonlinear state-space equations of which special cases are the examples of DEDS and related applications studied in [3,9,5]:

$$\begin{aligned} \boldsymbol{x}(k+1) &= (\boldsymbol{A} \boxplus \boldsymbol{x}(k)) \vee (\boldsymbol{B} \boxplus \boldsymbol{u}(k)) \\ \boldsymbol{y}(k) &= (\boldsymbol{C} \boxplus \boldsymbol{x}(k)) \vee (\boldsymbol{D} \boxplus \boldsymbol{u}(k)) \end{aligned} \tag{1}$$

where $\vee$ denotes pointwise maximum (or supremum), $\boxplus$ is a max-sum matrix 'product' defined later, $\boldsymbol{x} = [x_1, x_2, ..., x_n]^T$ with $(\cdot)^T$ denoting transpose is a n-dim state vector, $\boldsymbol{u}$ is a r-dim input or control vector, and $\boldsymbol{y}$ is a m-dim output vector. We shall refer to (1) as *max-sum control* systems. In applications to DEDS, the ith component $x_i(k)$ of the state vector may represent the earliest start-up or completion time of the kth cycle of machine i, the input $\boldsymbol{u}$ represents earliest availability times of parts, the output $\boldsymbol{y}$ represents

[1] P. Maragos' research work was partially supported by the US National Science Foundation under Grant MIP–9421677. This chapter was written while P.M. was with the Institute for Language & Speech Processing in Athens. A summary of this work was presented at EURISCON-98, Athens, June 1998.

earliest exit times, and the elements of the matrices $\boldsymbol{A}, \boldsymbol{B}, \boldsymbol{C}, \boldsymbol{D}$ represent service/delay times or activity durations. The components of the vector $\boldsymbol{x}$, of the vectors (or scalars) $\boldsymbol{u}, \boldsymbol{y}$ and of the matrices $\boldsymbol{A} \in \overline{\mathbb{R}}^{n\times n}$, $\boldsymbol{B} \in \overline{\mathbb{R}}^{n\times r}$, $\boldsymbol{C} \in \overline{\mathbb{R}}^{m\times n}$ and $\boldsymbol{D} \in \overline{\mathbb{R}}^{m\times r}$ are from the set $\overline{\mathbb{R}} = \mathbb{R} \cup \{-\infty, \infty\}$ of extended real numbers. In general, the *max-sum matrix 'product'* $\boxplus$ of an arbitrary $m \times n$ matrix $\boldsymbol{A} = [a_{ij}]$ with an arbitrary $n \times p$ matrix $\boldsymbol{B} = [b_{ij}]$ is the $m \times p$ matrix $\boldsymbol{M} = [m_{ij}]$ defined as

$$\boldsymbol{M} = \boldsymbol{A} \boxplus \boldsymbol{B} \quad , \quad m_{ij} = \bigvee_{k=1}^{n} a_{ik} + b_{kj} \tag{2}$$

with $a + (-\infty) = -\infty$ for any $a \in \overline{\mathbb{R}}$.

Replacing (in applications to DEDS) the meaning of the state, input, or output vectors as the latest possible values of the corresponding timing events leads to a *dual* algebraic model for control, which is obtained by replacing in (1) maximum ($\vee$) with minimum ($\wedge$) and the max-sum ($\boxplus$) with a min-sum matrix product ($\boxplus'$):

$$\begin{aligned} \boldsymbol{x}(k+1) &= \big(\boldsymbol{A} \boxplus' \boldsymbol{x}(k)\big) \wedge \big(\boldsymbol{B} \boxplus' \boldsymbol{u}(k)\big) \\ \boldsymbol{y}(k) &= \big(\boldsymbol{C} \boxplus' \boldsymbol{x}(k)\big) \wedge \big(\boldsymbol{D} \boxplus' \boldsymbol{u}(k)\big) \end{aligned} \tag{3}$$

where the *min-sum matrix 'product'* $\boxplus'$ of an arbitrary $m \times n$ matrix $\boldsymbol{A} = [a_{ij}]$ with a $n \times p$ matrix $\boldsymbol{B} = [b_{ij}]$ is the $m \times p$ matrix $\boldsymbol{M} = [m_{ij}]$ defined as

$$\boldsymbol{M} = \boldsymbol{A} \boxplus' \boldsymbol{B} \quad , \quad m_{ij} = \bigwedge_{k=1}^{n} a_{ik} +' b_{kj} \tag{4}$$

with $+'$ being regular addition extended by the rule $a +' (\infty) = \infty$ for any $a \in \overline{\mathbb{R}}$. We shall refer to (3) as *min-sum control* systems.

In our work we view the nonlinear state-space equations (1) and (3) as describing a large class of nonlinear dynamical systems whose system-theoretic and control aspects we call *max-min control.* This work deals with the theory of max-min control. We begin with some observations from its applications to DEDS and nonlinear filtering (of the morphological type). Then, our contributions include the following: 1) Representation of the matrix-based vector transformations and the signal transformations induced by the system in terms of nonlinear operators acting on vector or signal lattices. 2) Derivation of the complete solution of (1) in the discrete time domain and decomposition of this solution into two parts due, respectively, to the initial state and to the input. 3) Description of the latter part of the system's response in terms of nonlinear convolutions (of the morphological type) of the input signal with an appropriately defined impulse response of the system. 4) Study of stability, both via the system's impulse response and via its eigenvalues as defined in minimax algebra. 5) Study of controllability and observability by developing solutions based on minimax algebra and lattice operators. 6) Study of feedback.

2 Applications

2.1 Modeling DEDS

Max-min control systems have been used in [9,5] to model the dynamics of the start-up and completion times of machines connected in serial and/or parallel production lines. Figure 1 shows a serial production line of two machines and one buffer. Assume that the service/process times of the two machines are

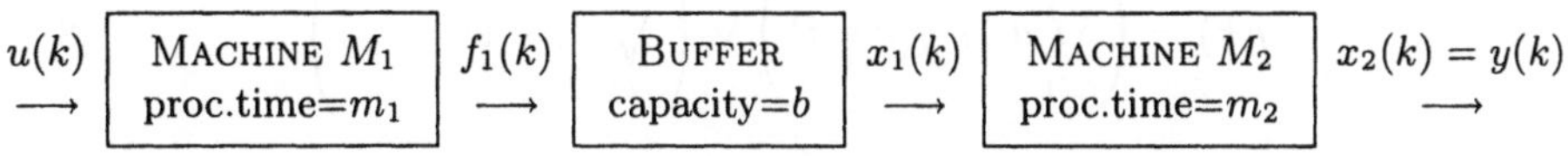

Fig. 1. A serial (assembly) production line.

constants m_1, m_2, the capacity of the buffer is an integer $b \geq 1$, and the transition times to transport machined parts from machine to buffer or from buffer to machine are zero. Let us define by $u(k)$ the arrival time of the materials needed for the kth performance cycle of machine M_1, by $y(k)$ the exit (output) time of the kth product from machine M_2, by $s_i(k)$ the start-up time of the ith machine, by $f_i(k)$ the completion time of the ith machine, and by $x_i(k)$ the output time from the ith machine for its kth performance cycle. Note that $f_i(k) = s_i(k) + m_i$, $i = 1, 2$. The first cycle starts at $k = 0$ and assume that $u(k) = y(k) = s_i(k) = f_i(k) = x_i(k) = -\infty$ for all $k < 0$ and the buffer was initially empty. If the buffer has infinite capacity ($b = \infty$), then there is no blocking and $x_1(k) = f_1(k)\ \forall k$. But if $b < \infty$, there is blocking and $x_1(k) \neq f_1(k)$; i.e.,

$$b < \infty \Longrightarrow x_1(k) = \max[f_1(k), x_2(k-b)] \tag{5}$$

Thus, M_1 cannot output its machined part at cycle k before M_2 has finished its task at cycle $k - b$. But the output time of M_2 is identical with its completion time. Combining all the above leads to the following state equations for the output times of the two machines in Fig. 1 (valid for two extreme values of b: $b = 1$ or $b = \infty$):

$$\begin{bmatrix} x_1(k+1) \\ x_2(k+1) \end{bmatrix} = \begin{bmatrix} m_1 & \theta \\ m_1 + m_2 & m_2 \end{bmatrix} \boxplus \begin{bmatrix} x_1(k) \\ x_2(k) \end{bmatrix} \vee \left(\begin{bmatrix} m_1 \\ m_1 + m_2 \end{bmatrix} + u(k+1) \right)$$

$$y(k) = [-\infty, 0] \boxplus \begin{bmatrix} x_1(k) \\ x_2(k) \end{bmatrix} \tag{6}$$

where $\theta = 0$ if $b = 1$ and $\theta = -\infty$ if $b = \infty$.

2.2 State-Space Models of Nonlinear Filters

A very large class of discrete linear time-invariant systems used in control and signal processing [2,8,13] can be described via the following class of linear difference equations

$$y(k) = \sum_{i=1}^{n} a_i y(k-i) + \sum_{j=0}^{p} b_j u(k-j) \tag{7}$$

Replacing sum with maximum and multiplication with addition gives us the following nonlinear *max-sum difference equation* [12]

$$y(k) = \left(\bigvee_{i=1}^{n} a_i + y(k-i) \right) \vee \left(\bigvee_{j=0}^{p} b_j + u(k-j) \right) \tag{8}$$

capable of modeling a large class of morphological systems used in nonlinear signal processing and image analysis [10,11,15].

The nonlinear state equations of (1) (or of (3)) can also model the dynamics of nonlinear discrete-time filters of the morphological type described by the above max-sum (or min-sum) difference equation. Specifically, if $p = 0$, setting $x_1(k) = y(k-n)$, $x_2(k) = y(k-n+1)$, ..., $x_n(k) = y(k-1)$ and choosing matrices

$$\boldsymbol{A} = \begin{bmatrix} -\infty & 0 & -\infty & \dots & -\infty \\ -\infty & -\infty & 0 & \dots & -\infty \\ \vdots & \vdots & & & \vdots \\ -\infty & -\infty & -\infty & \dots & 0 \\ a_n & a_{n-1} & a_{n-2} & \dots & a_1 \end{bmatrix}, \ \boldsymbol{B} = \begin{bmatrix} -\infty \\ -\infty \\ \dots \\ -\infty \\ b_0 \end{bmatrix}, \ \boldsymbol{C} = [a_n \dots a_1]^T, \ D = b_0$$

models (8) as a special case of (1).

3 Representation of Matrix and Signal Operations by Lattice Dilations and Erosions

Our set of scalars will be $\overline{\mathbb{R}}$. The triple $(\overline{\mathbb{R}}, \vee, \wedge)$ is a complete distributive lattice with partial order the standard relation $\leq$ among extended real numbers.

3.1 Vector Lattice

Consider now the vector space $\mathcal{L} = \overline{\mathbb{R}}^n$, equipped with the following 'supremum' and 'infimum' operation. Given two vectors $\boldsymbol{x}, \boldsymbol{y} \in \mathcal{L}$, their supremum is defined as the vector $\boldsymbol{z} = \boldsymbol{x} \vee \boldsymbol{y}$ with[2] $z_i = x_i \vee y_i$, $i = 1, ..., n$. Similarly we define the 'infimum' of vectors by replacing $\vee$ with $\wedge$. Then, $(\mathcal{L}, \vee, \wedge)$ is also a complete distributive lattice with an induced partial order

[2] *Notation:* If $\boldsymbol{M} = [m_{ij}]$ is a matrix, its (i, j)th element is also denoted as $\{\boldsymbol{M}\}_{ij} = m_{ij}$. Similarly, if $\boldsymbol{x}$ is a vector, its ith element is denoted as $\{\boldsymbol{x}\}_i$ or simply x_i.

defined componentwise; i.e., $\boldsymbol{x} \leq \boldsymbol{y}$ means $x_i \leq y_i \ \forall i$. Of interest are operators ψ on $\mathcal{L}$ (i.e., vector transformations) that are *increasing*, i.e., $\boldsymbol{x} \leq \boldsymbol{y}$ implies $\psi(\boldsymbol{x}) \leq \psi(\boldsymbol{y})$. Elementary increasing operators are the *translations* $\tau_v(\boldsymbol{x}) = v + \boldsymbol{x}$, which add a scalar v to a vector $\boldsymbol{x}$ componentwise. Actually, the set $\mathbb{T} = \{\tau_v : v \in \mathbb{R}\}$ of translations forms a commutative group under composition $\tau_v \tau_r(\boldsymbol{x}) = \tau_v(\tau_r(\boldsymbol{x})) = \tau_{v+r}(\boldsymbol{x})$. If we define the elementary vectors

$$\boldsymbol{e}_i \equiv [-\infty, ..., -\infty, 0, -\infty, ..., -\infty]^T$$

with a zero at the ith position and $-\infty$ elsewhere, then each vector $\boldsymbol{x} = [x_1, ..., x_n]^T$ can be represented as a max of translated elementary vectors:

$$\boldsymbol{x} = \bigvee_{i=1}^{n} x_i + \boldsymbol{e}_i = \bigvee_{i=1}^{n} \tau_{x_i}(\boldsymbol{e}_i) \tag{9}$$

Two very important types of increasing operators are the *dilations* δ and the *erosions* ε which are defined [16,6] as operators that distribute over sup and inf, respectively:

$$\delta(\bigvee_i \boldsymbol{x}_i) = \bigvee_i \delta(\boldsymbol{x}_i), \quad \varepsilon(\bigwedge_i \boldsymbol{x}_i) = \bigwedge_i \varepsilon(\boldsymbol{x}_i) \tag{10}$$

Two special examples of dilation (δ_M) and erosion (ε_M) are, respectively, the max-sum and min-sum 'product' of a matrix $\boldsymbol{M}$ with an input vector:

$$\delta_M(\boldsymbol{x}) \equiv \boldsymbol{M} \boxplus \boldsymbol{x}, \quad \varepsilon_M(\boldsymbol{x}) \equiv \boldsymbol{M} \boxplus' \boldsymbol{x} \tag{11}$$

An operator ψ on $\mathcal{L}$ is called *translation-invariant* or simply a $\mathbb{T}$-operator iff it commutes with any translation, i.e., iff $\tau\psi = \psi\tau \ \forall \tau \in \mathbb{T}$. The following theorem establishes a two-way correspondence between the max-sum and min-sum matrix-based vector transformations and the $\mathbb{T}$-dilations and $\mathbb{T}$-erosions.

Theorem 1. *(a) Any translation-invariant dilation δ on $\mathcal{L} = \overline{\mathbb{R}}^n$ can be represented as a matrix-based dilation δ_M where $\boldsymbol{M} = [m_{ij}]$ with $m_{ij} = \{\delta(\boldsymbol{e}_j)\}_i$, and vice-versa.*
(b) Any translation-invariant erosion ε on $\mathcal{L}$ can be represented as a matrix-based erosion ε_M where $\boldsymbol{M} = [m_{ij}]$ with $m_{ij} = \{\varepsilon(\boldsymbol{e}_j)\}_i$, and vice-versa.

Proof: (a) Let δ be a $\mathbb{T}$-dilation on $\mathcal{L}$. For any $\boldsymbol{x} = [x_1, ..., x_n]^T \in \mathcal{L}$ we have $\boldsymbol{x} = \bigvee_{j=1}^{n} \tau_{x_j}(\boldsymbol{e}_j)$. Hence,

$$\delta(\boldsymbol{x}) = \delta\left(\bigvee_{j=1}^{n} \tau_{x_j}(\boldsymbol{e}_j)\right) = \bigvee_{j=1}^{n} \delta(\tau_{x_j}(\boldsymbol{e}_j)) = \bigvee_{j=1}^{n} \tau_{x_j}(\delta(\boldsymbol{e}_j)) = \bigvee_{j=1}^{n} \delta(\boldsymbol{e}_j) + x_j$$

Hence, by defining the matrix $\boldsymbol{M} = [m_{ij}] = [\delta(\boldsymbol{e}_1), ..., \delta(\boldsymbol{e}_n)]$ with $m_{ij} = \{\delta(\boldsymbol{e}_j)\}_i$, we obtain

$$\{\delta(\boldsymbol{x})\}_i = \bigvee_{j=1}^{n} m_{ij} + x_j$$

which implies that $\delta(\boldsymbol{x}) = \boldsymbol{M} \boxplus \boldsymbol{x}$. Conversely: Consider an operator δ on $\mathcal{L}$ defined by $\delta(\boldsymbol{x}) = \boldsymbol{M} \boxplus \boldsymbol{x}$. Then, if $\boldsymbol{x}_k = [x_{k1}, ..., x_{kn}]^T$ is an indexed set of vectors, for each $i = 1, ..., n$ we have

$$\{\delta(\bigvee_k \boldsymbol{x}_k)\}_i = \bigvee_{j=1}^{n} m_{ij} + (\bigvee_k x_{ki}) = \bigvee_k \bigvee_{j=1}^{n} m_{ij} + x_{ki} = \{\bigvee_k \delta(\boldsymbol{x}_k)\}_i$$

Thus, $\delta(\bigvee_k \boldsymbol{x}_k) = \bigvee_k \delta(\boldsymbol{x}_k)$ and hence δ is a dilation. Further, for each translation τ_v we have for each $i = 1, ..., n$:

$$\{\tau_v(\delta(\boldsymbol{x}))\}_i = \left(\bigvee_{j=1}^{n} m_{ij} + x_j\right) + v = \bigvee_{j=1}^{n} m_{ij} + x_j + v = \{\delta(\tau_v(\boldsymbol{x}))\}_i$$

Thus, $\tau_v(\delta(\boldsymbol{x})) = \delta(\tau_v(\boldsymbol{x}))$ and hence δ is also translation-invariant.
(b) The proof is identical to that of (a) by replacing dilation with erosion and $\vee$ with $\wedge$. Q.E.D.

3.2 Signal Lattice

Consider the set $\text{Fun}(\mathbb{Z}, \overline{\mathbb{R}})$ of all discrete-time signals $f : \mathbb{Z} \to \overline{\mathbb{R}}$ with values form $\overline{\mathbb{R}}$. Equipped with pointwise sup $\vee$ and inf $\wedge$, this becomes a complete distributive lattice $\mathcal{L}$ with partial order the pointwise signal relation $\leq$. The signal translations are the operators $\tau_{i,v}(f)(k) = v + f(k - i)$, where $(i, v) \in \mathbb{Z} \times \mathbb{R}$ and $f(k)$ is an arbitrary input signal. An operator on $\mathcal{L}$ is called *translation-invariant* iff it commutes with any such translation. Consider now two elementary signals, called *upper impulse* ($q_\vee$) and *lower impulse* ($q_\wedge$):

$$q_\vee(k) \equiv \begin{cases} 0, & k = 0 \\ -\infty, & k \neq 0 \end{cases}, \quad q_\wedge(k) \equiv \begin{cases} 0, & k = 0 \\ +\infty, & k \neq 0 \end{cases}$$

then every signal f can be represented [12] as a sup of translated upper impulses or as inf of translated lower impulses:

$$f(k) = \bigvee_i f(i) + q_\vee(k - i) = \bigwedge_i f(i) +' q_\wedge(k - i)$$

Consider now operators on $\mathcal{L}$ that are *dilations* (resp. *erosions*), i.e., systems that distribute over any sup (resp. inf) of input signals. Special cases of dilation and erosion systems are, respectively, the supremal convolution $\oplus$ and the infimal convolution $\oplus'$ of two signals f and g defined by

$$(f \oplus g)(k) \equiv \bigvee_i f(i) + g(k - i), \quad (f \oplus' g)(k) \equiv \bigwedge_i f(i) +' g(k - i)$$

These two nonlinear convolutions are known in optimization [1], convex analysis [14], and especially in morphological signal processing where they (under the names of signal dilation and erosion) have found numerous applications in nonlinear filtering and computer vision [15,10,11]. The following theorem from [12] characterizes all translation-invariant dilation or erosion systems as nonlinear convolutions of the morphological type between the input signal and the system's impulse response.

Theorem 2. [12] *(a) An operator Δ on the signal lattice* $\mathrm{Fun}(\mathbb{Z}, \overline{\mathbb{R}})$ *is a translation-invariant dilation system iff it can be represented as the supremal convolution of the input signal with the system's upper impulse response* $h_\vee = \Delta(q_\vee)$.

(b) An operator E on the signal lattice $\mathrm{Fun}(\mathbb{Z}, \overline{\mathbb{R}})$ *is a translation-invariant erosion system iff it can be represented as the infimal convolution of the input signal with the system's lower impulse response* $h_\wedge = E(q_\wedge)$.

4 State and Output Responses

Based on the above lattice representations of max-sum matrix operations, the basic state-space model of a max-sum control system can now be represented via matrix-based dilations:

$$\begin{aligned} \boldsymbol{x}(k+1) &= \delta_A[\boldsymbol{x}(k)] \ \vee \ \delta_B[\boldsymbol{u}(k)] \\ \boldsymbol{y}(k) &= \delta_C[\boldsymbol{x}(k)] \ \vee \ \delta_D[\boldsymbol{u}(k)] \end{aligned} \tag{12}$$

Solving the state-space equations by using induction on k yields the *state response*:

$$\begin{aligned} \boldsymbol{x}(k) &= \boldsymbol{A}^{(k)} \boxplus \boldsymbol{x}(0) \vee \left(\bigvee_{i=0}^{k-1} \boldsymbol{A}^{(k-1-i)} \boxplus \boldsymbol{B} \boxplus \boldsymbol{u}(i) \right) \\ &= \delta_A^k[\boldsymbol{x}(0)] \ \vee \ \left(\bigvee_{i=0}^{k-1} \delta_A^{k-1-i} \delta_B[\boldsymbol{u}(i)] \right) \end{aligned} \tag{13}$$

where $\boldsymbol{A}^{(k)}$ denotes the k-fold max-sum matrix product of $\boldsymbol{A}$ with itself for $k \geq 1$ and $\boldsymbol{A}^{(0)} = \boldsymbol{I}_n$ where $\boldsymbol{I}_n$ is the $n \times n$ identity matrix I in max algebra consisting of zeros on its diagonal and $-\infty$ elsewhere.

The above result yields in turn the *output response:*

$$\boldsymbol{y}(k) = \underbrace{\delta_C \delta_A^k[\boldsymbol{x}(0)]}_{\text{'zero'-input resp.}} \ \vee \ \underbrace{\left(\bigvee_{i=0}^{k-1} \delta_C \delta_A^{k-1-i} \delta_B[\boldsymbol{u}(i)] \right) \vee \delta_D[\boldsymbol{u}(k)]}_{\boldsymbol{y}_{zs}(k) \equiv \text{'zero'-state resp.}} \tag{14}$$

Thus, the output response is found to consist of two parts: (i) the 'zero'-input response which is due only to the initial conditions $\boldsymbol{x}(0)$ and assumes an input equal to $-\infty$, and (ii) the 'zero'-state response which is due only to the input $\boldsymbol{x}(0)$ and assumes initial conditions equal to $-\infty$.

For *single-input single-output* systems the mapping $u(k) \mapsto y_{zs}(k)$ can be viewed as translation-invariant dilation system Δ. Hence, the 'zero'-state response can be found as the supremal convolution of the input with the impulse response $h = \Delta(q_{\vee})$ of Δ:

$$y_{zs}(k) = \Delta(u)(k) = \bigvee_i u(i) + h(k-i) \tag{15}$$

Assuming the system is initially at rest, its impulse response is found to be

$$h(k) = \begin{cases} -\infty, & k < 0 \\ D, & k = 0 \\ \boldsymbol{C} \boxplus \boldsymbol{A}^{(k-1)} \boxplus \boldsymbol{B}, & k > 0 \end{cases} \tag{16}$$

The last two results can be easily extended to multi-input multi-output systems.

5 Elements from Max Algebra

In this section we summarize some results of minimax algebra, mainly from [4], which are useful for our subsequent analysis.

Solving Max-Sum Equations: Let $\boldsymbol{A} \in \overline{\mathbb{R}}^{m \times n}$ and $\boldsymbol{b} \in \overline{\mathbb{R}}^{m}$. The set of solutions of

$$\boldsymbol{A} \boxplus \boldsymbol{x} = \boldsymbol{b} \tag{17}$$

over $\overline{\mathbb{R}}$ is either empty or forms a commutative semigroup under vector $\vee$. In [4] necessary and sufficient conditions are given for the existence and uniqueness of such solutions. One such result important for our analysis is given next, by using the conjugate matrix $\boldsymbol{A}^*$ where $\{\boldsymbol{A}^*\}_{ij} = -\{\boldsymbol{A}\}_{ji}$ for all i, j.

Theorem 3. [4] *Equation (17) has at least one solution iff $\boldsymbol{x} = \boldsymbol{A}^* \boxplus' \boldsymbol{b}$ is a solution; and $\boldsymbol{x} = \boldsymbol{A}^* \boxplus' \boldsymbol{b}$ is then the greatest solution.*

Vector Independence: Eq. (17) can also be written as

$$\bigvee_{j=1}^{n} \boldsymbol{a}(j) + x_j = \boldsymbol{b} \tag{18}$$

where $\boldsymbol{a}(j) \in \overline{\mathbb{R}}^{m}$, $j = 1, ..., n$, are the n consecutive columns of $\boldsymbol{A}$. If $x_j > -\infty \ \forall j$, we say that $\boldsymbol{b}$ is *max-sum dependent* on all the $\boldsymbol{a}(j)$, ..., $\boldsymbol{a}(n)$. By negation of max-sum dependence, the vectors $\boldsymbol{a}(j)$, ..., $\boldsymbol{a}(n)$ are called *max-sum independent* iff none of them is max-sum dependent on the others. A more useful (for our analysis) type of independence is the following. The vectors $\boldsymbol{a}(j)$, ..., $\boldsymbol{a}(n)$ are called *strongly max-sum independent (SMI)* iff there exists a finite $\boldsymbol{b} \in \mathbb{R}^m$ that has a unique expression of the form (18) with all x_j finite and the max of each row and column of $\boldsymbol{A}$ is a finite real.

Theorem 4. [4] *The vectors $\boldsymbol{a}(j)$, ..., $\boldsymbol{a}(n)$ are SMI iff there exists a finite $\boldsymbol{b} \in \mathbb{R}^m$ such that (17) is uniquely soluble.*

Matrix Rank: If we can find r columns of $\boldsymbol{A}$ ($1 \le r \le n$), but no more, that are SMI, then $\boldsymbol{A}$ is said to have column-rank equal to r.

Graph of a Matrix: Each square matrix $\boldsymbol{A} = [a_{ij}] \in \overline{\mathbb{R}}^{n\times n}$ can be represented by a *directed weighted graph* $\mathrm{Gr}(\boldsymbol{A})$ that has n nodes, is strongly complete, i.e., for each pair of nodes there is a corresponding directed graph branch (arc) joining them, and the weight of each arc joining a pair of nodes (i, j) is equal to a_{ij}. Consider a *path* on the graph, i.e., a sequence of nodes $P = (i_0, i_1, ..., i_t)$; its *length* $L(P)$ and weight $W(P)$ are defined, respectively, by:

$$L(P) \equiv \#\text{ arcs on } P = t, \quad W(P) \equiv a_{i_0 i_1} + ... + a_{i_{t-1} i_t}$$

A path is called a *circuit* if $i_0 = i_t$; the circuit is *elementary* if the nodes $i_0, ..., i_{t-1}$ are pairwise distinct. For any circuit P we can define its *average weight* by $W(P)/L(P)$. Let

$$\lambda(\boldsymbol{A}) \equiv \bigvee_{\text{all circuits } P \text{ of } \boldsymbol{A}} \frac{W(P)}{L(P)} \tag{19}$$

be the *maximum average circuit weight* in $\mathrm{Gr}(\boldsymbol{A})$. Since $\mathrm{Gr}(\boldsymbol{A})$ has n nodes, only elementary circuits (with lenth $\le n$) need be considered in (19). There is also at least one circuit whose average weight coincides with the maximum value (19); such a circuit is called *critical.*

Definite and Metric Matrices: A matrix $\boldsymbol{A}$ is called *definite* if every circuit in its graph has weight ≤ 0 and at least one such circuit has weight $= 0$. The *metric matrix generated by a matrix* $\boldsymbol{A}$ is defined by

$$\Gamma(\boldsymbol{A}) \equiv \boldsymbol{A} \vee \boldsymbol{A}^{(2)} \vee ... \vee \boldsymbol{A}^{(n)} \tag{20}$$

Eigenvalues, Eigenvectors: Given a square matrix $\boldsymbol{A} = [a_{ij}] \in \overline{\mathbb{R}}^{n\times n}$, we say that $\boldsymbol{x} \in \overline{\mathbb{R}}^{n}$ is an *eigenvector* of $\boldsymbol{A}$ and $\lambda \in \overline{\mathbb{R}}$ a corresponding *eigenvalue* of $\boldsymbol{A}$ if

$$\boldsymbol{A} \boxplus \boldsymbol{x} = \lambda + \boldsymbol{x} \tag{21}$$

If we can find finite λ and $\boldsymbol{x}$ satisfying (21), then we say that the eigenproblem is *finitely soluble* for $\boldsymbol{A}$. If $\boldsymbol{A}$ is definite, its associated graph $\mathrm{Gr}(\boldsymbol{A})$ contains at least one circuit with zero weight. An *eigennode* is any node on such a circuit.

Theorem 5. [4] *Let $\boldsymbol{A}$ be definite. Then:*
(a) j is an eigennode of $\mathrm{Gr}(\boldsymbol{A})$ iff $\{\Gamma(\boldsymbol{A})\}_{jj} = 0$.
(b) If j is an eigennode of $\mathrm{Gr}(\boldsymbol{A})$, then the jth column of $\Gamma(\boldsymbol{A})$ is an eigenvector of $\boldsymbol{A}$ whose corresponding eigenvalue is zero.

Thus, columns of $\Gamma(\boldsymbol{A})$ that correspond to eigennodes provide eigenvectors for $\boldsymbol{A}$, which are called *fundamental eigenvectors.* Two such eigenvectors are called *equivalent* if their corresponding eigennodes belong to the same critical circuit. Max-sum combinations of non-equivalent fundamental eigenvectors generate the *eigenspace* of $\boldsymbol{A}$, whose elements are eigenvectors of $\boldsymbol{A}$ with corresponding eigenvalue $= 0$.

Theorem 6. [4] *(a) If the eigenproblem for $\boldsymbol{A}$ is finitely soluble, then every finite eigenvector has the same unique finite eigenvalue, called the principal eigenvalue, which is equal to the maximum average circuit weight of $\boldsymbol{A}$ defined in (19). All finite eigenvectors of $\boldsymbol{A}$ lie in the eigenspace of the definite matrix $\boldsymbol{A} - \lambda(\boldsymbol{A})$. The non-equivalent fundamental eigenvectors which generate this space are SMI.*
(b) The eigenproblem for $\boldsymbol{A}$ is finitely soluble iff $\lambda(\boldsymbol{A})$ is finite and $\Phi(\boldsymbol{A} - \lambda(\boldsymbol{A}))$ has rows and columns whose maxima are finite, where $\Phi(\boldsymbol{A} - \lambda(\boldsymbol{A}))$ is any matrix whose columns form a maximal set of non-equivalent fundamental eigenvectors for the definite matrix $\boldsymbol{A} - \lambda(\boldsymbol{A})$.
(c) If $\boldsymbol{A}$ is finite, then the eigenproblem for $\boldsymbol{A}$ is finitely soluble.

Periodicity: A definite matrix $\boldsymbol{A}$ has zero principal eigenvalue. Such a matrix is called *order-d-periodical* if there is an integer k_0 such that $\boldsymbol{A}^{(k+d)} = \boldsymbol{A}^{(k)}\ \forall k \geq k_0$.

Theorem 7. [3] *If $\boldsymbol{A}$ has zero principal eigenvalue and the corresponding critical circuit is unique and has length d, then $\boldsymbol{A}$ is order-d-periodical.*

6 Causality, Stability

A max control system initially at rest can be viewed as a translation-invariant dilation system Δ mapping the input u to the output y. (Assume for brievity single-input single-output systems.) Let $h = \Delta(q_\vee)$ be the impulse response of Δ. A useful bound for signals $f(k)$ processed by such systems is their max absolute value over their support:

$$B_f \equiv \sup\{|f(k)| : f(k) > -\infty\}$$

Such systems are called bounded-input bounded-output (BIBO) *stable* iff a bounded input yields a bounded output, i.e., if $B_u < +\infty \Longrightarrow B_y < +\infty$. The following theorem provides us with simple algebraic criteria for checking the causality and stability of max control systems based on their impulse response.

Theorem 8. [12] *Consider a max-sum control system Δ initially at rest and let $h = \Delta(q_\vee)$ be its impulse response. (a) The system is causal iff $h(k) = -\infty$ for all $k < 0$. (b) The system is BIBO stable iff $B_h < +\infty$.*

In standard linear control the stability of the linear system can also be expressed via the eigenvalues of the matrix $\boldsymbol{A}$. We have derived a conceptually similar result (stated next) that links the stability of a max-sum control system with the principal eigenvalue of $\boldsymbol{A}$. (For simplicity, the theorem assumes a unique critical circuit, but it can extended to more general cases.)

Theorem 9. *Consider a max-sum control system whose matrices $\boldsymbol{A}, \boldsymbol{B}, \boldsymbol{C}, \boldsymbol{D}$ do not contain any $+\infty$ elements. Assume that there is a unique critical circuit of length d corresponding to the finite principal eigenvalue $\lambda(\boldsymbol{A})$ of $\boldsymbol{A}$. Then:*
(a) If $\lambda(\boldsymbol{A}) = 0$, the impulse response of the system is periodic with period equal to d.
(b) The system is BIBO stable iff $\lambda(\boldsymbol{A}) = 0$.

Proof: (a) The zero principal eigenvalue makes the matrix $\boldsymbol{A}$ order-d-periodical. Hence, by (16), there exists k_0 such that $h(k+d) = h(k)$ for all $k \geq k_0$. (b) If λ is the principal eigenvalue of $\boldsymbol{A}$, then (a) implies that $\boldsymbol{A}^{(k+d)} = d\lambda + \boldsymbol{A}^{(k)}$ for all $k \geq k_0$. Hence,

$$h(k+d) = d\lambda + h(k), \quad \forall k \geq k_0 \tag{22}$$

Further, the absence of $+\infty$ values in the system's matrices guarantees that $h(k)$ does not have any such values. Now, if $\lambda = 0$, then $h(k+d) = h(k)$ $\forall k \geq k_0$ and hence $B_h < +\infty$. In contrast, if $\lambda \neq 0$, then (22) will drive asymptotically (as $k \to \infty$) the values of $|h(k)|$ unbounded, and hence $B_h = +\infty$. Q.E.D.

Example 1 (DEDS): The DEDS with state-space equations (6) has principal eigenvalue $\lambda(\boldsymbol{A}) = \max(m_1, m_2)$ which is positive since $m_1, m_2 > 0$. Thus, this system is BIBO unstable. Of course, if the impulse response of the system (as is the case with this example) becomes unbounded only asympotically, then our definition of BIBO stability only affects the system if we let it run for an infinite time. In contrast, if we run the system only for a finite time, then its impulse response and output signal remain finite.

Example 2 (Recursive Nonlinear Filter): The max control system corresponding to the recursive nonlinear filter described by (8) has a principal eigenvalue equal to $\lambda(\boldsymbol{A}) = \bigvee_{k=1}^{n} a_k/k$. Hence, for this system to be stable, all the coefficients a_k must be nonpositive and at least one of them must be zero.

7 Controllability, Observability

A max-sum control system is *controllable* if the following system of nonlinear equations can be solved and provide the vector $\boldsymbol{u} = [u(0), u(1), ..., u(n-1)]^T$ of input values required to drive the system from the initial state $\boldsymbol{x}(0)$ to any desired state $\boldsymbol{x}(n)$ in n steps:

$$\boldsymbol{x}(n) = \boldsymbol{A}^{(n)}\boldsymbol{x}(0) \vee \left(\underbrace{[\boldsymbol{A}^{(n-1)} \boxplus \boldsymbol{B}, \cdots, \boldsymbol{B}]}_{\mathcal{C}} \boxplus \begin{bmatrix} u(0) \\ \vdots \\ u(n-1) \end{bmatrix} \right) \tag{23}$$

Assuming that the input is dominating the initial conditions, i.e., the second term is greater than the first term of the right hand side, we can rewrite the above as

$$\mathcal{C} \boxplus \boldsymbol{u} = \boldsymbol{x} \tag{24}$$

where $\boldsymbol{x} = \boldsymbol{x}(n)$ is the desired state. This is a system of max-sum equations which may be solved using the methods of minimax algebra. A necessary condition to find a unique solution of (24), i.e., a unique input vector to drive the system to the state $\boldsymbol{x}$ in n steps is the rank of $\mathcal{C}$ to be equal to n, i.e., the columns of $\mathcal{C}$ must be SMI. However, in certain applications Eq. (24) may be too strong of a condition and it may be sufficient to solve an approximate controllability problem that has some optimality aspects. Specifically, consider the problem of finding an optimal input vector $\boldsymbol{u}$ as solution to the following optimization problem:

$$\begin{array}{c}\text{Minimize } ||\boldsymbol{x} - \mathcal{C} \boxplus \boldsymbol{u}|| \\ \text{subject to } \mathcal{C} \boxplus \boldsymbol{u} \leq \boldsymbol{x}\end{array} \tag{25}$$

where the norm $||\cdot||$ is either the ℓ_∞ or the ℓ_1 norm. Namely, we wish to minimize a norm of the earliness subject to zero lateness. It follows [4] that the solution to the above optimization problem is

$$\boldsymbol{u} = \mathcal{C}^* \boxplus' \boldsymbol{x} \tag{26}$$

We note that the optimal controllability solution (26) is actually a lattice erosion, and its optimality can be proven simply by using properties of lattice dilations and erosions. Specifically, to the dilation $\delta(\boldsymbol{y}) = \mathcal{C} \boxplus \boldsymbol{y}$ there corresponds a unique erosion $\varepsilon(\boldsymbol{y}) = \mathcal{C}^* \boxplus' \boldsymbol{y}$ such that the operator pair (ε, δ) forms a lattice adjunction [6]. One property of such adjunctions is that the composition $\delta\varepsilon$, known as the *opening* operator, is antiextensive. Thus, $\delta(\varepsilon(\boldsymbol{x})) \leq \boldsymbol{x}$ and $\boldsymbol{u} = \varepsilon(\boldsymbol{x})$ is the largest solution with $\delta(\boldsymbol{u}) \leq \boldsymbol{x}$.

The above ideas on the controllability problem can also be applied to the observability problem. A max-sum control system is *observable* if we can estimate the initial state by observing a sequence of output values. This can be done if the following system of nonlinear equations can be solved:

$$\begin{bmatrix} y(0) \\ \vdots \\ y(n-1) \end{bmatrix} = \underbrace{\begin{bmatrix} \boldsymbol{C} \\ \vdots \\ \boldsymbol{C} \boxplus \boldsymbol{A}^{(n-1)} \end{bmatrix}}_{\mathcal{O}} \boxplus \boldsymbol{x}(0) \vee [h(n-1), \cdots, h(0)] \boxplus \begin{bmatrix} u(0) \\ \vdots \\ u(n-1) \end{bmatrix} \tag{27}$$

Assuming that the first term of the right hand side containing the initial state dominates the second term that contains the input, we can rewrite the above as

$$\mathcal{O} \boxplus \boldsymbol{x}(0) = \boldsymbol{y} = [y(0), ..., y(n-1)]^T \tag{28}$$

This max-sum matrix equation can be solved either exactly or approximately by using the same methods as for the controllability equation. The general solution is then

$$\hat{\boldsymbol{x}}(0) = \mathcal{O}^* \boxplus' \boldsymbol{y} \tag{29}$$

and has the property that it is the largest solution with $\mathcal{O} \boxplus \hat{\boldsymbol{x}}(0) \leq \boldsymbol{y}$.

8 Feedback

Consider a max control system described by the state space equations (12) and let its input signal $\boldsymbol{u}(k)$ be the max of a feedback of the state and a reference signal $\boldsymbol{w}(k)$. The state feedback law considered is

$$\boldsymbol{u}(k) = \boldsymbol{w}(k) \vee (\boldsymbol{F} \boxplus \boldsymbol{x}(k)) \tag{30}$$

where $\boldsymbol{F}$ is the feedback matrix. Then the new state space equations will be

$$\begin{aligned} \boldsymbol{x}(k+1) &= [\boldsymbol{A} \vee (\boldsymbol{B} \boxplus \boldsymbol{F})] \boxplus \boldsymbol{x}(k) \ \vee \ \boldsymbol{B} \boxplus \boldsymbol{w}(k) \\ \boldsymbol{y}(k) &= [\boldsymbol{C} \vee (\boldsymbol{D} \boxplus \boldsymbol{F})] \boxplus \boldsymbol{x}(k) \ \vee \ \boldsymbol{D} \boxplus \boldsymbol{w}(k) \end{aligned} \tag{31}$$

Thus, the use of feedback can affect the principal eigenvalue of the system, as well as its controllability and observability.

We illustrate a possible effect of feedback on the system's eigenvalue via the example of the DEDS of Fig. 1. Suppose as in [9] that, we want to design this system to have cyclic behavior, i.e.,

$$\boldsymbol{x}(k+1) = \lambda + \boldsymbol{x}(k) \tag{32}$$

where λ is a constant positive number (the cycle time). The original state-space equations (6) are

$$\boldsymbol{x}(k+1) = [\boldsymbol{A} \boxplus \boldsymbol{x}(k)] \vee [\boldsymbol{B} + u(k+1)] \tag{33}$$

If we use a state feedback of the form

$$u(k+1) = \boldsymbol{F} \boxplus \boldsymbol{x}(k) \ , \quad \boldsymbol{F} = [r_1, r_2] \tag{34}$$

then from (33) and (34) we obtain

$$\lambda + \boldsymbol{x}(k) = [\boldsymbol{A} \vee (\boldsymbol{B} \boxplus \boldsymbol{F})] \boxplus \boldsymbol{x}(k) \tag{35}$$

Thus the problem reduces to finding the principal eigenvalue of

$$\boldsymbol{Q} = \boldsymbol{A} \vee (\boldsymbol{B} \boxplus \boldsymbol{F}) = \begin{bmatrix} m_1 & \theta \\ m_1 + m_2 & m_2 \end{bmatrix} \vee \begin{bmatrix} m_1 + r_1 & m_1 + r_2 \\ m_1 + m_2 + r_1 & m_1 + m_2 + r_2 \end{bmatrix} \tag{36}$$

and the corresponding eigenvector $\boldsymbol{x}(0)$. Setting $r_1 = r_2 = 0$ yields the following matrix $\boldsymbol{Q}$ and its eigenvalue and eigenvector:

$$\boldsymbol{Q} = \begin{bmatrix} m_1 & m_1 \\ m_1 + m_2 & m_1 + m_2 \end{bmatrix} \Longrightarrow \lambda(\boldsymbol{Q}) = m_1 + m_2, \ \ \boldsymbol{x}(0) = \begin{bmatrix} m_1 \\ m_1 + m_2 \end{bmatrix} \tag{37}$$

Since $x_1(0) = m_1 + u(0)$ and $x_2(0) = m_1 + m_2 + u(0)$, it follows that the choice of feedback $\boldsymbol{F} = [0, 0]$ and initial input $u(0) = 0$ achieves the above cyclic behavior with cycle time $\lambda = m_1 + m_2$.

References

1. R. Bellman and W. Karush, "On the Maximum Transform", *J. Math. Anal. Appl.*, 6, pp. 67–74, 1963.
2. W. L. Brogan, *Modern Control Theory*, Quantum Publ., New York, 1974.
3. G. Cohen, D. Dubois, J.P. Quadrat and M. Viot, "A Linear System Theoretic View of Discrete Event Processes and Its Use for Performance Evaluation in Manufacturing", *IEEE Trans. Autom. Control*, vol. 30, pp. 210-220, 1985.
4. R. Cuninghame-Green, *Minimax Algebra*, Springer-Verlag, New York, 1979.
5. A. Doustmohammadi, "Modeling of Discrete Event Dynamic Systems based on Petri Nets and Minimax Algebra Approach", PhD thesis proposal, Georgia Inst. Tech., Atlanta, 1993.
6. H.J.A.M. Heijmans, *Morphological Image Operators*, Acad. Press, Boston, 1994.
7. Y.-C. Ho, Ed., "Special Issue on Dynamics of Discrete Event Systems", *Proc. IEEE*, vol. 77, Jan. 1989.
8. T. Kailath, *Linear Systems*, Prentice-Hall, N.J., 1980.
9. E. W. Kamen, "An Equation-Based Approach to the Control of Discrete Even Systems with Applications to Manufacturing", *Proc. Int'l Conf. on Control Theory & its Applications*, Jerusalem, Israel, Oct. 1993.
10. P. Maragos and R. W. Schafer, "Morphological Filters - Parts I & II", *IEEE Trans. Acoust., Speech, Signal Processing*, vol. 35, pp. 1153-1184, Aug. 1987. *ibid*, 37:597, Apr. 1989.
11. P. Maragos and R. W. Schafer, "Morphological Systems for Multidimensional Signal Processing", *Proc. IEEE*, vol. 78, pp. 690-710, Apr. 1990.
12. P. Maragos, "Morphological Systems: Slope Transforms and Max–Min Difference and Differential Equations", *Signal Processing*, vol. 38, pp. 57-77, 1994.
13. A. V. Oppenheim and R. W. Schafer, *Discrete-time Signal Processing*, Prentice-Hall, N.J., 1988.
14. R. T. Rockafellar, *Convex Analysis*, Princeton Univ. Press, Princeton, 1972.
15. J. Serra, *Image Analysis and Mathematical Morphology*, NY: Acad. Press, 1982.
16. J. Serra, Ed., *Image Analysis and Mathematical Morphology, Vol.2: Theoretical Advances*, NY: Acad. Press, 1988.

PART IV

ROBOTIC AND TRAVELING CRANE SYSTEMS

22 Force Sensing in Microrobot Systems : An Overview

S. Fahlbusch, S. Fatikow and K. Santa

1 Introduction

The progressing development of production methods in microsystem technology has led to a significant decrease in the size of microparts and an increase in their variety of materials, thus resulting in more and more complex microsystems. However, as miniaturisation proceeds, human beings increasingly experience performance limits imposed by the demands of the microassembly process. To a large extent these limits are even reached by modern precision mechanical products. Therefore, the flexible automation of microassembly processes, based on complex sensor feedback, offers considerable potential for rationalisation.

As a result of this, smart micromanipulation systems for high-precision manipulations are gaining great interest to overcome the restricted capabilities of the human hand – not only in industrial applications, but also in medicine. In medical research e.g. it is often necessary to pick up a single cell and to carry it to an apparatus used for further analysis of the cell. Therefore, the cell has to be cut loose from the surrounding cells, has to be picked up and carried away.

Nowadays, micromanipulations are performed using either mobile microrobots [1] or a precise positioning device under control of an optical or scanning electron microscope. Various tools for manipulating microparts and assembling microsystems have been developed and integrated into the above mentioned systems. During the initial state of development, special emphasis was laid onto the design of different microhandling tools applicable for appointed operations. Often sensor feedback is only given by means of optical sensors, thus leading to a lack of information about interaction forces between the endeffector and the microcomponents. In order to avoid breaking or damaging objects during the manipulation process, force reflection has to be assured as an essential component within the control structure of micromanipulation and microassembly systems. This paper summarises basic force sensing methods of and gives an overview about the current research activities within force sensing in microrobotic systems.

2 Fundamental Methods in Force Sensing

Mechanical stresses are characterised by the influence of force and torsion onto the component parts. They result in deformation and mechanical stresses and can be analysed with force, strain and tension measurement techniques. Forces can not be measured directly, but it is possible to determine them by investigating the actuated effects, e.g. changes in length or strain.

2.1 Strain Gauges

Mechanical stresses on a component part (length l_0, cross-section A) induced by a force F result in a strain $\varepsilon = \Delta l/l_0$, a mechanical stress $\sigma = F/A$ and, if the deformation is linear elastic (i.e. reversible), a proportional relation between stress and strain $\sigma = E\varepsilon$, where E defines the Young modulus. Strain measurement techniques provide information about deformation and stress conditions of components and allow the realisation of force sensors based on strain gauges.

Strain gauges are made of meander-shaped measuring lattices embedded in a thin carrier layer; they transform strains into changes of the electrical resistance ΔR:

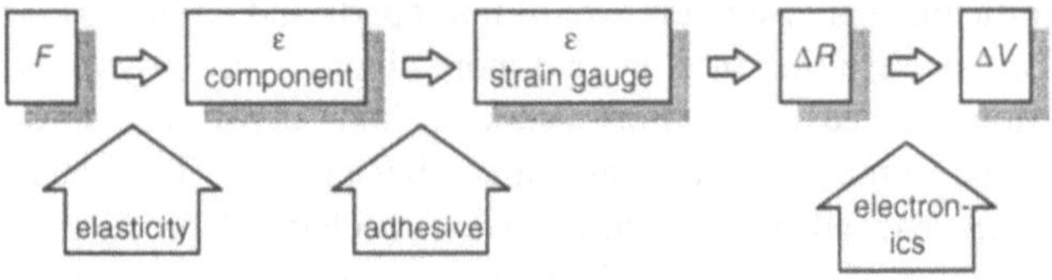

The electrical resistance R of a wire and its change by an infinitive variation of diameter D, length l and specific resistance ρ are given by:

$$R = \frac{4\rho l}{\pi D^2}, \frac{dR}{R} = \frac{d\rho}{\rho} + \frac{dl}{l} - 2\frac{dD}{D} \quad (1)$$

Inserting $\varepsilon = dl/l$ and the Poisson number $\mu = -(dD/D)/(dl/l)$ leads to:

$$\frac{\Delta R}{R} = \left(1 + 2\mu + \frac{\Delta\rho / \rho}{\varepsilon}\right)\varepsilon = k\varepsilon \quad (2)$$

where k is the gauge factor. It has a value of about 2 for metal-foil strain gauges and up to 200 for semiconductor types (e.g. silicon with piezoresistive effect, see below). The change of the electrical resistance is usually measured with a Wheatstone bridge, thus providing a compensation of mechanical and thermal influences.

Strain gauges are either made out of metal-foils or of semiconductor materials using the piezoresistive effect. One disadvantage of force sensors based on this effect is the time dependency of the measurement. The charge is only generated when there is a change in the applied force. For static loads it will leak away across the input resistance of the measuring device. A quasi-static force sensing can be obtained by a very high input impedance of the charge amplifier [2].

2.2 Piezoelectric ("Self-sensing") Effect

Ideally, a technical system should be capable of adapting its function to external influences like temperature, mechanical loads, moisture or electrical fields. The availability of smart actuators fulfils an important requirement for realising self-adapting systems. An actuator consists of the series connection of an electronic amplifier and an electromechanical energy transducer. Interesting materials include piezoelectric ceramics as well as magnetostrictive materials such as Terfenol-D.

A change in the mechanical output load on piezoelectric ceramics can be detected and evaluated at the electrical transducer input, making the simultaneous use of the energy transducer as a sensor possible [3, 4]. The transducer voltage is not affected by the load, but the current signal is superimposed by a disturbance. Integration of the current signal with disturbance gives the charging response of the piezoceramic. Measurement of the control voltage and the charge makes it possible to determine a sensor signal representative of the mechanical load. Inaccuracy of measurements results in small deviations in the identification of the load.

2.3 Piezomagnetic Effect

Piezomagnetic materials are a subset of ferromagnetic materials, which have a number of characteristic features arising from the coupling between the local magnetic moments and the lattice. In an applied magnetic field they undergo a dimensional change (Joule effect). The inverse of magnetostriction is that the application of mechanical strain may change the state of magnetisation within the sample (Villari effect). This is usually seen as a change in permeability.

There are a number of basic characteristics which are required for a high performance piezomagnetic material. These essentially reduce to the requirement for a well defined domain structure dominated by uniaxial anisotropy. For maximum dimensional change the magnetisation should coherently rotate through 90° on application of an external field. For maximum piezomagnetic response in sensors, the strain should be uniaxial and in a direction orthogonal to the magnetisation in zero field [5].

2.4 Capacitive Sensors

Capacitive sensors make use of the change in capacitance between two metal plates. The membrane deflects when pressure or force are applied, which causes the distance between the two electrodes to be changed. Hereby the capacitance increases or decreases. From the amount of membrane deflection, the capacitance change is measured and the value of pressure or force can be calculated. Compared to piezoresistive signal transformers, capacitive sensors have no hysteresis, a better long-term stability and a higher sensitivity.

3 Force Sensing in Microrobotic Systems

A profound study in literature and industrial applications about force sensing in microrobotic systems showed that there is a distinct demand for sensors capable of measuring forces in the range of mN or even μN. In regard to its importance for the grasping of microobjects, there are not many suitable solutions offered at present, thus requiring increasing efforts within this scope in future research works.

Strain gauges offer a well known method for force sensing in microrobotics. Semiconductor types are available in geometrical sizes below 1 mm and can be applied in many microhandling applications. Combined with an intelligent mechanical design of the force sensor, they offer suitable solutions in many micromanipulation tasks. Other methods for strain measurement are also explored, e.g. a capacitive force sensor [6] and a Laser Raman Spectrophotometer proposed by [7], but strain gauge techniques seem to offer a comparable performance in terms of accuracy, speed and represent one of the cheapest solutions. Some of the proposed force sensors use the piezoresistive [8, 9] or the piezoelectric effect in piezoactuators [10, 11, 12, 13], but in most cases poor sensitivity is reached with such systems [11].

In general, few works are presented in literature containing a comprehensive description of force sensors for micromanipulation in terms of accuracy, precision, linearity and resolution of the sensor system. These data are of fundamental importance for a proper grasping functionality in microrobotic systems with force feedback. Sensing of microforces with high precision is not trivial, and calibration and stability of such sensors is a very delicate item.

3.1 Force Sensing With Strain Gauges

Force sensing with strain gauges, either metal-foil or semiconductor type, is the most common technique used in the macro world. The investigations already gained within this field can be transferred into the micro world up to a certain extent. Although alternative methods are being investigated successfully, the use of strain gauges seems to be the most widely used approach, because of its good performance in terms of cost, speed and accuracy of measurement. One disadvantage is the geometrical constraint that has to be followed in the design of endeffectors for micromanipulators, as fabrication of strain gauge has miniaturisation limits.

A lot of research work in force measurement exerted by microactuators deals with measuring the strains induced on passive structures that are part of the actuation system [14, 15, 16, 17, 18, 19, 20, 21]. One important reason therefore is that the advantage of force sensing based on stress information rises against the one based on displacement when the structure is miniaturised. Scaling analysis in case of a force applied at the tip of a cantilever [7] shows that maximum stress is proportional to L^{-2} (where L is the scale unit of length), while maximum displacement is proportional to L. On the other hand, it is difficult to fabricate strain gauges when the structure is miniaturised. Spacing and positioning for gauge installation should be taken into account in the mechanical design of the grasping tool or endeffector.

3.1.1 Metal-foil Strain Gauges

Although most force sensing applications used in microrobotics are based on semiconductor strain gauges, research is also partially done with metal-foil types. The teleoperated microgripper for microsurgery developed by [17] is capable of measuring gripping forces with metal-foil strain gauges. Each gauge forms one arm of a lead-wire temperature compensated quarter-bridge arrangement. In order to further reduce the affects of temperature on the strain measurements, the bridge excitation is pulsed, i.e. using a positive square wave with a small duty cycle.

3.1.2 Semiconductor Strain Gauges

Due to their advantages compared to metal-foil gauges, semiconductor strain gauges are mostly preferred when small forces have to be measured in microrobotic applications. In their work, [15] describe an electrostatic microactuator capable of force sensing. They measured the force generated at the tip of the actuator's arm using a semiconductor strain gauge. This strain gauge was attached at the root of the arm with thermohardening type adhesive (Figure 3.1).

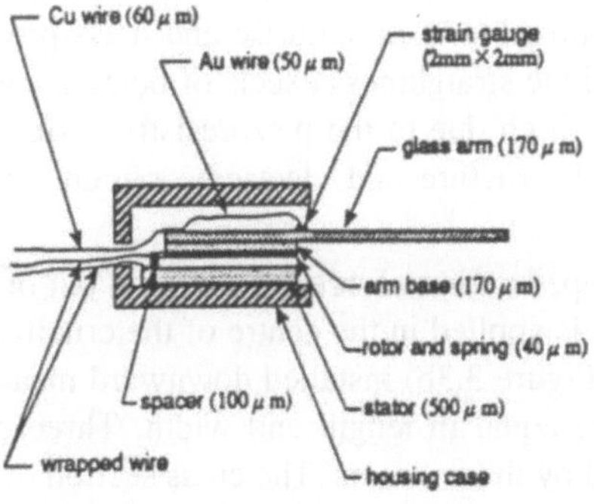

Figure 3.1: Cross section of the microactuator

During their experiments, two different cases of the object to be handled have been considerd. One is the case with a "hard" object, the form of which is not transformed. The other case includes a "soft" object, the shape of which is transformed when in contact with the tip of the endeffector. The measured forces were compared to calculated values and showed only for the second case an acceptable conformity.

[20] describes a piezoelectric bimorph microgripper capable of force sensing (Figure 3.2). The configuration of the actuator and the sensors enables high accuracy position and force sensing at the same time. A rigid-actuator soft-sensor pair is realised on a single cantilever plate. Due to the deformation characteristics of piezoelectric elements, no strain is induced by the actuator where there is no actuator layer. The first sensor pair, positioned at the most sensitive location regarding external forces, measures only the tip force induced strain free from the actuator induced strains. The second sensor pair, which is located on the actuator layer, mainly measures the actuator induced strains and is used to compensate creep and hysteresis effects of the piezoceramics.

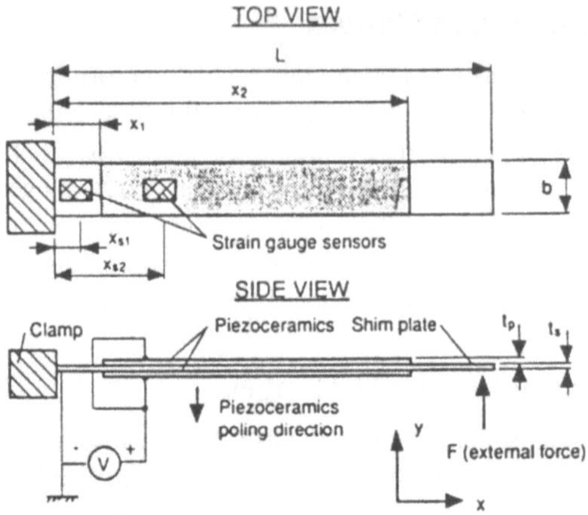

Figure 3.2: Structure of the cantilever plate

Silicon is a suitable material for micro force sensors because of several advantages: 1) Semiconductor manufacturing technologies, e.g. lithography and etching, can be used. Therefore, it is possible to miniaturise and mass-produce the sensors. 2) The hysteresis is small and the straightness excels of being a completely elastic body. 3) The sensitivity is very high due to the piezoresistive effect of silicon. 4) A system integrating mechanical structure and electronic circuit on a silicon wafer can be achieved.

Hence, [14] developed a micro force sensor made out of silicon, shown in Figure 3.3a. The acting force is applied in the centre of the cruciform beam. Four semiconductor strain gauges (Figure 3.3b) installed downward measure the strains caused in four beams, which are equal in length and width. Three components of the force vector can be obtained by these strains. The cross section of the beam is a trapezoid.

a) 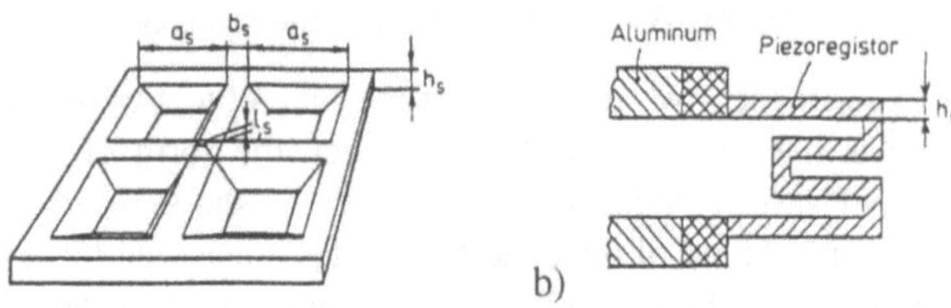

b)

Figure 3.3: a) Shape type, b) Micro force sensor

The force sensor was manufactured using bulk and micromachining techniques. The size is 4×4 mm^2 including the aluminium wiring on the back. The size of the four beams, where the force is detected, is 1.15×1.15 mm^2. By improvement of the force sensor's design, it can be manufactured by placing two or more sensors on a matrix.

A LIGA microtesting system with integrated strain gauges was presented by [19]. In this system, strain gauges are integrated in the force transmitting unit of the actuator for direct measurement of the forces transmitted to the specimen. The system was manufactured using the standard LIGA process combined with the sacrificial layer technique to realise the movable parts of the structure. The measurements showed a linear relation between the signal and the applied forces as expected.

A force sensor was realised by [18] using a combination of a parallel plate structure. Two rigid blocks are connected with two thin plates on the roots of which strain gauges are glued. When a force perpendicular to the thin plate is applied to the parallel plate structure, a strain proportional to the applied force can be measured. As the structure is very simple, it is easy to make a force sensor with any rated value by changing thickness, length, width and the number of the thin plates. In addition to this, a multi-axis sensor can be build by piling up the structure serially.

3.2 Piezoresistive Effect

The force sensor for microhandling purposes developed by [8] exploits the piezoresistive effect in silicon. Four piezoresistors are connected into a Wheatstone bridge configuration. The aim is to obtain an output signal from the bridge that is linearly dependent to the applied force. The active part of the sensor consists of a thin bridge structure, which is clamped on both sides, with a thicker part (so called boss) in the centre of the bridge (Figure 3.4). The boss is the contact surface between the sensor and the micropart. Four transversally oriented piezoresistors are connected into the Wheatstone bridge in order to amplify the output signal and to reduce the temperature influence on the output signal due to the change of the nominal values of resistors. In order to get a sensor with higher sensitivity and lower nonlinearity of the output signal, the boss is integrated into the bridge.

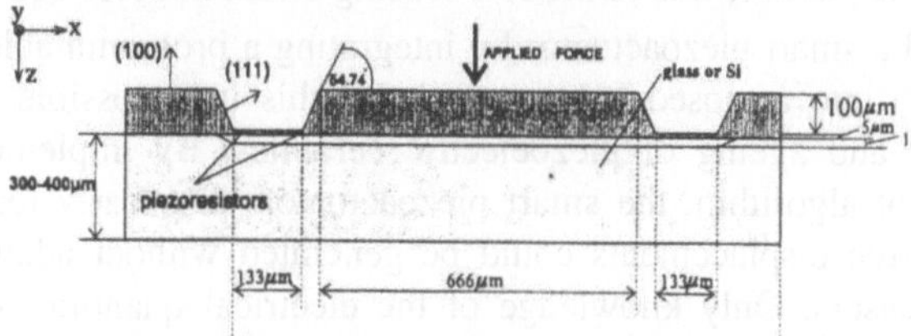

Figure 3.4: Sensor design

During the process of gripping, the applied force is mostly dislocated from the centre of the boss. This influences the measured signal. Hence, special attention was paid to an error estimation due to the positioning of the microparts between the endeffectors and several FEM simulation have been performed by the authors. The sensor dimensions were optimised using a structural simulation tool in order to obtain the highest possible sensitivity of the sensor.

A micromachined tactile gripper system with a thermal bimorph actuator and a piezoresistive force sensor was presented by [9]. This gripper is used for handling and assembly tasks with a nanorobot. For a first order temperature drift compensation the piezoresistors are arranged as a Wheatstone bridge. In order to measure not only the mechanical stress in the beam but also to obtain information about the gripping force and its position, a second Wheatstone bridge is located at the middle of the sensor finger.

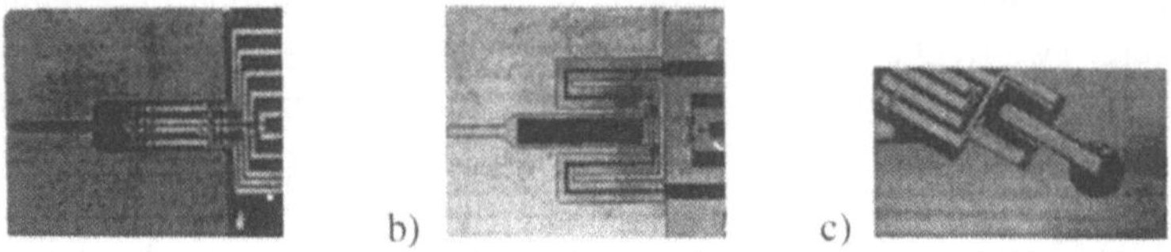

a) b) c)

Figure 3.5: a) Sensor finger, b) Actuator finger, c) Microgripper

The approach of the authors consists of a combined system with integrated actuation and force sensing. A sensor finger (Figure 3.5a) and an actuator finger (Figure 3.5b) are assembled to the complete microgripper, thus minimising the crosstalk between them (Figure 3.5c). The experimental results show a linear variation of the voltage difference in the bridge circuit with the applied force. The sensitivity of the sensor is about 70 mV/mN, the resolution is 0.2 μN.

3.3 Piezoelectric ("Self-sensing") Effect

Energy transducers with multifunctional materials such as piezoelectric PZT ceramics fulfil the requirements for a compact and inexpensive construction by combining actuator and sensor capacities. They enable load dependent control without implementing extra force or displacement sensors. The realisation of smart piezoelectric actuators is possible due to the self-sensing effect as described in section 2.1.

[11] realised a smart piezoactuator by integrating a programmable digital signal processor (DSP) into a closed loop control. By this it is possible to compensate hysteresis, drift and ageing of piezoelectric ceramics. By implementing a DSP-supported control algorithm, the smart piezoactuator was nearly load-independent and hysteresis free displacements could be generated without additional force or displacement sensors. Only knowledge of the electrical quantities of voltage and current at the transducer input are necessary.

[12] proposed a micromachined force sensor based on PZT and PNNZT thin films. After these films have been deposited on substrates, free-standing piezoelectric microcantilevers with dimensions of $125 \times 50 \times 3.7\ \mu m^3$ were sculpted from the substrates by a microfabrication process. A new excitation and detection mechanism has additionally been developed for force sensing microcantilevers used in scanning force microscopes by measuring the deflection of the cantilever.

Due to the above described advantages and possibilities offered by the piezoelectric effect, other projects within this research area have recently started [13]. Although being in an initial state of development, they show some interesting concepts to integrate sensors and actuators into one component. Within their work, [10] developed monolithic flexible structures for micromanipulation cut directly into piezoelectric ceramics. A microgripper with an integrated force sensor using the self-sensing effect has been tested.

3.4 Piezomagnetic Effect

[5] have started to develop piezomagnetic materials in thin film form for the use in microelectromechanical systems (MEMS). The feasibility of introducing such materials in MEMS was demonstrated by a pressure sensor prototype based on a piezomagnetically coated membrane. The magnetic material chosen was an amorphous Fe-Si-B-C alloy. The prototype device was fabricated using an undoped semi-insulating GaAs wafer as the substrate. The results obtained within this work offer promising starting points for the use of piezomagnetic thin films concerning force sensing in microrobotic systems in further developments [22].

3.5 Capacitive Force Sensor

A capacitive sensor is shown in Figure 6 [6]. The electrodes are made up of a planar comb structure. Here, the applied force is exerted parallel to the sensor surface. In force sensors which use membranes, the force is usually applied perpendicular to the sensor surface. In this case, nonlinearity and cross-sensitivity may cause problems. In the device described, the sensor element mainly consists of two parts: first, a movable elastic structure which transforms a force into a displacement, and, second, a transformation unit consisting of the electrodes which transform the displacement into a measurable change of capacitance.

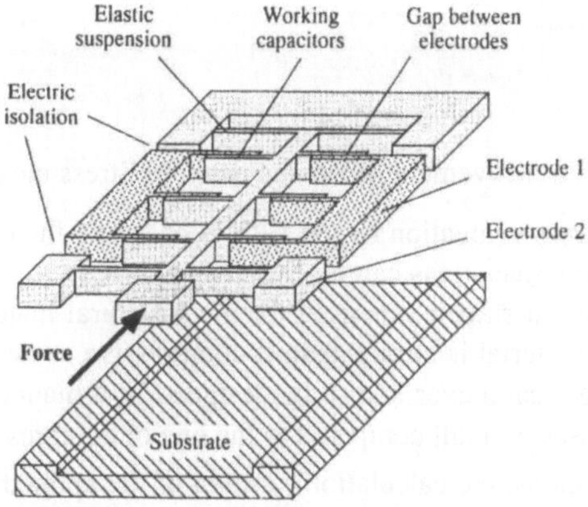

Figure 3.6: Capacitive force sensor made of silicon

The displacement is stored by an elastic suspension beam. The capacitors consist of two electrically insulated thin electrodes with a very narrow gap between them. They are placed on both sides of the sensor chip, making the capacitance on one side increase and decrease on the other side. By the separate measurement of the capacitance changes on both sides a high linearity and sensitivity is obtained. The sensor unit is made by anisotropically etching (110) silicon and then fastening it to a pyrex substrate through anodic bonding. With the proposed sensor it was possible to measure very small forces with a resolution of 20 nm in a range between 0.01 N and 10 N.

3.6 Laser Raman Spectrophotometer

A method of force sensing based on a Laser Raman Spectrophotometer (LRS) was proposed by [7]. The LRS has been found capable of measuring crystallinity, stress and crystal-axis orientation over areas in the size of 1 μm. When light strikes a solid object, the majority of the scattered light is of the same frequency as the incident light (Rayleigh scattering). In addition to Rayleigh scattering, a very small fraction of the incident light is disseminated at lower or higher frequencies, due to an inelastic interaction between the photons and atoms of the target. This phenomena is known as Raman scattering. The peak frequency and bandwidth of the Raman scattering light provides quantitative information about the internal stress and the crystallinity of the target. The difference in energy between the incident and the scattered light corresponds to the lattice-vibration energy of the object. So the Raman scattering frequency is decreased by tensile and increased by compressive stresses (Figures 3.7a, 3.7b).

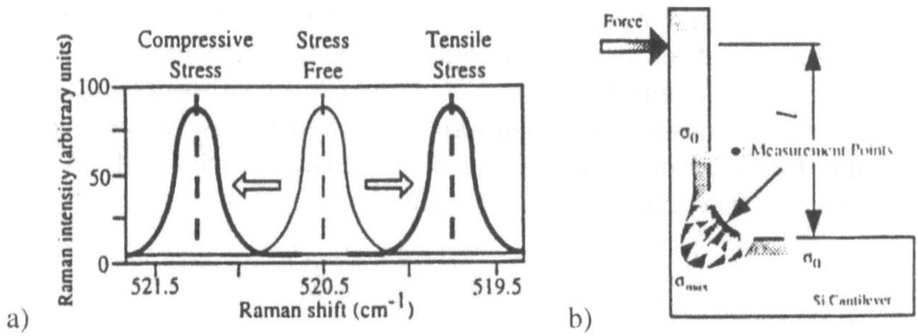

Figure 3.7: a) Movement of Raman peak, b) Stress measurement points

The principle of stress evaluation by the LRS is based on the observation of lattice-vibration energy. Internal stress can be easily measured in a single crystal material. Hence, a single crystal silicon was used for the structural material of the endeffector. Although this material is strong against compressive stresses, it is weak against tensile stresses. So a cantilever had to be designed in a manner that the concentration of tensile stresses is small compared to the one of compressive stresses.

With simplifications, the calculation formula of the applied force F can be written as (α: stress concentration factor, σ_{max}: stress, t: thickness, b: beam width):

$$F = \frac{bt^2 \sigma_{\max}}{6l\alpha}$$

The force measurement results obtained by LRS were compared to forces measured by strain gauges. The measurement range was 0 to 0.15 N. Both results showed a good linearity, but did not match perfectly. This may have several reasons: 1) neglection of the anisotropic character of Si; 2) use of two-dimensional stress evaluation for three-dimensional deformation; 3) errors in stress measurement and the production of the silicon cantilever. One disadvantage of the LRS is the time needed to measure stress for each point, which is about 5 to 90 seconds.

4 Conclusions

A brief overview about force sensing in microrobotic systems has been given. Fundamental methods of force measurement were presented and the current research work in the field of micro force sensors was proposed. The most common technique used for force sensing in micromanipulation is that of strain gauges. Some of the proposed force sensors use the piezoelectric effect in piezoactuators, but the sensitivity obtained is often poor. Other methods for strain measurement were also explored but at present strain gauges offer a comparable performance in terms of accuracy, speed and prize. In general, only few work is present in literature providing a basic description of force sensors for micromanipulation tasks regarding accuracy, precision, linearity and resolution of the sensor system. These data are of fundamental importance for a proper functionality of microrobots capable of grasping with force feedback. Hence, a lot of work has to be done within the field of micro force sensing. The use of piezoresistive and piezoelectric effects offers promising solutions for future developments, especially in regard to an integration of force sensors into piezoceramic actuators.

Acknowledgement

This research work has been performed at the Institute for Process Control and Robotics, Prof. Dr.-Ing. H. Woern, Prof. Dr.-Ing. R. Dillmann and em. Prof. Dr.-Ing. U. Rembold, Department of Computer Science, University of Karlsruhe, and has been funded by the VDI/VDE-Technologiezentrum Informationstechnik GmbH within the programme "Mikrosystemtechnik 1994-1999".

References

1. Rembold, U., Fatikow, S. (1997): *Autonomous Microrobots.* Journal of Intelligent and Robotic Systems 19, Kluwer Academic Publishers, pp. 375-391.
2. Gardner, J.W. (1994): *Microsensors: Principles and Applications.* John Wiley & Sons Ltd, Chichester, England, pp. 177-190.
3. Hurlbut, B.J., Regelbugge, M.E. (1993): *Development of a self-sensing multilayer piezoceramic actuator for structural damping applications.* 4th Int. Conf. on Adapt. Struct., Cologne, Germany, pp. 29-41.
4. Jendritza, D., Janocha, H., Scheer, P. (1996): *Smart actuators with piezoelectrical materials.* 3rd European Conf. on Smart Structures and Materials, Lyon, France.
5. Gibbs, M.R.J., Watts, R., Karl, W., Powell, A.L., Yates, R.B. (1997): *Microstructures containing piezomagnetic elements.* Sensors and Actuators A 59, pp. 229-235.
6. Despont, M. et al. (1993): *New Design of Micromachined Capacitive Force Sensor.* Journal of Micromechanics and Microengineering, Vol. 3, pp. 239-242.
7. Arai, F., Nonoda, Y., Fukuda, T., Oota, T. (1996): *New Force Measurement and Micro*

Grasping Method Using Laser Raman Spectrophotometer. Proc. of the 1996 IEEE Int. Conf. on Robotics and Automation, Minneapolis, Minnesota, pp. 2220-2225.

8. Delic, N., Detter, H., Brenner, W., Popovic, G., Vujanic, A., Hannenheim, W. (1996): *Development of Force Sensor for Microhandling Purposses.* Micro System Technologies 96, 5th Int. Conf. On Micro Electro, Opto, Mechanical Systems and Components, Potsdam, Germany, pp. 363-368.
9. Greitmann, G., Buser, R. (1996): *A Micromachined Tactile Gripper System.* Microrobots: Components and Applications, SPIE Proceedings Series, Vol. 2906, Boston, pp. 2-12.
10. Breguet, J.-M., Henein, S., Mericio, R., Clavel, R. (1997): *Monolithic Piezoceramic Flexible Structures for Micromanipulation.* Progress in Precision Engineering and Nanotechnologies, Proc. of the 9th Int. Precision Engineering Seminar, Braunschweig, Germany, Vol. 2, pp. 397-400.
11. Janocha, H., Jendritza, D., Scheer, P., Stephan, P. (1996): *Principle of smart piezoactuators.* Actuator 96, 5th Int. Conf. on New Actuators, Bremen, Germany, pp. 148-151.
12. Lee, C., Itoh, T., Maeda, R., Suga, T. (1996): *Micromachined Piezoelectric Force Sensors Based on PZT and PNNZT Thin Films.* Micro System Technologies 96, 5th Int. Conf. On Micro Electro, Opto, Mechanical Systems and Components, Potsdam, Germany, pp. 357-362.
13. Spanner, K., Wolny, W.W. (1996): *Trends and Challenges in New Piezoelectric Actuator Applications (Review).* Actuator 96, 5th Int. Conf. on New Actuators, Bremen, Germany, pp. 140-145.
14. Horie, M., Funabashi, H., Ikegami, K. (1995): *A study on micro force sensors for microhandling systems.* Microsystem Technologies, Vol. 1, Nr. 3, pp. 105-110.
15. Kimura, M., Fujiyoshi, M. (1991): *Force Measurement and Control of Electrostatic Micro-actuator.* Proc. 2nd Int. Symp. on Micro Machine and Human Science, Nagoya, Japan, pp. 119-124.
16. Kobayashi, H., Iijima, S. (1991): *Micro Robotic Manipulation.* Proc. of 2nd International Symposium on Micro Machine and Human Science, Nagoya, Japan.
17. Ku, S., Salcudean, S.E. (1996): *Design and Control of a Teleoperated Microgripper for Microsurgery.* Proc. of the 1996 IEEE Int. Conf. on Robotics and Automation, Minneapolis, Minnesota, pp. 889-894.
18. Morishita, H., Hatamura, Y. (1993): *Development of ultra precise manipulator system for future nanotechnology.* Proc. of the 1st IARP Workshop on Micro Robotics and Systems, Karlsruhe, Germany, pp. 34-42.
19. Ruther, P., Bacher, W., Feit, K., Menz. W. (1997): *LIGA-Microtesting system with integrated strain gauges for force measurement.* Proc. of the 10th Annual Int. Workshop on Micro Electro Mechanical Systems, Nagoya, Japan, pp. 541-545.
20. Seki, H. (1992): *Modelling and Impedance Control of a Piezoelectric Bimorph Microgripper.* Proc. 1992 IEEE/RSJ Int. Conf. Intelligent Robotics and Systems.
21. Seki, H. (1992): *Piezoelectric bimorph microgripper capable of force sensing and compliance adjustment.* ASME, Japan/USA Symposium on Flexible Automation, Vol. 1, pp. 707-713.
22. Kleinke, D.K., Uras, H.M. (1994): *A magnetostrictive force sensor.* Rev. Sci. Instrum. 65 (5), pp. 1699-1710.

23
Vision-Based Control Approaches for Trajectory Correction of Microrobots

K. Santa, U. Rembold and H. Woern

1 Introduction

For automated micromanipulation station (Figure 1) developed at the University of Karlsruhe several piezoelectric microrobots have been built, they are capable of moving over long distances and of manipulating in a range of a few nm [1].

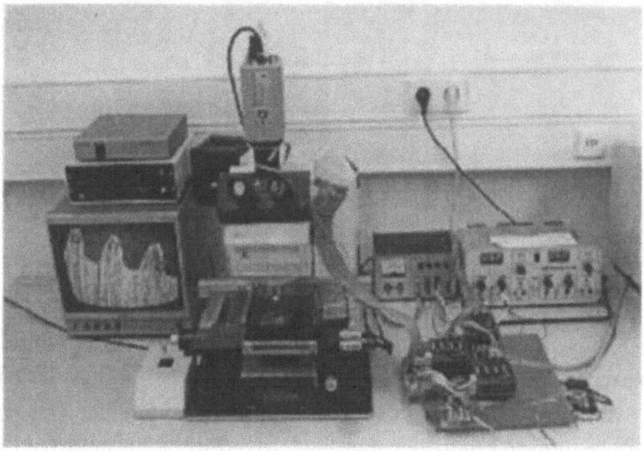

Figure 1: The automatized micromanipulation station (University of Karlsruhe)

The microrobots operate on a glass plate under an automated light microscope (Figure 2). In order to automatically control the manipulation process in the station, there must be sensor feedback. The microrobots have no integrated sensors. The light microscope equipped with a CCD-camera serves as a local sensor, which supplies visual information about fine manipulations of the robot endeffectors and the location of the microobjects. The actual position of the robots travelling on the glass plate is detected by another CCD-camera serving as a global sensor. The microrobots' motions on the working plate and their manipulations under the objective of the microscope are carried out by using only information obtained from the visual system. A controller is implemented on a PC and on a parallel computer system. A micromanipulation process in the station may be exposed to the disturbing influence of the environment such as dust, vibration, humidity, etc. For this reason, the microrobots used within the station have to be able to adapt themselves to the process requirements, i.e. to operate in a partially defined environment and to guarantee a reasonable behaviour in unpredicted situations.

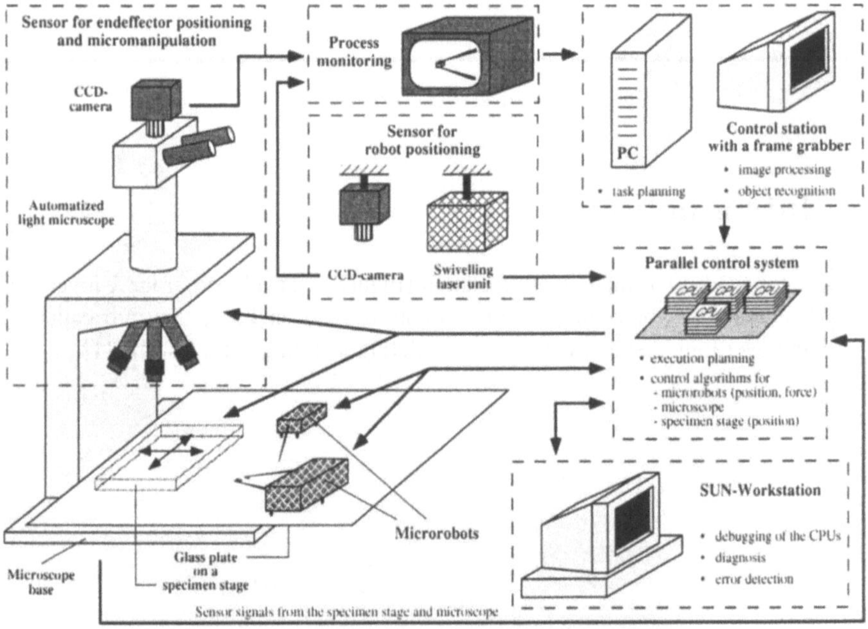

Figure 2: Design of the micromanipulation station

A behaviour-based control methods was developed, which allows a good compromise between real-time processing and obtained quality. It leads to defined decisions, even if only vague sensor information is available. The paper presents a neural adaptive motion controller that has proved to be an improved solution over a conventional one.

2 Aspects of the microrobots' mobility

As with conventional mobile robots, a flexible microrobot consists of a manipulating unit equipped with an endeffector which is integrated into a mobile platform. A specific feature of microrobots - compared to large mobile robots - is the fact that requirements to the motion performance depends on the task to be solved. Firstly, it may be the task of transportation of grasped microobjects over a relatively long distance (coarse motion or positioning). This requires a high speed of the microrobot. Secondly, the robot may perform fine manipulations under microscopic supervision (fine motion or manipulating). When performing fine manipulations, the robot movements may consist of that of the platform movements and of that of the manipulating unit.

To provide a microrobot with both positioning and micromanipulation capabilities, various designs of piezoelectric actuators can be used. Our microrobots use a platform with three triangularly arranged piezo legs (Figure 3). The legs are piezoceramic tubes which are equipped with an inner and four outer electrodes. The achievable speed is in the range of several mm/sec. The platform has a very high motion accuracy of 10 nm.

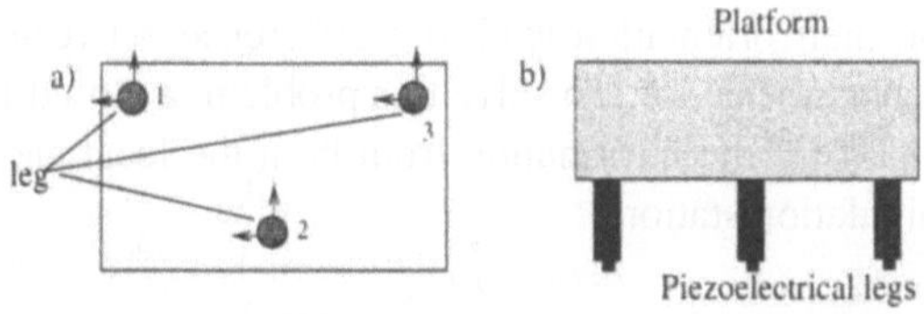

Figure 3: Robotplatform: a) bottom view; b) side view

A microrobot must be able to operate in an environment with many uncertainties. There are inaccuracies associated with the sensory system and manufacturing tolerances of the robot. Additionally, friction effects, influence of the connecting cable or a slightly varying slope of the working plate may considerably influence the behaviour of the microrobots.

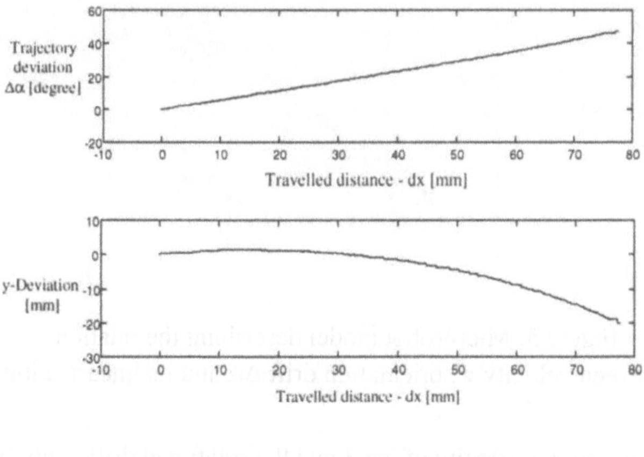

Figure 4: Uncontrolled microrobot: measured trajectory (orientation and position deviation)

For this, a control concept is needed, which is able to adapt the robot's behaviour to changing environmental conditions and to correct deviations from the desired behaviour in real time. Due to the nonlinear characteristics of both the sensors and actuators. The controller must be able to handle the behavior. Additionally, a good compromise between the real-time control and the quality of assembly has to be found. The most important problem is the inexact motion of the microrobot along the desired trajectory (Figure 4) [2].

Although the robot is directed to walk along a calculated trajectory, the tip of its endeffector does not allways arrive at the target position under the microscope. Typically without the controller the microrobot has a 45° angular deviation and a 20 mm deviation of the trajectory along a traveled distance of 80 mm (Figure 4).

This may has different causes. Firstly, the control system gets inexact input data from the global CCD-camera: its accuracy is only 0.5-1mm. Secondly, there is friction between the robot legs and the glass plate. Thirdly, there are some inaccuracies in the assembly of the microrobot itself. The smallest deviations in the legs' position and their orientation lead to a difference between the calculated and that of the real robot's behavior. To solve this problem, a closed-loop control system was developed that uses the information from both the local and the global sensors of the micromanipulation station.

3 Development of a neural controller

To study the control behavior, a model of the microrobot was conceived by standardizing the deviation of orientation versus the covered distance, depending on the velocity parameter of leg 3. The achieved nonlinear function is presented in Figure 5.

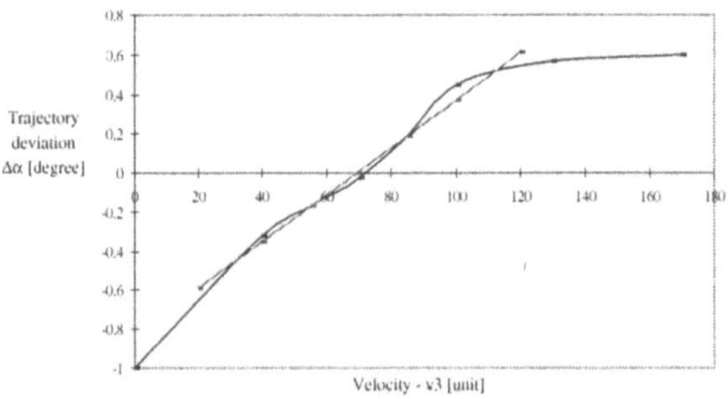

Figure 5: Microrobot model describing the relation between velocity v_3, orientation drift $\Delta\alpha$ and its linearization

The relation between the velocity of leg 3 and the rotational drift is nonlinear.

The translational drift can be reduced by using a nonlinear controller. It will be presented a learning control approach, based on a neural reference model for an adaptiv control method (MRAC).

The basic principle of the MRAC-based control method is to define the desired behavior of the neural controller in terms of the output of a reference model which is called the reference behavior. The output of the reference model is then used to adjust the controller's parameters in such a way that the output signal of the neural controller approaches the output of the reference model. The reference model and the neural controller receive the same reference input data. Both output signals are evaluated and compared referring to the input date [3,4].

The result is the error between the desired and the real output signal. This error is backpropagated through a neural network, and its parameters (weights) are adapted to minimize the error. This procedure is repeated until the error for an epoch falls below a user defined limit.

For the definition of the reference signal, the dynamic of the process must be known in order to keep the manipulated variable in a desired bandwidth. To improve the controller's action a reference signal was chosen to obtain the same results of the linear controller (PDT2-behavior), however, with a drastically reduced amplitude of 0.5 degree and a faster settling distance of 5 mm at a starting velocity parameter of 170 units. The arithmetical Equation of the reference signal is than

$$f_{REF}(k)=A*\exp(-d*k)*\sin(2\pi f*k)$$

with an amplitude of A=15, a damping value of d=0.5 and a frequency of f=0.08.

To be able to backpropagate through the plant and to adapt the controller's parameters, the nonlinear function in Figure 5 representing the robot's behavior has to be reproduced in a neural network. As a first step, a neural model of the plant behaviour has to be learned.

Because of the fact that the network needs to adapt to the complete dynamic nonlinear relationship between the manipulated signal, the delayed controlled signal by one sampled time step and the actual controlled signal (Figure 5), the training data were chosen as equal distributed over the interval [0; 170] units. These data were generated by using the Matlab's Simulink toolbox.

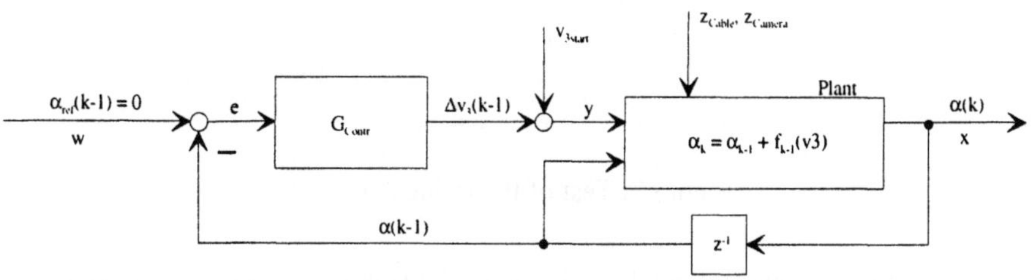

Figure 6: General structure of the closed-loop

The closed-loop structure in Figure 6 shows both the inputs of the neural model and the outputs and also the external feedback to generate a dynamic behavior. The number of neurons in the hidden layer depends on the task that has to be solved. Because of the nonlinearity of the function that needs to be learned and the fact that the neurons in the hidden layer represent the knowledge of the network, the *tanh* activation function was assigned to this layer. With this activation function it is also possible to produce negative outputs that are needed for the nonlinear function without using negative weights. The output layer summarizes the evaluated signals in the hidden layer and therefore its activation function is the linear function.

Using the generated training data and the plant's structure, from Figure 6 several neural networks with a different number of neurons in the hidden layer were trained. The best results were achieved using a 2-10-1 network structure. Increasing the number of hidden units did not significantly improve the approximation performance. The network was trained for 1000 epochs with 1500 training samples using the Levenberg-Marquard learning algorithm. Some test data were presented to the trained network to check its functionality and to avoid overfitting effects. Here, the sinusoidal test function that covers the entire input area was created, with an amplitude of a = 85 units and a frequency of f = 1. The systems output of the nonlinear function in Figure 5 of the trained neural network and the difference between both outputs are shown in Figure 7.

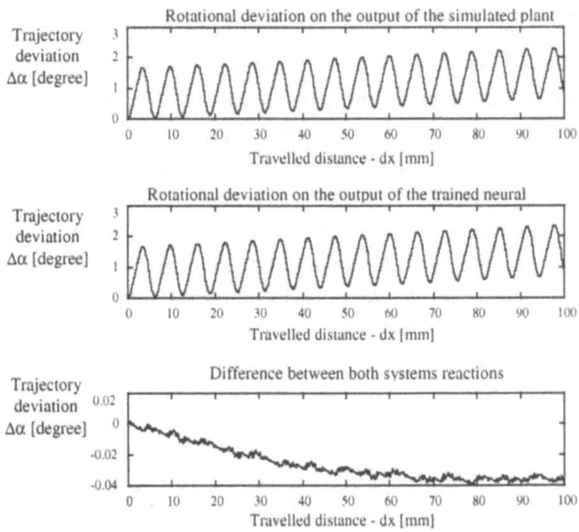

Figure 7: Test of the trained network

In each test point there is a small deviation of the neural network's output compared to the output of the nonlinear function but the achieved accuracy turned out to be sufficient to learn an appropriate controller. With the trained neural network representing the microrobot's behavior it is now possible to create a neural network controller with the MRAC-method.

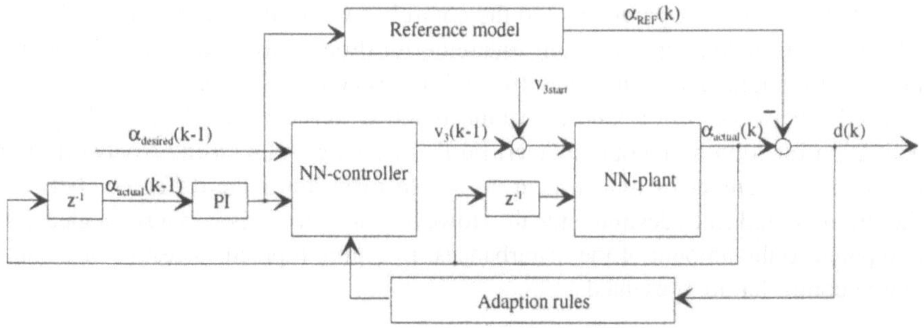

Figure 8: The MRAC neural control structure for the microrobot orientation control

The basic MRAC-structure has to be adapted to the microrobot system. The overall neural control system for the orientation control of the microrobot is shown in Figure 8. The PI indicates the back coupled orientation drift and its integrated signal.

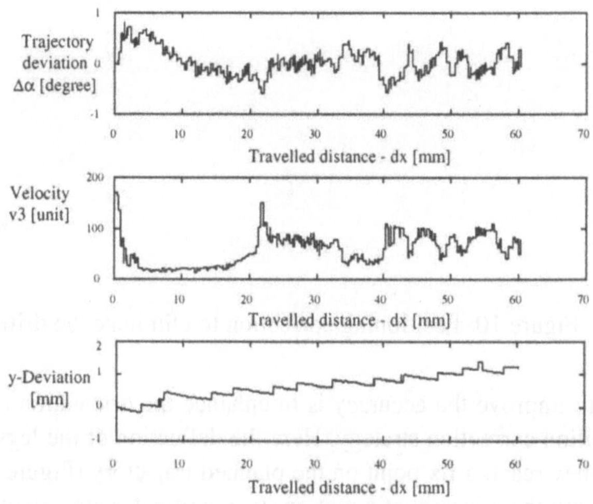

Figure 9: Test results by using a neural controller

By using the modified MRAC-method and the reference signal in Figure 9, the parameters of the neural network controller can be adapted. At the beginning they are randomly initialized. In the first step the velocity parameter causes a rotatoric deviation at the output of the neural controller. This deviation represents the controller's inputs in the second step. The generated output of the controller is beeing superpositioned with the velocity parameter. This signal and the deviation in the first step are the plant's inputs and cause a new rotational deviation.

This is now compared with the desired output from the reference model. The difference between both signals is the error that is backpropagated through the plant with its fixed network parameters to the controller where the parameters are beeing adapted. This is done until the sum-squared error drops below a user defined bound. For the investigated system it was 10^{-3} and it was achieved by using a 2-5-1 neural controller network with a *tanh* activation function in the hidden layer and a linear activation function in the output layer. Further training and by using a network with an increased number of hidden neurons would even lower the difference between the trained and the desired signal. However, the error is already neglectable compared to the amount of the disturbances. In Figure 8 practical results of using a neural controller are presented.

The rotational deviation in Figure 9 is kept at a bandwidth of ±0.8° (compared to 1.5° for the linear PI controller). The drift in the x-direction is reduced to 1.2mm over a distance of 60 mm (about 2.1 mm for the linear controller). The nonlinear neural controller thus clearly outperforms the PI controller in this application. By further training the neural controller with the real system a further improvement of accuracy might be expected.

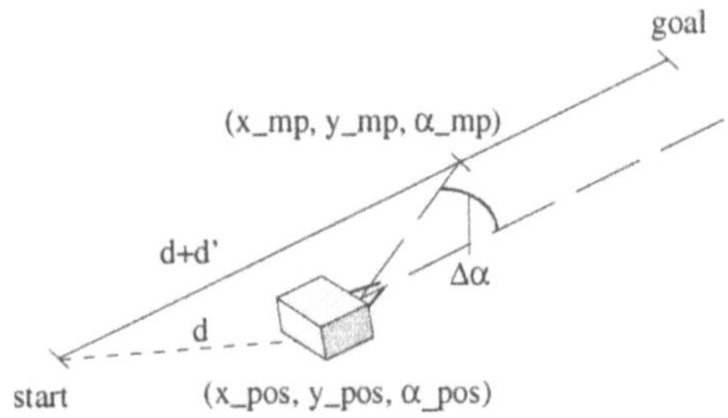

Figure 10: Positioning correction to eliminate the drift

Another way to improve the accuracy is to enhance the orientation controller by an additional position correction strategy. Here the deflection of the legs (amplitude) is done, so that they reach a fix point on the planned trajectory (Figure 10). This point moves along with the microrobot's motion. Its coordinates are calculated depending on the covered distance by the microrobot and an additional buffer that is needed for a smooth motion.

If the distance between the start position and the moving point is bigger than the distance between the start and the goal position, the microrobot will bent its legs towards the goal position. Figure 11 shows the performance of the entire control system consisting of the neural orientation controller and the additional position control algorithm.

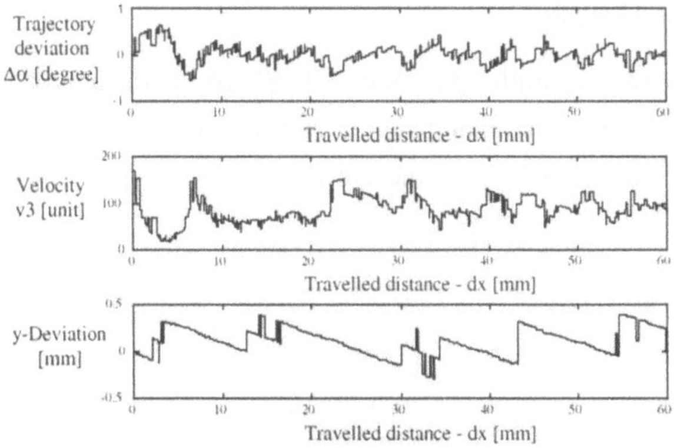

Figure 11: Practical results with rotational (neural controller) and positional correction

The microrobot can move over a long distance with a high accuracy (smaller than 0.5mm over 60mm distance).

4 Conclusion

An intelligent neural controller for positioning a microrobot is described in this article. It operates under the guidance of a powerful sensor system for positioning a microrobot on a work platform and for performing micromanipulations. The neural controller does not need a mathematical model of the controlled system. The control tasks that can be solved with the help of neural networks are presented. This control system will be used to build up a fully automated microrobot-based micromanipulation desktop-station.

References

[1] Th. Fischer, K. Santa, S. Fatikow (1996): „Sensor system and powerful computer system for controlling a microrobot-based micromanipulation station", Proc. of the 7th Micromechanics Europe 1996, Barcelona, Spain, pp. 283-286.

[2] K. Santa, H. Woern (1997): „Intelligente Ansteuerung von autonomen Mikrorobotern in einer Mikromanipulationsstation", Stuttgart, Germany, Proc. of the 13th Autonome Mobile Systeme 1997, pp. 199-209.

[3] K. Santa, S. Fatikow, G. Felso (1997): „Control of microassembly-robots by fuzzy-logic and neural networks", Advanced Summer Institute 1997,

Budapest, Hungary.

[4] K. Santa, M. Riedmiller, M. Mews (1998): „A Neural approach for the control of piezoelectric micromanipulation robots“, Journal of Intelligent & Robotic Systems, Netherlands, 1998 (accepted).

24

Exploiting a Human Arm Model for Fast, Intuitive and Time-Delays-Robust Telemanipulation

P. Prokopiou, W.S. Harwin and S.G. Tzafestas

1 Introduction

Although master-slave systems have been investigated since the 1940s, in addition to the traditional application fields (e.g. space, underwater, hazardous industrial ones), new fruitful areas have arisen, including the entertainment and health care industry [1]. However, current telemanipulator designs [2],[3],[4],[5],[6], are often accused of providing an obscure feel of the telemanipulated objects. This in part results from the poor modelling of the operator arm and the brain decision tactics, which due to their complexity and the ambiguity of early models, are either neglected or overly simplified.

The authors have explored various ways to employ a human arm model in the telemanipulation area. A key feature of our research is the use of the neural input to the human arm (NI, its measurement termed electroneurograph - ENG) as an observable variable, either measured or estimated through an inverse model of the arm. We consider an "extended teleoperator system", including the manipulators and the operator arm (with its reflex and local neural loops), whose input is the NI rather than the "macroscopic" variables of position and force. Given the current level of research, this is a realistic advance [7]. Previous research has regularly thought of the NI (often termed "intentional force") as a system starting point, but only for stability analysis. Note that the ENG has to be rectified and further processed to be suitable for use within a model (see [8] and references therein). In this chapter, we will employ the acronym NI to refer to this processed signal.

Another key feature of our designs is the prediction of the operator's arm movement as a response to the NI, which, as outlined in Section 3, enables the compensation of time delays in the communication channel. Three main groups of techniques robustifying against time delays have appeared up to date (refer to [9] for a thorough survey). The first is based on the use of predictive displays for the slave and the remote environment. The second group entails the use of wave variables to passivate the communications channel (e.g. [5]). These provide stability but alter the force fed back. A final solution is to use supervisory control [9], which however leads to a non continuous form of teleoperation, i.e. it changes the basic specification of manipulating as close to physical as possible. The scheme proposed in this chapter, effectively uses a prediction of the master state *only*, which can be much more accurate than predicting the slave and the remote environment, and incorporates this in a stable force-feedback scheme. Also, it cancels the computational burden of visually representing the slave future state, needed in predictive displays: the slave-side cameras' image is enough for optical feedback, since its timing is correct (Fig. 1). This is to the best of the authors' knowledge an unexplored approach. In [9] two early efforts predicting the control input along with the rest of the system state are reported for non-telemanipulation tasks, but are judged inadequate for telemanipulation.

In this chapter a novel "from scratch" teleoperator design, named "neuropredictive teleoperation" (NPT) scheme, attempting to exploit the opportunities offered by the peculiarities of the arm physiology, will be presented. An effort to use a human arm model in a conservative way [10], by providing a classic scheme [2] with robustness to time delay, confirmed that the combination of employing a predictor module and the NI, renders classical designs obsolete: through a "traditional" design process, it was shown that the mechanical part of the teleoperator reduces under perfect prediction to an open-loop system. This convinced us to abandon the obsolete bilateral constraint that the two robots are simultaneously compliant to each other's state (equal positions and forces in both sides), thus obtaining a succession of only locally controlled modules and a much simpler control problem. The compensation of time delays of up to 500 msec was also made possible. Our designs were inspired by well established physiological evidence that the brain, rather than controlling the movement on-line, "programs" the arm with an action plan of a complete movement, which is then executed largely in open loop, regulated only by local reflex loops [11]. It acts as a supervisory controller of the "semiautonomous" arm, in the sense of [9].

As a model of the human arm we employed the Stark model [12],[13] (also refer to the Appendix) adapted for the arm with parameters from [14], because it is well-established and, being anthropometric, can give valuable insight. The control schemes outlined in the sequence are however valid for *any* human arm model.

This chapter is organized as follows: In the next Section models of the human arm and their stability are discussed. In Section 3 the proposed design is outlined, whereas in the next Section its stability is analyzed. After presenting simulation results in Section 5 and concluding remarks in Section 6, in the Appendix the Stark model of the human arm and the modifications we propose, are outlined.

2 Human Arm Models in Current Designs

Although first attempts to model the biological muscles date in the 30s, an increasing amount of models of human and animal limbs continue to appear, the latter easily adaptable to the former. These can be divided in two broad groups: anthropometric models [13],[14],[15] which represent each part of the arm physiology separately, and neural networks or other 'black-box' mathematical models [16],[17],[18],[19]. The first type facilitates the studying of physiology and the invention of cures or ways to imitate it, whereas the second is suitable to use within control systems, as well as more adaptive towards injury, illness or other natural variations. Inputs of the models are mostly the electromyograph (EMG) or the NI. Measurement devices for both, e.g. surface or implanted electrodes, are available, although (especially for the latter) in need of improvement. Implanted electrodes have been successfully used on humans for both stimulation and neural signal measurement over a large period (see e.g. [8]).

The simpler an arm model, the easier it is to be exploited in a teleoperator implementation. Bobet et al [18] proposed one of the most useful models: a simple 2nd order linear model but with time varying parameters. Slightly more complex is the discrete-time model of Bernotas et al. [19]. The parameters are in both papers bounded, with small variations, and the system poles are always stable. If these parameters are considered independent of the muscle input and output, BIBO-stability is guaranteed by applying theorems of "almost time-invariant systems" [7],[20]. If not, total stability theorems are expected to yield this result. More accurate is a Hammerstein model recently proposed by Efranian et al. [16]. This is also BIBO-stable, as verified with similar arguments. Gollee et al. [17] employed LMN neural networks, comprising of a set of linear submodels with regions of application determined by a gaussian function. Since the linear submodels are stable and the gaussian function bounded, the network is BIBO stable. This model, or another neural networks one, is ideal for engineering applications.

A literature survey of teleoperation-related research [7], [21] has revealed that the human arm dynamics and the decision making process of the human brain are neglected or excessively simplified. Although it is acknowledged that the system is in reality more complex, most papers opt for simulations and stability analysis based on a simple, time invariant, spring-damper-mass model for the human arm (exceptions include [3],[4]). Stability and passivity results are based on a (usually underlying) assumption that "the operator's input τ_{op} is independent of the state of the master-slave system" [2]. As argued in [11], [21], this is correct "most of the time" and "the best we can assume" for the brain, since it neither *can* nor *has to* be analyzed for teleoperation. Thus current designs ensure that the mechanical part of the system is stable or passive and rely on the operator to use it successfully, like a 'low-tech' passive tool such as a hammer.

3 Neuropredictive Teleoperation

A basic feature of our designs is the use of a module that predicts the operator's arm movement as a response to the NI. This is used to direct the slave along the desired trajectory *ahead* of the master. The slave's response is fed back to the master appropriately delayed, so that the operator feels the results of his actions simultaneously with his actions, exactly as in physical "by hand" operation. This is illustrated in Fig. 1.

It is preferable to make a prediction of the NI / EMG signal, rather than the model *output*, since it has some well studied characteristics (e.g. the three pulses' sequence for fast movements, modelled by varying the amplitude *or* the period, roughly square NI, roughly sinusoidal rectified EMG, 1st order activation [12]). We can predict the output by simulation ("running the model forward") or analytical calculations. Depending on the desired prediction accuracy and the available hardware, a lead ($=T_t/2$) of 100 to 500 msec is achievable [21]. Considering typical values [9] we "erase" time delay for many ground applications (e.g. telesurgery with dedicated communication lines or Internet link), underwater ones for depth of 400m, or even single-link earth-to-orbit ones. Employment of hints about the intended movement or combination with existing techniques could increase those limits.

For contact tasks, the prediction can be made with the nominal arm parameters, i.e. without load, and then rely on the slave controller to provide compliance. However, since the brain calculates the NI expecting a certain load, a model of the manipulated object was also included in the predictor. If the human arm model and

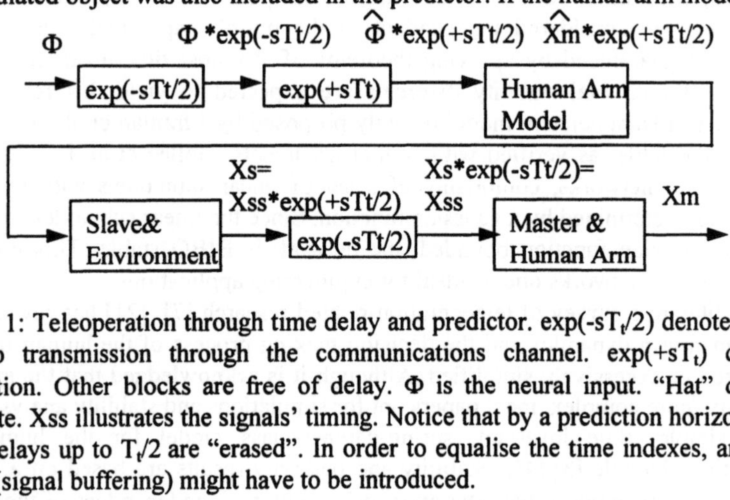

Figure 1: Teleoperation through time delay and predictor. $\exp(-sT_t/2)$ denotes delay due to transmission through the communications channel. $\exp(+sT_t)$ denotes prediction. Other blocks are free of delay. Φ is the neural input. "Hat" denotes estimate. Xss illustrates the signals' timing. Notice that by a prediction horizon of T_t time delays up to $T_t/2$ are "erased". In order to equalise the time indexes, artificial delay (signal buffering) might have to be introduced.

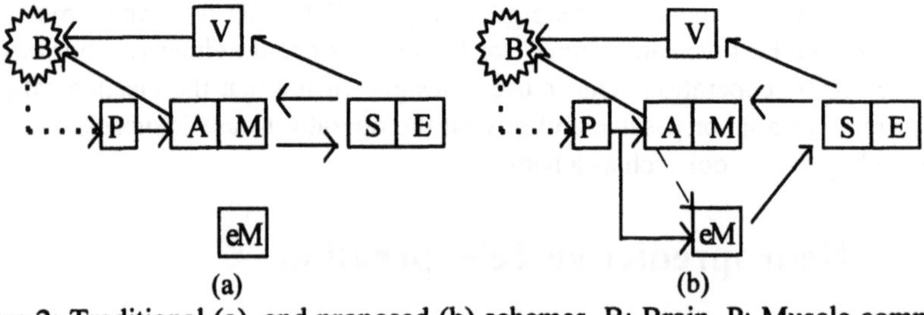

Figure 2: Traditional (a), and proposed (b) schemes. B: Brain, P: Muscle command program buffer V:Screen, A:Arm, M:Master, eM: Arm Model and Predictor, S: Slave, E: Environment, Dashed lines should be ignored,unless NI is estimated through an inverse arm model or when the arm model is tuned online. The dotted connection is open most of the time.

predictor modules can be placed in the slave side, an object representation does not induce heavy calculations. If it were placed in the master side, a prediction of the NI and the environment's status would be needed all the time, whereas in the slave side the prediction of just the NI is adequate: measured variables will determine the environment's behavior. An environment model is only needed when a new prediction cycle is initiated, and, due to the short prediction horizon, it can be local and thus simple. Moreover, choosing the slave side is conceptually equivalent to placing a virtual human arm there. For linear arm model both cases are mathematically equivalent.

By employing a model of the human arm, the "forward" signal path (brain to slave) can be decoupled from the "backward" one (slave to brain), so that a simpler, open loop, system is obtained. In classic schemes (Fig. 2a) there are two basic control loops in series: the "hardware" one (master-slave-master) and the "proprioceptive" one (brain- operator arm -brain), which are nested inside the "visual" loop (slave- brain- operator arm-master-slave). However, since the two latter loops are most of the time open (only close while new "command programs" are downloaded), the "hardware" loop is dominant, from a designer's point of view. To further complicate things, traditionally the bilateral constraint that the two robots should be simultaneously compliant to each other's state (equal positions and forces) is imposed, so that the operator can correctly "feel" the slave state. Such a multiplicity of feedbacks, although dictated by the need to use easily measured "macroscopic" variables, complicates the analysis. Of course, there are also local control loops inside the operator arm, the master and the slave.

With neuropredictive teleoperation (NPT) (Fig. 2b), we break up the control loop master-slave-master. We attempt to close the loop in the human brain, mimicking natural human movements. The slave is directed by the predictor output. The master is still directed by the slave trajectory, so that the operator senses the results of his actions, but it is not compared against the predictor output. Thus a succession of only locally controlled modules is obtained, each one of which follows (through suitable local control) the command supplied by the previous one. I.e. we are left with an open-loop system, a much simpler control problem. Notice also that the aim of teleoperator design is redefined to imposing to the slave tracking of the intentional movement, as this is encoded in the NI, instead of tracking the master. The latter is now an intermediate performance specification, enforced by local control. NPT is sensitive to the quality of the operator arm model and the predictor. Prediction corrections can be done without affecting the open-loop nature of the scheme, whereas online adaptation of the model or NI estimation by inverting, will spoil it.

The dynamics of the system will be described in the sequence. The formalism is compatible to that of [3].

Master: $$\boldsymbol{f}_{\mathrm{am}} + \boldsymbol{f}_{\mathrm{em}} = \boldsymbol{M}_{\mathrm{am}}\ddot{\boldsymbol{x}}_{\mathrm{m}} + \boldsymbol{U}_{\mathrm{bm}} \quad (1)$$

Operator: $$\boldsymbol{f}_{\mathrm{h}} - \boldsymbol{f}_{\mathrm{em}} = \boldsymbol{M}_{\mathrm{o}}\ddot{\boldsymbol{x}}_{\mathrm{m}} + \boldsymbol{B}_{\mathrm{o}}\dot{\boldsymbol{x}}_{\mathrm{m}} + \boldsymbol{K}_{\mathrm{o}}\boldsymbol{x}_{\mathrm{m}} \quad (2)$$

Coupled: $$\boldsymbol{f}_{\mathrm{am}} + \boldsymbol{f}_{\mathrm{h}} = (\boldsymbol{M}_{\mathrm{am}} + \boldsymbol{M}_{\mathrm{o}})\ddot{\boldsymbol{x}}_{\mathrm{m}} + \boldsymbol{B}_{\mathrm{o}}\dot{\boldsymbol{x}}_{\mathrm{m}} + \boldsymbol{K}_{\mathrm{o}}\boldsymbol{x}_{\mathrm{m}} + \boldsymbol{U}_{\mathrm{bm}} \quad (3)$$

Slave: $$\boldsymbol{f}_{\mathrm{as}} + \boldsymbol{f}_{\mathrm{es}} = \boldsymbol{M}_{\mathrm{as}}\ddot{\boldsymbol{x}}_{\mathrm{s}} + \boldsymbol{U}_{\mathrm{bs}} \quad (4)$$

where $\boldsymbol{f}$ denotes forces, $\boldsymbol{x}$ positions, subscripts m the master, s the slave, h and o the operator, a actuator forces. M_o, B_o, K_o, are constants, modelling the passive part of the human arm dynamics. In (1) and (4), M represents the effective inertia matrix and U the bias force (related to gravity and centripetal / Coriolis force). f_h is generally a function of x_m and the neural input. By linearizing, i.e. setting:

$$f_{am} = f_{ddm} + M_{am}\ddot{x}_m + U_{bm} \quad (5a)$$

$$f_{as} = f_{dds} + M_{as}\ddot{x}_s + U_{bs} \quad (5b)$$

we get:

$$f_{ddm} + f_h = M_o\ddot{x}_m + B_o\dot{x}_m + K_o x_m \quad (6a)$$

$$f_{dds} + f_{es} = 0 \quad (6b)$$

To avoid acceleration measurements, we can solve the equation specifying f_{ddm}, f_{dds} (to be specified later) to the acceleration, substitute in (3),(4) and solve for f_{am}, f_{as}.

The environment force f_{es} is assumed locally linear (x_e is the object's position):

$$f_{es} = M_e(\ddot{x}_e - \ddot{x}_s) + B_e(\dot{x}_e - \dot{x}_s) + K_e(x_e - x_s) \quad (7)$$

The above general system set-up, will be adapted for our scheme by specifying suitable f_{ddm}, f_{dds}. The manipulation will be as close to physical as possible, if we choose f_{dds} so as to completely replicate the dynamics of the human operator:

$$f_{dds}(t) = f_h(t + T_t/2) - (M_o\ddot{x}_s(t) + B_o\dot{x}_s(t) + K_o x_s(t)) \quad (8)$$

$$\Leftrightarrow f_{es}(t) + f_h(t + T_t/2) = M_o\ddot{x}_s(t) + B_o\dot{x}_s(t) + K_o x_s(t) \quad (9)$$

In this case f_h is clearly an estimate (prediction). If we set to the master:

$$f_{ddm}(t) = f_{es}(t - T_t/2) + M_o(\ddot{x}_m(t) - \ddot{x}_s(t - T_t/2)) + B_o(\dot{x}_m(t) - \dot{x}_s(t - T_t/2)) + K_o(x_m(t) - x_s(t - T_t/2)) \quad (10)$$

we get:

$$f_{es}(t - T_t/2) + f_h(t) = M_o\ddot{x}_s(t - T_t/2) + B_o\dot{x}_s(t - T_t/2) + K_o x_s(t - T_t/2) \quad (11)$$

which shows that the master replicates the slave dynamics, delayed by $T_t/2$ secs.

Another option is to use simple force feedback, so that:

$$f_{ddm}(t) = f_{es}(t - T_t/2) \quad (12)$$

and

$$f_{es}(t - T_t/2) + f_h(t) = M_o\ddot{x}_m(t) + B_o\dot{x}_m(t) + K_o x_m(t) \quad (13)$$

If $x_m(t) \approx x_s(t-T_t/2)$ - a reasonable assumption, since master and slave have become dynamically equivalent and with perfect prediction receive the same input, then:

$$f_{es}(t - T_t/2) + f_h(t) \approx M_o\ddot{x}_s(t - T_t/2) + B_o\dot{x}_s(t - T_t/2) + K_o x_s(t - T_t/2) \quad (14)$$

which approximately replicates the situation at the slave side.

Since the above rely on good knowledge of the dynamics, a 'safer' approach is using impedance or PID control laws, ensuring trajectory following. With suitable f_{ddm}, f_{dds} we e.g. obtain (PD rather than PID control shown, as used in the sequence):

$$f_h(t) + f_{es}(t - T_t/2) + K_p(\hat{x}_m(t) - x_m(t)) + K_d(\hat{\dot{x}}_m(t) - \dot{x}_m(t)) = M_o\ddot{x}_m(t) + B_o\dot{x}_m(t) + K_o x_m(t) \quad (15)$$

$$f_{es}(t) + f_h(t + T_t/2) + K_p(\hat{x}_m(t + T_t/2) - x_s(t)) + K_d(\hat{\dot{x}}_m(t + T_t/2) - \dot{x}_s(t)) = M_o\ddot{x}_s(t) + B_o\dot{x}_s(t) + K_o x_s(t) \quad (16)$$

$$f_{es}(t) = M_{ds}\ddot{x}_s(t) + B_{ds}(\dot{x}_s(t) - \hat{\dot{x}}_m(t + T_t/2)) + K_{ds}(x_s(t) - \hat{x}_m(t + T_t/2)) \quad (17)$$

where hat denotes estimate and eq (15) reflects the master and eqs. (16),(17) the slave. In our simulations PID and impedance schemes were equally successful.

4 Stability Analysis

To facilitate analysis, the whole system is divided in two parts: the "human" part, including the physical arm and its model, and the "mechanical" part, encompassing robots, controllers and the environment. These two main blocks are in series or in feedback connection, depending on the control scheme used. In the latter case the resulting loop is in a suitable form to derive BIBO-stability and passivity, by applying the small gain or the passivity theorem [20]. Partitioning also enables a universal analysis of variations of our scheme: by modifying their contents, various local controllers, predictors and human arm models are studied within the same framework. As output of the above system the master's state x_m, f_m is considered, rather than the slave's x_s, f_s. A model of the contacted object (eq. (7)) is also included in the mechanical part. To analyze for a range of environments, various parameter values, or "worst case" ones can be used. Although this setup is dictated by stability analysis, it elucidates the consideration of the teleoperator and the remote environment interfaced to the operator as a generalized, non-physical object, a "teleobject" (not a virtual one, since it materially exists). Input to the system is the NI and output the position and force on the human arm, just as in direct manipulation of a physical object. The impression to the human brain is more important than the tracking between slave and master [3],[21].

With all the control schemes of Section 3, the "mechanical part" is BIBO-stable for a wide range of parameter values [7]. In this Section representative results are reported.

First, let's assume the prediction is nearly perfect, so that there are no robustifying terms in the master. A system of this type is described by eqs (13) and (17) and the block diagram of Fig. 3a, and is an open loop one. It is obvious that if G,M, and eM are stable, the whole system is stable too, which is indeed the case for various parameter values and models. In the s-domain, G represents (1-DOF):

$$F_{ddm} = F_s e^{-sT_t/2} = \frac{(B_{ds}s + K_{ds})(-M_e s^2 - B_e s - K_e)}{(M_{ds} + M_e)s^2 + (B_{ds} + B_e)s + (K_{ds} + K_e)} \hat{X}_m$$

Notice that all exponentials have been canceled out, so that the system is causal. Stability conditions for its poles are easily derived. Typical parameters used in our simulations are: m_e=b_e=0, K_e =40, M_{ds}=1, B_{ds}=4000, K_{ds}=5000, which give stable poles. In fact for this set of parameters the system is stable for any K_e. Employment

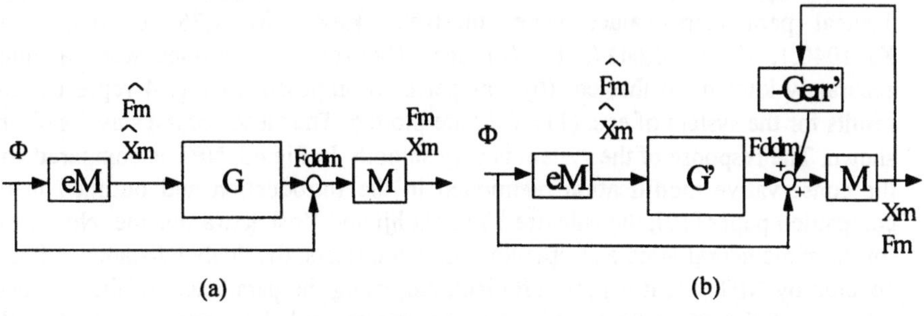

Figure 3: (a) System of (13),(17), (b) System of (15),(16). G,G': "mechanical" part.

of a PD instead of impedance controller in the slave yields equivalent results.

To be on the safe side, terms relative to the prediction error or providing better tracking can be included in the master control law. By e.g. implementing PD tracking controllers (eq. (15), (16)), the transfer function of the system becomes:

$$F_{ddm} = \frac{\left[-M_e s^2 + (K_{dt} - B_e)s + (K_{pt} - K_e)\right]\left[\hat{F}_m + (K_d s + K_p)\hat{X}_m\right]}{(M_o + M_e)s^2 + (B_o + B_e + K_d)s + (K_o + K_e + K_p)} - (K_{dt}s + K_{pt})X_m$$

Once again the system is open loop and the stability analysis easy. The non-fractional part of the above equation (Gerr') is represented in Fig. 3b as a separate block,encircling the human arm model. This is the only feedback connection in this family of schemes. The stability of the loop can be analysed by choosing a particular human arm model. Since this is generally nonlinear, a general method would involve the application of the small gain theorem [20], according to which the loop is BIBO stable if both blocks M and Gerr are stable and $\gamma_M * \gamma_{Gerr} < 1$, where γ_S, is the L∞ gain of system S. γ_M is not easily calculated by the models. However, since the human arm physiology has to be respected, an upper limit for it is $\gamma_M=1$ (for angle expressed in rad), although our simulations of the Stark model point towards $\gamma_M \approx 0.5$. K_{dt}, K_{pt} satisfying the above condition can always be specified, although not any arbitrarily strong feedback is possible. Breach of the condition does not necessarily implies instability, and that the rest of the system can tolerate bigger K_{dt}, K_{pt}. Similar conclusions are reached when a term relative to prediction error is included.

Instead of using an object model in the predictor, f_{ddm} can be issued as input to it. f_{ddm} equals f_{es} (eq. (12)) or f_{es} plus some correcting terms, which are normally much smaller (eq (10), (15)). By predicting f_{ddm} with simple techniques such as a Taylor extrapolation, a less accurate but also less complicated predictor is obtained.

If f_{ddm} is used in the prediction, or if NI is estimated with an inverse model, rather than measured, the loop closes. The small gain theorem can be employed to derive stability conditions [7], which impose trivial constraints to simple systems like the ones of eqs. (13),(17), and more demanding to systems similar to those of eqs (15) and (16). Passivity analysis is similar to the stability one of this chapter [7],but will be omitted due to lack of space.

5 Simulation and Comparison with Other Schemes

Neuropredictive teleoperation (NPT) was tested through extensive simulation. Typical parameters values were: $m_e=b_e=0$, $k_e=40$, $M_o=1{,}75$, $B_o=0{,}4$; $K_o=5$, $K_d=1040{,}1$, $K_p = 50{,}6417$, $T_t = 0{,}1$ sec. The robots' dynamics were assumed accurately known, so that eq. (6) was perfectly imposed. In Fig. 4 representative results for the system of eqs. (11), (16) are plotted. The "lead" of the slave is clearly shown. The response of the system is very smooth. In Fig 4d, NPT is compared with the conservative modification mentioned in the Introduction and outlined in the companion paper [10], the enhanced Yokokohji and Yoshikawa scheme. Notice that for the same neural input and operator arm parameters, the double distance is finally covered by NPT, i.e. it is more efficient. Adjusting the parameters of the enhanced scheme only improves the tracking between master and slave, but *cannot cover this gap*, since it is due to the fact that the robots' position and force are pushed to their mean. Most, if not all, other classical teleoperator systems, directly or indirectly as in [3], also follow this bilateral constraint, which is here proven obsolete.

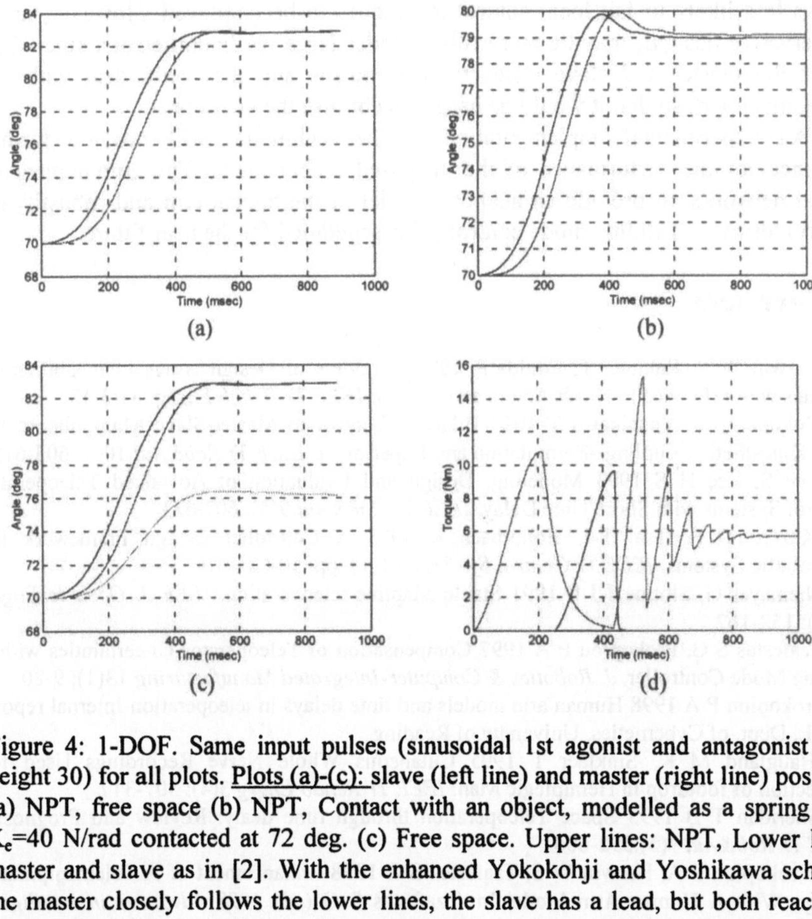

Figure 4: 1-DOF. Same input pulses (sinusoidal 1st agonist and antagonist with height 30) for all plots. Plots (a)-(c): slave (left line) and master (right line) position. (a) NPT, free space (b) NPT, Contact with an object, modelled as a spring with K_e=40 N/rad contacted at 72 deg. (c) Free space. Upper lines: NPT, Lower lines: master and slave as in [2]. With the enhanced Yokokohji and Yoshikawa scheme, the master closely follows the lower lines, the slave has a lead, but both reach the same final position (≈76 deg.-not plotted for clarity). Plot (d): Operator arm's extensor (left line) and flexor (central bell) force for the contact task of (b). During the movement forces are smooth (contact at around 200 msec), but due to our imperfect inverting algorithm, the clamping phase is oscillatory. In real life, it would be smoother.

6 Conclusions

In this chapter we described how the inclusion of a model of a human arm, combined with the measurement of its neural input and a predictor, can significantly simplify teleoperator design by delivering an open loop system, and provide robustness under time delay. Our trials gave clear indications of the superiority of the new scheme over traditional as well as the enhanced Yokokohji and Yoshikawa architectures. Its main advantages are: the *time-lead of the slave*, the *more efficient, and providing a more natural feeling manipulation,* and the fact that incorporating an operator arm model leads to *more credible stability* results. Finally, its *simplicity* allows less likely to fail local control techniques to be employed. However, more conservative designs, may be more robust under large prediction errors, since they direct the master and slave states to their means and they only deal with the mechanical part, so that it would be harder to confuse the operator.

An experimental implementation of the techniques will provide further evidence for the performance of the proposed architectures. The employment of neural networks, to provide an adaptive model of the human arm and robustifying control terms, and sliding mode control [6] is scheduled for the near future.

References

[1] Harwin W S, Rahman T, Foulds R 1995 A review of Design issues in rehabilitation robotics with reference to North American research. *IEEE Tr Rehab Engrg*, 31:3-13

[2] Yokokohji Y, Yoshikawa T 1994 Bilateral Control of Master-Slave Manipulators for Ideal Kinesthetic Coupling--Formulation and Experiment. *IEEE Tr Rob&Aut* 10(5):605-619

[3] Lee S, Lee H S 1993 Modeling, Design and Evaluation of Advanced Teleoperator Control Systems with Short Time Delay. *IEEE Tr Rob&Aut* 9(5): 607-623

[4] Kazerooni H, Tsai T-I, Hollerbach K 1993 A Controller Design Framework for Telerobotic Systems. *IEEE Tr Control Sys Techn* 1(1): pp. 50-62

[5] Niemeyer G, Slotine J-J E 1991 Stable adaptive teleoperation. *IEEE J. Oceanic Engrg* 16(1): 152-162

[6] Tzafestas S G, Prokopiou P A 1997 Compensation of Teleoperator Uncertainties with a Sliding Mode Controller. *J. Robotics & Computer-Integrated Manufacturing* 13(1): 9-20

[7] Prokopiou P A 1998 Human arm models and time delays in teleoperation Internal report, tHRIL, Dept. of Cybernetics, University of Reading

[8] Haugland M K, Sinkjaer T 1995 Cutaneous Whole Nerve Recordings Used for Correction of footdrop in Hemiplegic Man. *IEEE Tr Rehab Engrg* 3(4): 307-317

[9] Sheridan T B 1993 Space Teleoperation through time dealy: Review and Prognosis. *IEEE Tr Rob&Aut* 9(5):592-606

[10] Prokopiou P A, Harwin W S, Tzafestas S G 1998 Enhancement of A Telemanipulator Design With A Human Arm Model In: Tzafestas S G (ed) *1998 Advances in Intelligent Systems: Concepts and Applications*, Kluwer Academic Publishers (in preparation).

[11] Houk J C, Buckingham J T, Barto A G 1996 Models of the cerebellum and motor learning. *Behavioral and Brain Sciences* 19: 368-383

[12] Zangemeister W H, Lehman S, Stark L W 1981 Simulation of Head Movement Trajectories: Model and Fit to Main Sequence. *Biol Cybernetics* 41:19-32

[13] Ramos C F, Stark L W 1990 Postural maintenance during fast forward bending: a model simulation experiment determines a "reduced trajectory". *Exper Brain Research*, 82:651-657

[14] Gossett J H, Clymer B D, Hemami H 1994 Long and Short Delay Feedback on One-Link Nonlinear Forearm with Coactivation, *IEEE Tr SMC* 24(9):1317-1327

[15] Prochazka A, Gillard D, Bennett D J 1997 Positive Force Feedback Control of Muscles, J. Neurophysiology 77(6): 3226-3236

[16] Efranian A, Chizeck H J, Hashemi R M 1998 Using evoked EMG as a synthetic force sensor of isometric electrically stimulated muscle. *IEEE Tr Biomed Engrg* 45(2): 188-202

[17] Gollee H, Hunt K J 1997 Nonlinear modelling and control of electrically stimulated

muscle: a local model network approach. *Intl. J. Control* 68(6):1259-1288
[18] Bobet J, Stein R B, Oguztoreli M N 1993 A Linear Time-Varying Model of Force Generation in Skeletal Muscle. *IEEE Tr Biomed Engrg* 40(10):1000-1006
[19] Bernotas L, Crago P, Chizeck H J 1986 A Discrete-Time Model of Electrically Stimulated Muscle. *IEEE Tr Biomed Engrg* 33(9): pp.829-838
[20] Narendra K S, Annaswamy A M 1989 *Stable Adaptive Systems*. Prentice-Hall Intl.
[21] Prokopiou P A, Harwin W S, Tzafestas S G 1998 Fast, Intuitive And Time-Delays-Robust Telemanipulator Designs Using A Human Arm Model In: *Proceedings Symposium Intelligent Robotic Systems*, Edinburgh, UK 1998

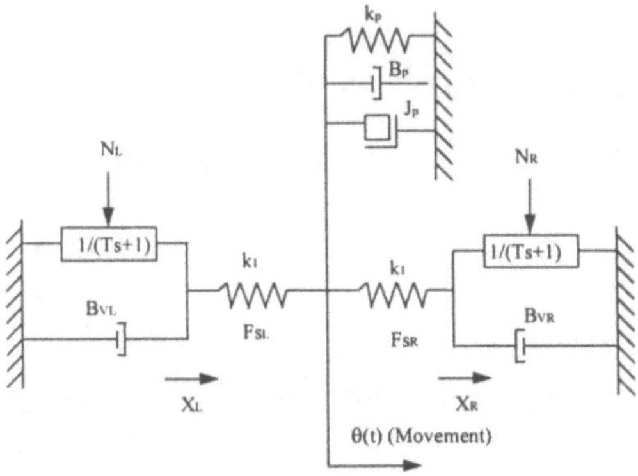

Figure 5: The (unmodified) Stark muscle model.

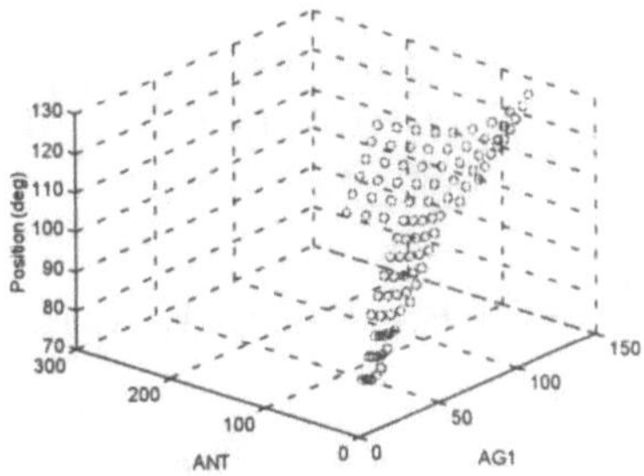

Figure 6: Stark muscle model. Final arm position for various 1st agonist (AG1) and antagonist (ANT) pulses' heights. The starting position was always 70 degrees.

Appendix: The Stark Model of Human Arm

The Stark model, as outlined in [12],[13], forms a good compromise between complexity and accuracy ([14] being more of both). However, it is difficult to use it for simulations, because of its nonlinear and not well defined mathematical description. Some modifications were considered necessary, which do not alter the model's core, but just expand it for cases the original model was unrealistic. (Fig. 5, Table 1).

The original model collapses when the muscle activation (HTL, HTR) approaches zero. Instead of considering a minimum muscle coactivation as in [13], we produced a simplified form of the original equations for the case of small activation by ignoring terms that at that point become too small. Such an elimination of terms leads to two possible choices for Fsl / Fsr: either set them to zero (due to the eq. (A.1)'), or define them from eq. (A.5). We chose to set them equal to HTL/HTR, which is equivalent to ignoring only the first term of (A.1)' related to HTL/HTR, which decreases faster. The switching between the two operating regions is smoothed by a sigmoid function. Also, in the original model X_L /X_R are not updated when the muscle receives zero input, unless the non natural hypothesis of a negative neural input holds. Thus subsequent activations of the muscle lead to increasingly high forces. In the modified scheme, X_L/X_R is updated when the muscle is inactive, taking into account the natural movement of the muscle inside a moving arm. This allows us to also drop the absolute from eq. (A.1)'. These modifications result in a smooth and realistic arm trajectory.

Table 1: Stark Model for the Human Arm - State Equations

Modified Model:

$$\dot{X}_L = \begin{cases} \dfrac{B_h(HTL-Fsl)}{0.25*HTL+Fsl}*\sigma_L + \left(v + \dfrac{H\dot{T}L}{k_2(HTL+1)}\right)*(1-\sigma_L) & \text{if } HTL \geq thres\_HTL \\ \dfrac{f_{xl}(\theta)-X_L}{\tau_L} & \text{if } HTL < thres\_HTL \end{cases} \quad \text{(A.1a)}$$

$$\dot{X}_R = \begin{cases} \dfrac{B_h(HTR-Fsr)}{0.25*HTR+Fsr}*\sigma_R + \left(v + \dfrac{H\dot{T}R}{k_2(HTR+1)}\right)*(1-\sigma_R) & \text{if } HTR \geq thres\_HTR \\ \dfrac{f_{xl}(\theta)-X_R}{\tau_R} & \text{if } HTR < thres\_HTR \end{cases} \quad \text{(A.1b)}$$

$$\dot{v} = J_p^{-1}(-B_p v - K_p\theta + F_e + F_{sl} - F_{sr}) \quad \text{(A.2)}$$

$$\dot{\theta} = v \quad \text{(A.3)}$$

$$\text{with: } \sigma_L = \left(1+e^{-1.5*(\frac{B_h(HTL-Fsl)}{0.25*HTL+Fsl}-4.5*dXlo)}\right)^{-1}, \sigma_R = \left(1+e^{-1.5*(\frac{B_h(HTR-Fsr)}{0.25*HTR+Fsr}-4.5*dXro)}\right)^{-1} \quad \text{(A.4a, b)}$$

$$Fsl = \max(0, k1(e^{k_2(X_L-\theta)}-1)), \quad Fsr = \max(0, k1(e^{k_2(X_R-\theta)}-1)), \quad \text{(A.5a, b)}$$

$$f_{XL}(\theta) = f_{XR}(\theta) = \theta - 0.1(\theta - X_{lo})^2 \quad \text{(A.6)}$$

$$H\dot{T}L = \frac{N_L - HTL}{T}, \quad H\dot{T}R = \frac{N_R - HTR}{T} \quad \text{(A.7)}$$

The **original Stark model** comprises of eqs. (A.2) , (A.3), (A.5), (A.7) and (A.1a,b)':

$$B_{VL}\dot{X}_L = \frac{1.25HTL}{B_h + |\dot{X}_L|}\dot{X}_L = HTL - Fsl, \qquad B_{VR}\dot{X}_R = \frac{1.25HTR}{B_h + |\dot{X}_R|}\dot{X}_R = HTR - Fsr \quad \text{(A.1a, b)'}$$

where: the subscripts l and r denote left and right muscle, θ,v position and velocity of the arm, X_L, X_R internal model variables, K_p, J_p, B_p, passive parameters of the arm (load for the muscles), B_h,T,k_1,k_2,X_{lo},dX_{lo}, X_{ro}, dX_{ro},thres_HTL, thres_HTR are constants, N_L, N_R the neural input, HTL, HTR activation levels ("hypothetical tension"), F_{sl}/F_{sr} the left/right muscle's force, F_e an external force.

Acknowledgement

We gratefully acknowledge the support of the MobiNet EU-TMR Research Network to the first author.

25

Intelligent Robotic Assembly and Disassembly : General Architecture and Implementation Case Studies

C.S. Tzafestas and *S.G. Tzafestas*

1 Introduction

Assembly is one of the sophisticated industrial processes and constitutes a challenging area for the application of artificial intelligence (AI), and intelligent robot control techniques and implementations [1-6]. Very broadly, the problem of robotic assembly consists in designing a computer – controlled robotic system that can assemble a product from its constituent parts. The two major drawbacks of conventional assembly systems are the high cost of reprogramming and the lack of flexibility. Both of these difficulties can be faced by using AI-expert system techniques and sensor-based intelligent control. Particular problems that have to be treated for designing a complete functional flexible robotic assembly system are: geometry description of assemblies [1-3], large data base design, computation of assembly times (feeding, handling, orienting and insertion times) using feature extraction methods, design for assembly [4-5], selection of assembly sequences, sensory feedback, and hierarchical intelligent control design [6,10,13-26].

A reliable robotic assembly system must be able to accommodate uncertainties in the environment. A first approach to deal with such uncertainties is to use passive instruments such as the *remote centre compliance* (RCC) [7], which can be mounted between the robot's wrist and the gripper to introduce compliance for assembly. Since RCC is only suitable for workpieces of a given shape and geometry, active instruments such as the *instrumented remote centre compliance* (IRCC) were used which can measure the force and moment encountered and enable the robot to compensate for them. A second approach is to use the *SCARA* (*Selective Compliance Assembly Robot Arm*) robot which is a horizontally articulated manipulator with a vertical insertion axis of the wrist end. SCARA – type robots are very stiff in the vertical direction but are relative compliant laterally, a feature very convenient for several assembly tasks. However, the produced leads of the components can easily be jammed during vertical insertion due to several reasons (bending, misalignment, gripper orientation error, error in the PCB hole position, etc). Thus it becomes necessary to monitor (sense) and control the insertion force in order to avoid damaging the assembled parts, the PCB or the robot. An alternative cheaper approach is to estimate via a proper observer the external reaction force during component insertion by using standard position and velocity information. Robots are characterized by *parametric* (*structured*) *uncertainties*, (e.g. mass / inertia and payload variations) which can be dealt with adaptive controllers, and *unstructured uncertainties* (e.g. unmodelled dynamics, nonlinear frictional torques) that need the application of robust controllers.

The purpose of this paper is to present a general architecture of intelligent robot – based assembly / disassembly systems which in produced by taking into account all functions and tasks required to be performed by the system. This architecture involves as special cases several systems developed throughout the years by different research and applications groups. For the convenience of the reader a brief review is provided of the available works since 1986. Then, a number of prototypes implemented for various levels of the hierarchy (organization level, coordination level, execution level), as well as an overall implementation of a car disassembly system are discussed.

2 Brief review of robotic assembly

A comprehensive review of the work in assembly carried out up to 1986 can be found in [8]. Here a brief review of some results produced afterwards will be provided. We start by first mentioning the works of [9-12] where useful general investigations of intelligent manufacturing systems are presented, and the works of [13-17] where important knowledge – based issues and approaches to robot – based assembly system design are provided. In particular, the application of the blackboard (BB) model to robotic systems with an assembly case study is presented in [18], and the design of a fuzzy – logic – based expert system for determining the optimum assembly sequence is proposed in [19].

In [20] a graph oriented simulation method was proposed which helps the assembly engineer to design in an objective manner alternative process sequences and structural forms of flexible assembly systems. The work in [20] also includes a detailed discussion of the three main obstacles in the wide use of simulation, namely: (i) great expense of modeling, (ii) great expense of experiment, and (iii) unsuitable representation and treatment of results.

In [21] the design of a workspace prediction and fast matching system is presented. This system exploits the special constraints on the identity and positioning of objects in a typical robotic assembly cell to perform fast analysis of stereo – vision data produced in such an environment where typically there is considerable information available *a priori* about objects. This quantitative geometric information is usually provided by a CAD-based model of the assembly workstation and the known position of the camera system used to view the 2D or 3D scene. To avoid redundant processing for constant scene features or for objects for which accurate estimates of current position are available and so maximize the operational efficiency of the vision system, this *a priori* knowledge should be carefully utilized. This is actually done in [21] with the beneficial result of reducing the computational cost of vision processing.

In [22] the star network model around a blackboard is employed to deal with the concurrent processes involved in a robot – based assembly system. The blackboard stores information of general value to all routines, such as the jobs already completed or the actual position of the robot. This information is updated regularly by a relevant module of the system. The blackboard increases the system's flexibility and offers the capability of adding more modules to the system. The control structure involves a supervisor (called motivator) and a local (executing) controller, and an error recovery unit is included to periodically check and restore the occurring errors. STRIPS is employed for the implementation of the task manager and concurrent execution of the rules, i.e. concurrent robot jobs are allowed for a faster overall operation.

In [23] an approach to assembly sequence planning is given defined in terms of mating and non – mating operations and based on a dynamic expansion of the assembly tree obtained from a knowledge base management system. The best – first search technique with forward chaining is employed to avoid considering all possible assembly sequences. Hard and soft constraints are introduced to manage the tree's dynamic expansion. Stability analysis under gravity is performed at each mounting of a single component, and valid subassemblies satisfying equilibrium under arbitrarily oriented forces are ensured. The paper proposes two complete sets of hard and soft constraints and a powerful scheme for managing the dynamic expansion of the assembly tree. The stability analysis is proposed as hard constraint in addition to the standard geometrical precedence, and the assembly sequences are defined in detail by adding several non – mating operations.

3 General structure of intelligent robotic assembly or disassembly systems

Intelligent control systems (ICS) are typically structured as 3-level hierarchical systems each level being decomposed in a number of layers according to the particular needs. The three levels are:

- The organization level (OL)
- The coordination level (VL)
- The execution level (EL)

The intelligence required in OL is the highest one and decreases as we go dawn to CL and EL, whereas for the required functional (operational) precision the opposite is true, i.e. the level with the highest precision is EL. This is known in ICS theory as the IPDI principle (Increasing Precision with Decreasing Intelligence principle) [24]. The time horizon over which each level operates is decreasing from the top (OL) to the bottom (EL) of the hierarchy.

In assembly / disassembly systems the organizer performs the overall organization of the assembly / disassembly process in its most general form taking into account general criteria such as cost, risk, etc. The coordinator supervises and programs the overall assembly / disassembly process by taking into account the actual task plans sent by the organizer, and transmits suitable commands for execution by the execution (control) level. The execution level accepts, manipulates, translates, and executes the commands of the coordinator, and detects and restores any malfunction or error. Of course such an error detecting and recovering ability must actually be available to all hierarchical levels.

The above hierarchical intelligent control structure of assembly / disassembly systems (Fig. 1) is generic and is used in all cases in an explicit or implicit form [5,6,15,19,25].

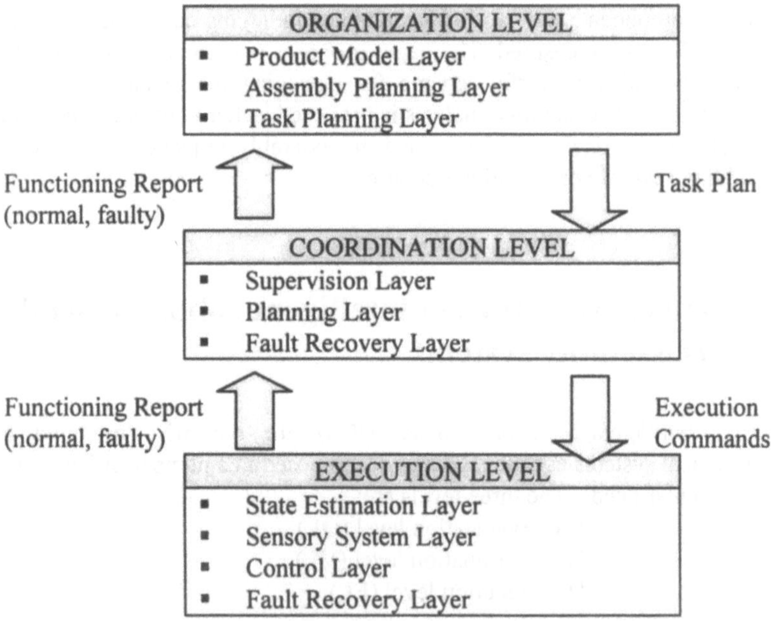

Fig. 1. General structure of intelligent assembly / disassembly systems

The *organization level* (sometimes called *strategic planning level*) typically involves three principal layers:

- Product model layer
- Assembly planning layer
- Task planning layer

These layers exchange data in a well defined way (data flow from a higher to a lower layer and only in case of an error / fault occurrence data are allowed to be transmitted from a lower to a higher layer). A subsystem (module, unit) of one layer receives its input (s) from its next higher neighboring subsystem and sends its output (s) to the immediate layer neighboring subsystem.

At the highest layer of OL there is a *product model*. Here the system is only interested in the product to be assembled / disassembled and its constituent components, and not in the way these can be handled. The next OL layer is the *assembly (disassembly) planning layer* (assembly / disassembly planner) which produces an assembly / disassembly plan for the product using geometrical and experiential knowledge about the parts (to be assembled or disassembled) and their spatial relationship. The generated plan can be represented in several ways (AND/OR graphs, precedence diagrams, etc.). The third layer of the organizer is the *task planning layer* which given a specific assembly / disassembly sequence and operations (plan) finds a sequence of *high level actions* of the robot (e.g. pick – up (component - i), test (component – i), transfer (component – i), etc), checks them for possible inconsistencies (e.g. unavailability of components, uncertainties, etc.) and splits them into elementary robot operations such as *move*, *approach*, *grip*, *dismantle*, *depart – with – load*, *approach – with – load*, *ungrip*, *depart*, and so on.

The coordination and execution (control) level together constitute the so-called *tactical level* system which is obliged to detail and execute the task plan commanded by the organizer. The tactical level system deals with short – and – medium term faults. The coordination level involves the following three principal layers:

- Supervision layer
- Planning layer
- Fault recovery layer

The *supervision layer* [26] receives the task plan from the OL's task planning layer as a sequence of elementary robot operations to be detailed and introduced for execution. Typical sub functions of the supervisor are: scheduling, coordination and task distribution; communication (with various units of the task planning layer and execution level), fault state management and world model management. The *planning layer* involves a variety of planners to perform trajectory planning, grasp planning, gross motion planning, fine motion planning and part mantle / dismantle planning. The *fault recovery layer* involves several units depending on the desired level and policy of fault detection and restoration. Typical units of this kind are: (i) *reflexive agents* which fire via alarm demons a predefined reflex (e.g. reverse or stop motor motion) when one or more of the monitored parameters appear to go to abnormal values, (ii) the trajectory *control manager* which initiates proper control actions to compensate small scale faults, (iii) the *state estimation manager* which undertakes the improvement of the quality of the state estimates (obtained by Kalman and other state estimators) and the task of initiating the measurement noise covariance matrix R determined from the experimentation, (iv) *the state fault*

detector which is responsible for detecting and isolating medium – term errors and failures, and (v) the *fault restoration manager* which determines and issues a sequence of corrective actions in order to bring the system back to normal operations. The faults may be recoverable or unrecoverable.
The execution level is responsible for the measurement of the system variables, it performs the necessary processing on them (filtering, estimation) and issues the proper low – level control actions in order to compensate small deviations of the state variables' values from the desired ones. The principal layers of the execution level are:

- State estimation layer
- Sensory system layer
- Control layer
- Fault recovery layer

The *state estimation layer* receives data from the sensory system, fuses them and estimates the relative position / orientation of the end – effector with respect to the part to be mantled or dismantled. The *sensory system layer* involves the sensors (encoders, contact force, proximity, range and vision sensors) which measure the various variables of the system that are involved in the gross motion control stage. In particular, the force sensor is used to provide the trajectory controller with the wrist – force information during the compliant motion stage. The control subsystem layer is responsible for all low – level control functions viz. position control, trajectory control and gripper control. Finally, the *fault recovery layer* is responsible for detecting, locating, isolating and repairing / restoring possible errors and faults taking place at the layers of the execution level. Several techniques are available for designing the fault recovery layer [27-29].

Although the final design goal is to fully automate the operation of the assembly / disassembly system, the human operator assistance / intervention will always be necessary at least for global perception and modeling of the robotic world and for fault detection / restoration purposes. Therefore an important component in any intelligent assembly / disassembly system is the human – system interface which has to be friendly, visible, transparent and complete as much as possible [31-32].

4 Prototype implementation examples

The general intelligent robot assembly/disassembly architecture investigated in Section 3 can be and has been implemented in several ways depending on the types of products assembled or disassembled on the particular technologies available for use. In this section we present three particular implementations designed and realized by the NTUA Intelligent Robotics and Automation Laboratory. The first two implementations concern particular levels of the hierarchy whereas the third concerns a full-scale prototype involving all levels (organization, coordination, and execution).

4.1 Fuzzy-Logic Based Implementation of the Assembly Organizer

The aim of the organizer's implementation to be described here was to deal with the uncertainty introduced by the user (operator) using fuzzy logic [19]. The operator enters the data (the product and the part description, and the assembly graph) and the organizer constructs the product model (the symbolic part) and determines the best assembly sequence, which is fed as input to the supervision layer. To satisfy the user-friendliness requirement and achieve a strong interactiveness, a Windows based approach for IBM compatible computers was adopted. The structure of this system (called OPASSEQ: *OP*timum *AS*sembly *SEQ*uence) is shown in Fig.2.

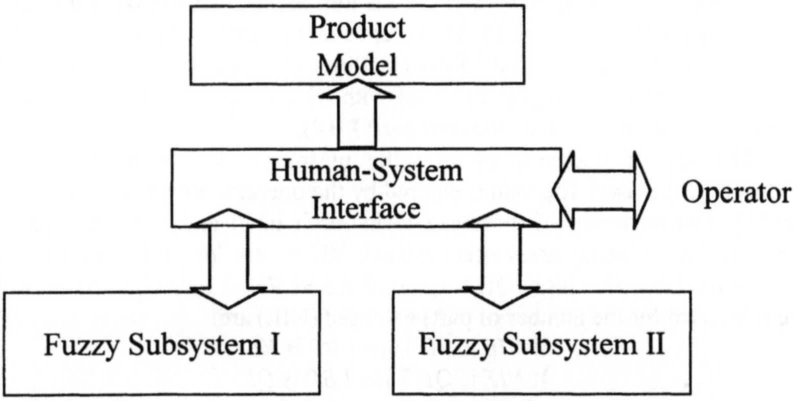

Fig. 2: Organization level: Expert system determining the best assembly sequence

The operator enters the assembly graph, a general description of the product and the parts that have to be assembled in order to produce this product. The system asks (interactively) for further characteristics of the parts and for some assembly features in order to find the proper base part and the sequences of the assembly process. The fuzzy subsystem I determines the appropriate base part and the fuzzy system II finds the best tenor given a starting mode of the assembly graph. Both fuzzy systems have the standard structure shown in Fig. 3 where the singleton method is used for fuzzification and center of gravity (COG) method for defuzzification. The fuzzy subsystem I has eight input, namely **NIE** (number of parts enclosed), **NIS** (number of parts surrounding it), **DCM** (distance of center of mass), **NPHS** (number of physical contacts), **M** (mass), **V** (volume), **NAD** (number of assembly directions).

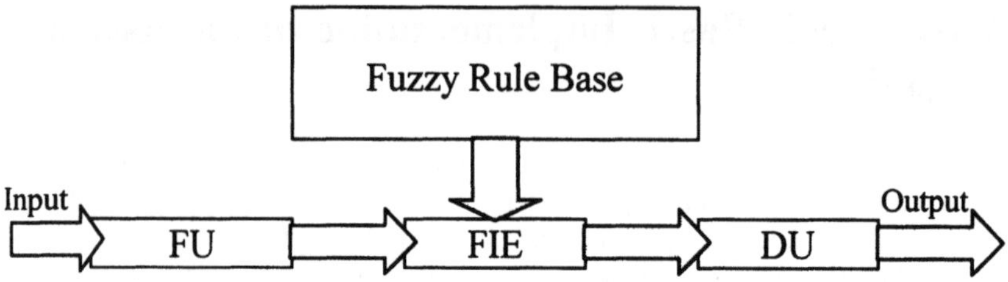

Fig. 3: Structure of fuzzy subsystems I and II (FI: Fuzzification Unit, FIE: Fuzzy Inference Engine, DU: Defuzzification Unit)

and **PS** (potentiality to stabilize), and one output which is the belief with which a certain assembly part should be selected as the base part. (i.e. the priority to be the base part: **PBS**). The fuzzy subsystem II has four inputs, namely: **DCAD** (degrees of change in assembly direction), **TCM** (transposition of center of mass), **HPA** (hinders other parts to be assembled) and **FPA** (facilities other parts to be assembled), and one output which is the degree of belief with which a certain assembly part is the next part (i.e. the priority to be the next part: **PBP**).

The use of fuzziness provide the means to deal with the uncertainty introduced by the user. The values entered by the operator are uniformly normalized in [0,12]. The universes of discourse U for each input and V for the output are divided in the following fuzzy sets (values): VL = very low, QL = quite low, L = low, M = medium, H = high, QH = quite high and VH = very high. As an example three rules used for the number of parts enclosed (NIE) are:

If *NIE* is *VL* Then *PBP* is *VL*
If *NIE* is QL Then *PBP* is *QL*
If *NIE* is *L* Then *PBP* is *L*

The set of rules that concern a certain input has a given weight of affecting the overall output, different from the weight of the rules of the other inputs. The OPASSEQ system was used to assemble successfully several components. Two representative component examples usually used for testing "best assembly sequence" systems are the *gear shaft* and the *drip-cleaner*. The results obtained by the present system in these examples are provided in [19].

4.2 Blackboard-based implementation of a 2-robot assembly system

The structure of Fig. 1 was followed in the implementation of a 2-robot assembly cell with principal component a blackboard system [18]. The cell consists of: 1) two single-armed robots: a stationary and a mobile robot, 2) a pallet-parts and subassemblies temporary storage space, 3) a worktable to which the stationary robot is docked, 4) a platform docked to the mobile robot, 5) an input stream where the parts are supposed to arrive, 6) an output stream where the assembled (ready) products are put, 7) a dustbin where the faulty parts and subassemblies are thrown, 8) a toolkit and 9) a set of sensors.

New parts arrive in any order and in any number (up to a certain limit), and the part arrival is detected in the input stream by an appropriate sensor (e.g. a weight sensor). The mobile robot moves from its safe position to a prespecified position where the input stream is within its reach. Using the sensors each new part arrived is recognized and grasped or not for assembly. The scanning of the input stream stops when either the platform is full or no more new parts to recognize exist. Any two objects (parts or subassemblies) in the pallet that fit are retrieved by the stationary robot and assembled together. After an assembly operation, new sensoring is performed to find out if the subassembly is faulty or if the ready product was reached. The new assembled object is then stored to the pallet, from where the mobile robot will carry it to the dustbin or to the output stream respectively. There exists one single and rare case of "stuck" which occurs when all storehouses are full, no assembly operation can be performed and new arrivals are detected.

The *blackboard* stores the various pieces of information including current state description and control data. The domain is splitted into four *knowledge sources* (**KS**), namely: arrival, access-pallet, assemble, user. The control mechanism (task planner) can also be regarded as a special purpose knowledge source, which is executed at predetermined moments during the process. All knowledge sources communicate and interact implicitly via the blackboard control structure as shown in Fig. 4. The task planner does not interact with the KSs but rather it schedules their execution. A suitable set of rules were selected and used in the task planner in order to support parallelism in execution and achieve efficiency and avoid deadlock situations.

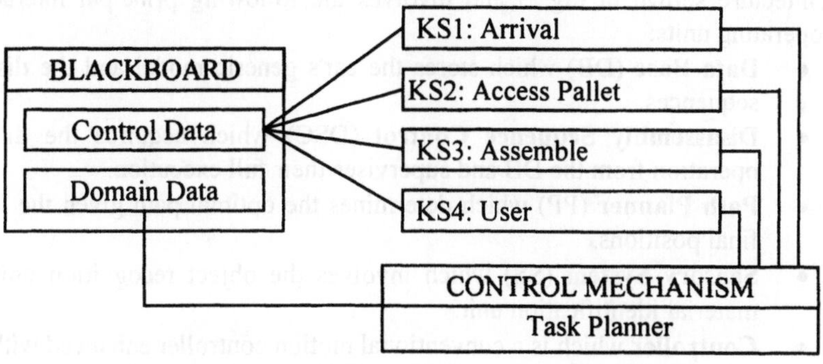

Fig.4: Blackboard implementation of the 2-robot assembly system

The system was developed in a SUN workstation under Unix 4.2BSD and was implemented in C-Prolog, which is very powerful knowledge-based systems development tool. It has the ability to handle simultaneously several types of products in several volumes of production (a part can be used by several product types and more than one time in the same product type). The state space of the system was represented by using the Petri net model which corresponds to a single direction graph and involves the concepts of state and transition. The advantage of this approach is the natural modeling of the assembly system where several subassemblies of several parts can be combined together in parallel.

The system was verified to possess integrity (correctness), consistency and robustness. It is robust in the sense that slightly different data lead to slightly different results, i.e. its performance does not deteriorate exponentially when the size of the data increase linearly. Also, the speed of the system in simulation form was proved to be satisfactory and can be further improved by using extended parallelism and reducing the communication overheads.

4.3 Implementation of an integrated car-disassembly system

Here, an integrated intelligent robotic car-disassembly system (**NEUROBOT**) designed and implemented with the philosophy described in section 3 will be outlined [33]. Car disassembly is performed at the cars' end-of-life for recycling purposes with many economic and social advantages. Here it is useful to mention the comprehensive *DELPHI* study [34] carried out by Delft University of Technology in order to chart the technical feasibility and the economic attractiveness of automated disassembly technologies that are currently under development for metals and plastics.

The Neurobot system was developed by an EU consortium [33] within the ESPRIT CIM stream of R&D projects. The overall system has the general architecture shown in Fig. 5 and involves the following principal interacting and cooperating units:

- **Data Base** (DB) which stores the car's generic model and the disassembly sequences.
- **Disassembly Sequence Control** (DSC) which acquires the disassembly operation from the DB and supervises their full execution.
- **Path Planner** (PP) which determines the optimal path given the initial and final positions.
- **Sensory System** (SS) which involves the object recognition unit and the material identification unit.
- **Controller** which is a conventional motion controller enhanced with a neural network module for inverse kinematic and dynamic modeling.

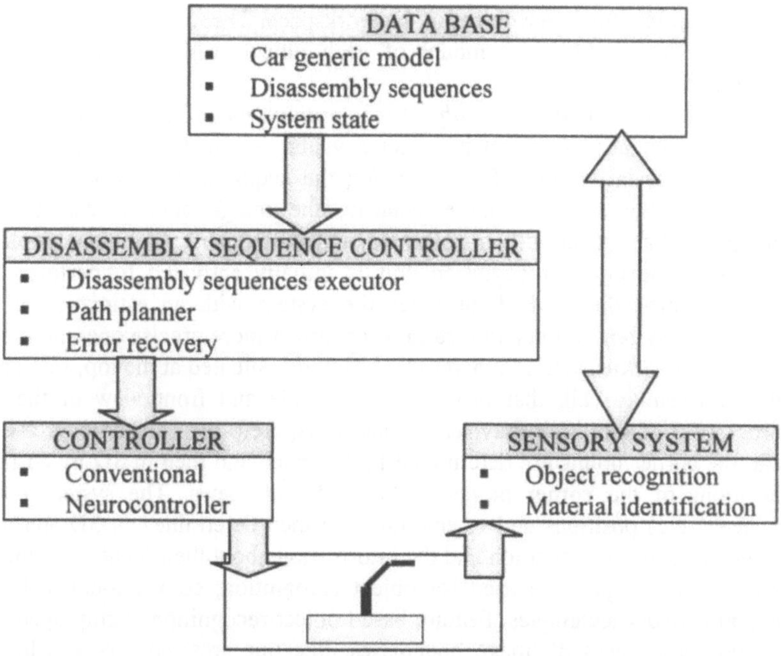

Fig. 5: Structure of intelligent car-disassembly system

The **DB** serves two purposes: a) to store the *generic car reference model* (GCRM) and the *disassembly sequences*, and b) to integrate the various subsystems at both the physical and logical level. The GCRM contains information about all car components of interest, about their rough shapes, sizes, materials and locations within certain limits. The disassembly sequences contain the exact steps the system has to follow in order to complete its task. For every car component to be disassembled there is a different sequence describing in detail the specific actions required.

The **DSC** is the heart of the system since it communicates with all subsystems and coordinates their operation using the disassembly sequences retrieved from the DB. Before each operation it receives a status signal from the responsible subsystem, and after the end of the operation it receives an acknowledgement. If an error occurs during the procedure, the subsystem sends an error message and the *Executor of Disassembly sequences* (EDS) activates the *Error Recovery Unit* (ERU). The DSG is also responsible for handling the system faults.

The **PP** determines the optimal route between the desired initial and final positions of the robotic manipulator. To this end, the n-dimensional configuration space is discretized to form a graph and the A^* algorithm is employed to find the path from the initial to the target node. Before implementation the workcell was simulated in a virtual environment using the *Telegrip* graphic simulation tool of DENEB Robotics, Inc. A 3D-sensor system (LCD camera) generates clouds of

250,000 points for every object in the workspace. Then, a triangulation technique is used to form a 3D-point model of each object, which is further filtered and smoothed.

The **SS** consists of the *object recognition unit* and the *material identification unit*. The object recognition unit uses a vision system for data acquisition and an image-processing module for processing the acquired images and identifying the objects in the scene. Complementarily, the image contour can be extracted, compressed and invariantly transformed, and subsequently fed into a Hopfield neural network. The object recognition unit is initially activated to perform the rough scanning procedure, which provides the system with an estimation of the car's coordinate system origin and locations for further more precise operations.

The *vision system* involves three cameras, situated at the top, side and front of the disassembly cell, that provide a plan, side and front view of the car. With processing of the three gray-scale images acquired, the car contours are extracted and the corner points are determined in 2D space and then in 3D by correlating the positions of the corner points in the different views. The system components' (wheels, etc) positions and orientations are then determined in 3D space from the estimate of the car's position and the information about their location in the DB. Two kinds of techniques are used for object recognition; conventional techniques and neural network techniques. Feature based object recognition is employed, where the features are pieces of image boundaries, like line arcs, corners and holes. In the neural based approach a Hopfield network is used, which serves as a non-linear or content addressable memory, the primary function of which is to retrieve a pattern (object) stored in memory, in response to the presentation of an incomplete or noisy version of that pattern. The image provided by the vision system is first *compressed*, and then represented by its *invariant features*. Then the Hopfield network based on the Zernike moments classifies the input image.

Knowledge of object materials is important for the recycling procedure and the *material identification* unit is performing this task. The approach followed is based on the fact that differences in properties characterize materials. A combination of sensors, each of which examines a different property, integrated in a robot-mounted head was the practical solution adopted. Four types of automotive materials can be identified, namely: a *ferrous metal* (e.g. steel), a *non-ferrous metal* (e.g. aluminum), *plastic* and *textile*. The sensors employed for the detection of properties of the above materials, which are mounted on a head as shown in Fig. 6, are:

- Standard inductive proximity sensor
- Inductive proximity sensor with selective features
- Capacity proximity sensor
- Ultrasonic sensor

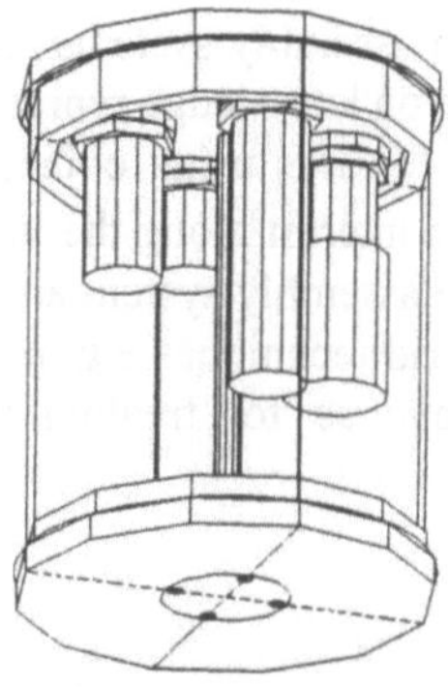

Fig. 6: Layout of multisensor head

Again two alternative approaches were used for material identification based on the readings of these sensors; *conventional approach* and *neural network approach* (using the multi-layer perceptron: MLP). The MLP is fed with the outputs of the three sensors (inductive, standard inductive, capacitive) acquired at 20 different distances (measured by the ultrasonic sensor). The network is trained by 40 measurements taken using the sensor head, which corresponds to 10 tests per material (steel, aluminum, plastic, and textile). For each material, five patterns are used for training and five for testing. In all cases, the classification accuracy for both training and testing patterns was 100%.

5 Conclusions

Automated robotic assembly has high potential for application since the labor costs of assembly operations are high. However, statistical studies showed that till now only a small percentage of industrial robots are employed for assembly/disassembly operations. Therefore a great deal of research is conducted during the recent years towards designing and implementing intelligent sensor-based robotic assembly systems, most recently including systems for car disassembly. The prospects are very good since the technology has by now developed accurate and robust components and methodologies. This paper is a step towards developing a unified framework for the design of robotic assembly/disassembly systems. After a brief survey of some research papers appeared since 1986, an effort is made to provide a general hierarchical architecture that covers all functions and tasks potentially needed for efficient and successful operation of any robot-based assembly/disassembly system including organization, coordination and execution issues. Then three particular case studies carried out by the authors' group are outlined to show the genericity of the proposed architecture and some of the specialized problems that are encountered in graphical designs. Much room is still available for work in the automated assembly/disassembly field in all areas (modeling, intelligence, sensory hardware and software).

Of central importance in assembly systems are the symbolic and knowledge-based techniques with emphasis on knowledge representation and control techniques. In latest years there is a trend towards using object-oriented methods. Fuzzy logic based methods are finding some interest along the line to handle uncertain situations due to both the operation of the assembly system and the robot environment. Here it is remarked that the use of hybrid techniques e.g. neuro-fuzzy, neuro-expert, neuro-fuzzy-expert is a promising avenue for treating more sophisticated assembly / disassembly problems.

References

1. Lee K, Gossard D C 1985 A hierarchical data structure for representing assemblies: Part I. *Computer Aided Design* 17(1):15-24.
2. Ambler A P, Barrow H G , Brown H G, Burstall C M, Crawford C F, Popplestone R J 1975 A versatile system for computer-controlled assembly. *Artificial Intelligence* 6(2):125-156.
3. Popplestone R J, Ambler A P, Bellos I M 1978 RAPT: A language for describing assemblies. *Industrial Robotics* 131-137.
4. Boothroyd G, Dewhurst P 1983 *Design for assembly: A designers handbook.* Mech. Engrg. Dept., Univ. of Massachusets, Amherst, MA.
5. Boothroyd G, Dewhurst P 1983 Computer aided design for assembly. *Assembly Engrg.* 26(2):18-22.
6. DeWinter D, Van Brussel H 1985 An expert system for flexible assembly system design. *Proc. 8th Ann. British Robot Association Conf.,* Birmingham, UK, pp 133-142.
7. DeFazio T L 1980 Displacement-state monitoring for the remote center compliance (RCC): rationalization and application. *Proc. 10th Intl. Symp. on Industrial Robots,* pp 559-570.
8. Cho H, Warnecke H, Gweon D 1987 Robotic assembly: a synthesizing overview. *Robotica* 5: 153-165.
9. Allen D K, Van Twelves W 1986 CAD in the CIM environment: where do we go from here? *Appl. Mech. Reviews* 39(9):1345-1349.
10. Stoll H W 1986 Design for manufacture: an overview. *Appl. Mech. Reviews* 39(9):1356-1364.
11. Heragu S S, Kusiak A 1987 Analysis of expert systems in manufacturing design. *IEEE Trans. Syst., Man, Cybern* 17(6):898-912.
12. Tzafestas S G 1990 AI techniques in computer-aided manufacturing. *In: Adeli H (ed), Knowledge Engineering, Vol II, McGraw-Hill, New York,* pp 161-212.
13. Billatos S B 1997 Application of axiomatic design in intelligent design for assembly environment. In: Tzafestas S G (ed.) *Knowledge Based Systems: Advanced Concepts, Techniques and Application,* World Scientific, Singapore, pp. 443-463.
14. Chang K-H, Wee W G 1988 A knowledge-based planning system for mechanical assembly using robots. *IEEE Expert* Spring: 18-30.

15. Almgren R 1989 On knowledge based planning and programming systems for flexible automatic assembly. *Thesis N. 176, Linkoping Studies in Science and Technology.*
16. Jain A, Donath M 1992 Knowledge representation systems for assembly using robots. *Tech. Report, Univ. of Minessota, Minneapolis, U.S.A.*
17. Irizary-Gaskins V M, Chang T-C 1989 Knowledge-based process planning for electronic assembly. *J Intell. And Robotic Syst.* 3:27-50.
18. Tzafestas S G, Tzafestas E S 1991 The blackboard architecture in knowledge-based robotic systems. *In: Jordanides T, Torby B (eds), Expert systems and Robotics*, Springer, Berlin, pp 285-317.
19. Tzafestas S G, Stamou G B 1997 Concerning automated assembly: knowledge-based issues and a fuzzy system for assembly under uncertainty. *Computer Integrated Manufacturing Systems* 10(3):183-192.
20. Binger G, Flemming V 1989 Graph oriented simulation of assembly process. *Syst. Anal. Model. Simul.* 6(10):781-786.
21. Aylett J T, Fisher R B, Fothergill A P 1988 Predictive computer vision for robotic assembly. *J. Intell. and Robotic Syst.* 1:185-201.
22. Shoureshi R, Momot M, Mitchell O R, Feddema J 1989 Vision-based intelligent control for automated assembly. *J. Intell. and Robotic Syst.* 2:153-170.
23. Roberto C, Enrico C 1997 Forward assembly planning based on stability. *J. Intell. and Robotic Syst.* 19:411-436.
24. Saridis G N 1983 Intelligent robotic control. *IEEE Trans. Autom. Control* 29:547-557.
25. Janabi-Sharifi F, Wilson WJ 1995 An intelligent assembly robotic system based on relative pose measurements. *J. Intell. and Robotic Syst.* 12:49-86.
26. Kelley R B 1991 Knowledge-based robot workstation: supervisor design. *In: C S George Lee (ed), Sensor-Based Robots: Algorithms and Architectures*, NATO ASI Series (F66), Springer, Berlin, pp 107-128.
27. Tzafestas S G, Watanabe K 1990 Modern approaches to system/sensor fault detection and diagnosis. *Journal A* 31(4): 42-57.
28. Watanabe K, Tzafestas S G 1990 Stochastic controls for systems with faulty sensors. *ASME J. Dyn. Syst. Meas. and Control* 112(1):143-147.
29. Tzafestas C S, Rigatos G G, Tzafestas S G 1998 Design of fault-tolerant control systems: passive and active approaches. *Proc. Systems Science XIII Conf.* Wroclaw, Poland, Sept.
30. Hörmann A, Meier W, Schloen J 1991 A control architecture for an advanced fault tolerant robot system. *J. Robotics and Autonomous Systems* 7(2-3):211-225.
31. Laengle T, Lueth T C, Herzog G, Stopp E, Kamstrup G 1995 KANTRA: A natural language interface for intelligent robots. *In: Rembold V, Dillmann R, Hertzberger L O, Kanade T (eds), Intelligent Autonomous Systems*, IOS Press, Amsterdam, pp 357-364.
32. Fukuda T, Sekiyama K, Ueyama T, Arai F 1993 Efficient communication method in the cellular robotic system. *Proc. IROS IEEE/RSJ Intl. Conf. on Intelligent Robots and Systems, Yokohama, Japan,* pp 1091-1096.

33. Tzafestas S G, Anthopoulos Y, Katevas N, Spyropoulou E 1997 Architecture and implementation of an autonomous car-disassembly system. *Syst. Anal. Modell. Simulation* 29:129-149.
34. Boks C B, Tempelman E 1997 Delphi study on future disassembly and recycling technology for electronic and automotive industry. *Final Report*, Delft Univ. of Technology, The Netherlands.

26
Integrated Design for Flexible Assembly System : Robotic Arc Welding Stations for Containers

A. Benmounah and M. Merzak

1 Introduction

There are two powerful forces acting upon the future development of manufacturing technology : the need for greater flexibility and the need for reduction of work-in-process inventories. Many of the limiting factors in improving flexibility and inventory have to do with time, cost and difficulty of set-ups [1]. The aim of flexibility with reference to the product range makes it essential to turn away from the organisational modes conventionally in production, such as workshop and assembly line production [2]. These are constantly being replaced by flexible assembly line or some specific cases by flexible assembly station as a new mode of rational organisation.
In the case of workshop production the parts are transported from one workshop to another in batches. In the case of assembly line production the parts are often rigidly tied into the conveyor system. The flow of material is therefore relatively simple, makes much heavier demands on the material flow system. Flexible systems that react quickly are required for flow of materials, they must reduce the time in which machines are out of action to a minimum without the necessity for large buffer stores that occupy valuable space that could be used for production. Fulfilment of theses demands necessitates the highest degree of automation in conveyor, storage, and transfer systems, computer-assisted distribution with on-line link to the production control systems and high availability of all components of the material flow systems [3, 4]. The assembly manner proposed in this chapter is a combination of flexible assembly-line productions, each for most of the different parts of the container, and of flexible assembly stations that permits most of the container parts to be transferred and prepared , mounted to form the container shape, and welded from inside and outside by welding robots.

2 Robotic Assembly Process for Container

The related proposed idea for studying container production ways that permits technological and industrial application integration, such as robot use, instigates capital interest. This in fact deals with an attractive process of fabrication and/or an important economical activity such as auto industries. Containers construction must obey preliminary requirements of a known and confirmed homologation by one of the international untitled control organisms such as : Loyds, Bureau verites and others. Generally, a 20 foot container type ICC is shaped like a cube (rectangular parallelepiped) with the following dimensions : Length L= 6058mm, Width W = 2438 mm, Height H = 2591 mm. This type of container is made up of: 01 wooden floor, 02 lateral walls, 01 end wall, 01 roof, 02 folding doors. In this case, the order of the different parts of container assembly and robotic, arc welding in three flexible stations, are discussed.

3 Flexible Station No. 1

3.1 Operation Mode Description

Using the details of Figure 1, which represents the container different parts, the chassis guide mark 1 is represented and fixed on an automotor cart (on rail) using a system fixation as seen on Figure. 2 [5]. One after the other and using semi-portico, the lateral walls 3 are walled up against a dispositive put previously inside. The structure of this dispositive is light steel made and must be positioned considering chassis axes. It is equipped with permanent magnets capable of maintaining the lateral walls vertical and perpendicularly to the chassis respecting a precise positioning of the different parts without any human intervention or material means. These magnets exist under different commercial forms depending on the application case. Bearing in mind the assembly plan, the same operation is followed concerning the fixation of the end wall 2 by walling it up against the mentioned dispositive. After raising up and positioning the folding doors frame 5, again they are walled up against the magnet. Finally, the wooden floor 6 is put and fixed on the chassis, and the operation consists of conveying the roof 4 from the store to the station, lifting it up and putting it over the container structure by adjusting it using the assembly plan. The different pre-assembly phases cited above are all not more than loading operations. Consequently, and because of the heavy-weight of the container parts (200 to 300 kg), these simple operations can not be fulfilled by robots. At this level, and before the transfer of the pre-assembled container it is recommended to proceed for the container diagonals control norms (Table. 1 and Figure. 3) using telescopic rulers that have on each side a fixation dispositive for the container corner pieces (top and bottom).

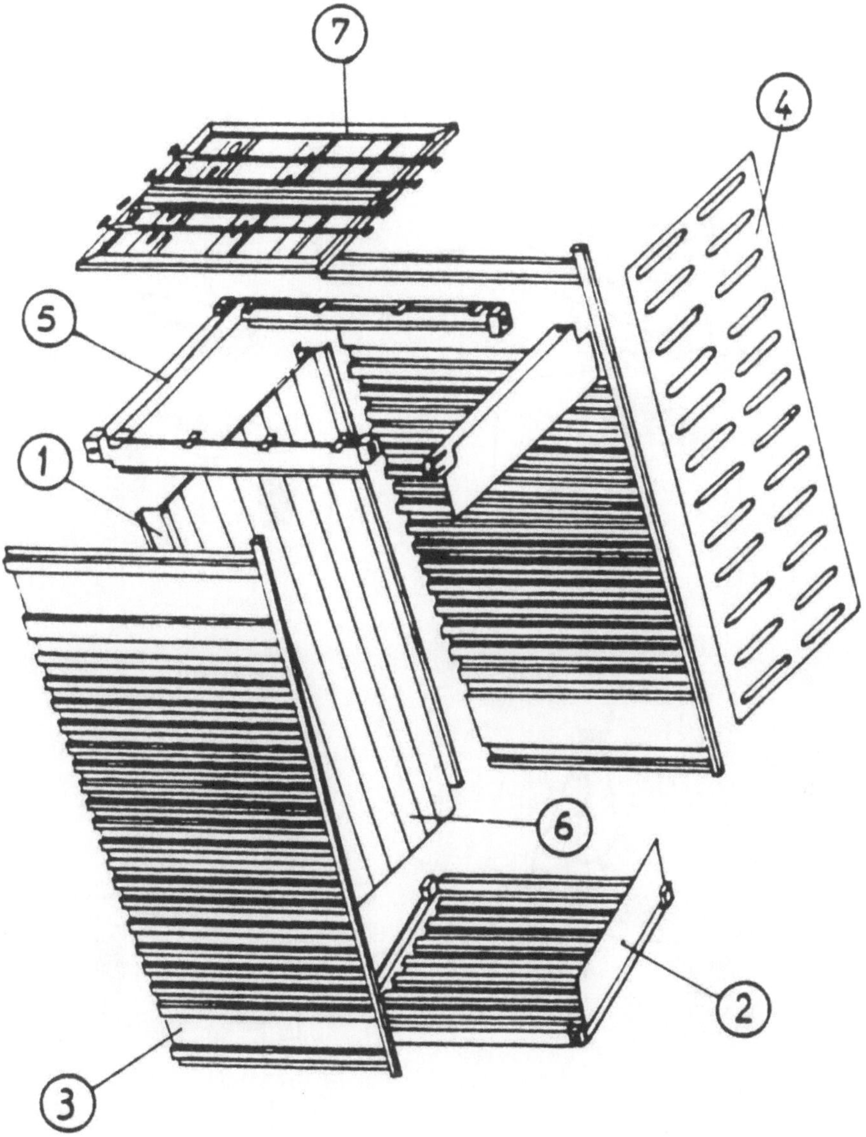

Fig. 1 General view of the container

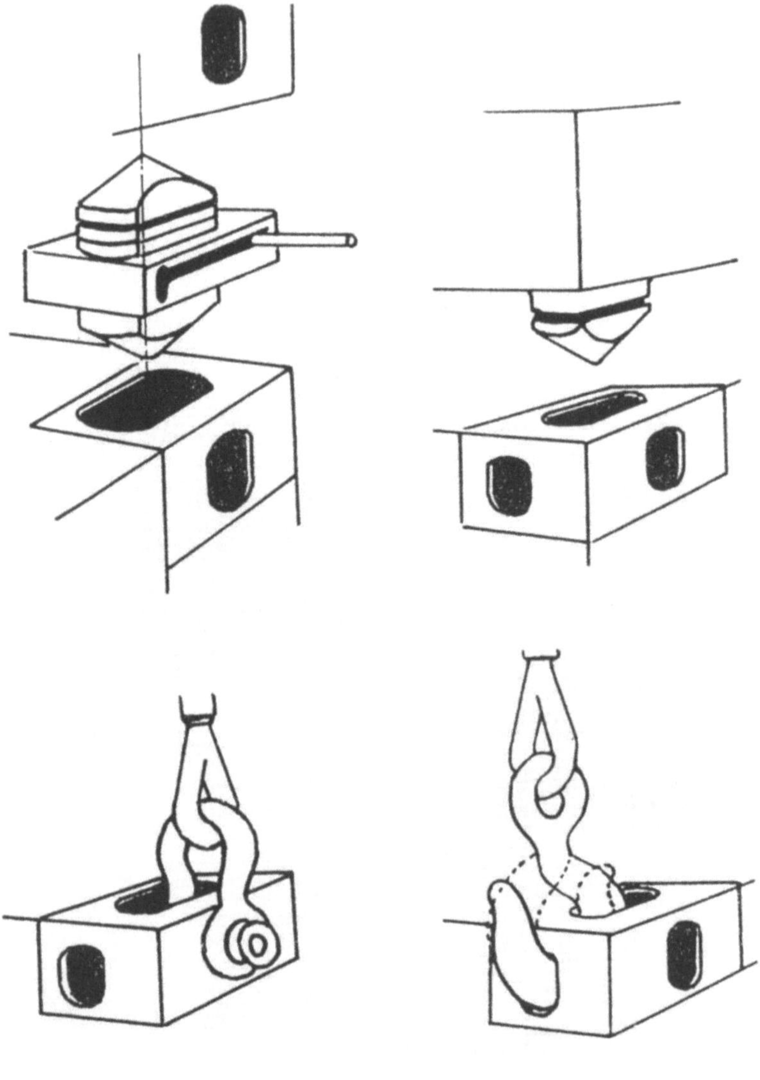

Fig. 2 Fixation dispositive for container pieces

Type of Container	Height (mm)		Width (mm)		Length (mm)		Mass (Ton)
	Outside	Outside	Outside	Outside	Outside	Outside	
1 AA	2591	2350	2438	2330	12192	11998	30,480
1 A	2438	2197	2438	2330	12198	11998	30480
1 BB	2591	2350	2438	2330	9125	8931	25,400
1 B	2438	2197	2438	2330	9125	8931	25,400
1 CC	2591	2350	2438	2330	6058	5867	20,320
1 C	2438	2197	2438	2330	6058	5867	20320
1 D	2438	2197	2438	2330	2991	2802	10,160
1 E	2438	2197	2438	2330	1968	1780	7,110
1 F	2438	2197	2438	2330	1460	1273	5,080

Table. 1 Container different dimensions

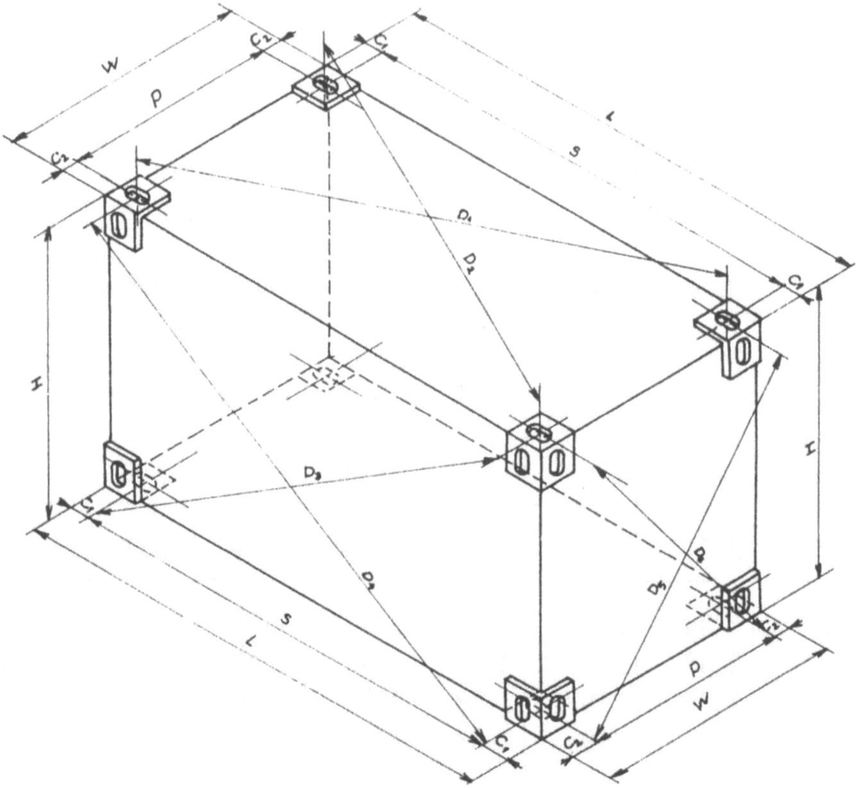

S- Length between the center holes of corner pieces
P- Width between the center holes of corner pieces
C1- 101.5mm
C2- 89.0mm
L- Length of the container
W- Width of the container
D- Distance between the center hole of the corner pieces diagonally in front gives six measures D1, D2, D3, D4, D5, D6
K1- Difference between D1 and D2 or D3 and D4
K2- Difference between D5 and D6
H- Height of the container

Fig.3 Container corner pieces-diagonales adjustment

4 Flexible Station No. 2

4.1 Operation Mode Description

This station must be equipped with a set of automatic welding that uses a Cartesian robot under portico of the type RP 23 (AFMA-ROBOTS) Figure. 4 . Using the details of Figure 4 [6], the field of application here is the case of external continuous rectilinear welding. The pre-Assembled container that is initially presented on the automotor cart is directly transferred from the flexible station No. 1 under portico for welding, Figure 5. The RP 23 robot model is equipped with a welding torch.

4.2 RP 23 Cartesian Portico Robot Characteristics

From Figure 4 (technical sheets of RP 23), rotations of the robot end (vertical or horizontal position) can be determined depending upon the case of geometrical application of the joints to be welded. The portico must be dimensioned compared to the size of the product considered (in our case containers 20 and 40 foot).

5 Flexible Station No. 3

5.1 Operation Mode Description

This station must be equipped with a set of automatic welding that uses a robot of the type FR 10 DAROS, Figure 6 [6]. Using the details of Figure 6, the field of application is the case of internal discontinuous welding. The transfer of the preliminary welded container from flexible station No.2 to flexible station No.3 using the automotor cart takes place, and before starting welding, the dismantling of the used permanent magnet assembled to the dispositive of gap adjustment for lateral and end walls must be done. First, one takes out the dispositive by means of mono-rail light made for this purpose. At this stage, teach the robot to repeat all the prescribed movements in order to execute discontinuous welding inside the container, over the inferior (chassis), superior (roof) walls, and the inside perimeter according to the plan. Extraction of the container from flexible station No. 3 must be done without any potential deformation. In other words, geometrical and dimensional control must be done for each container.

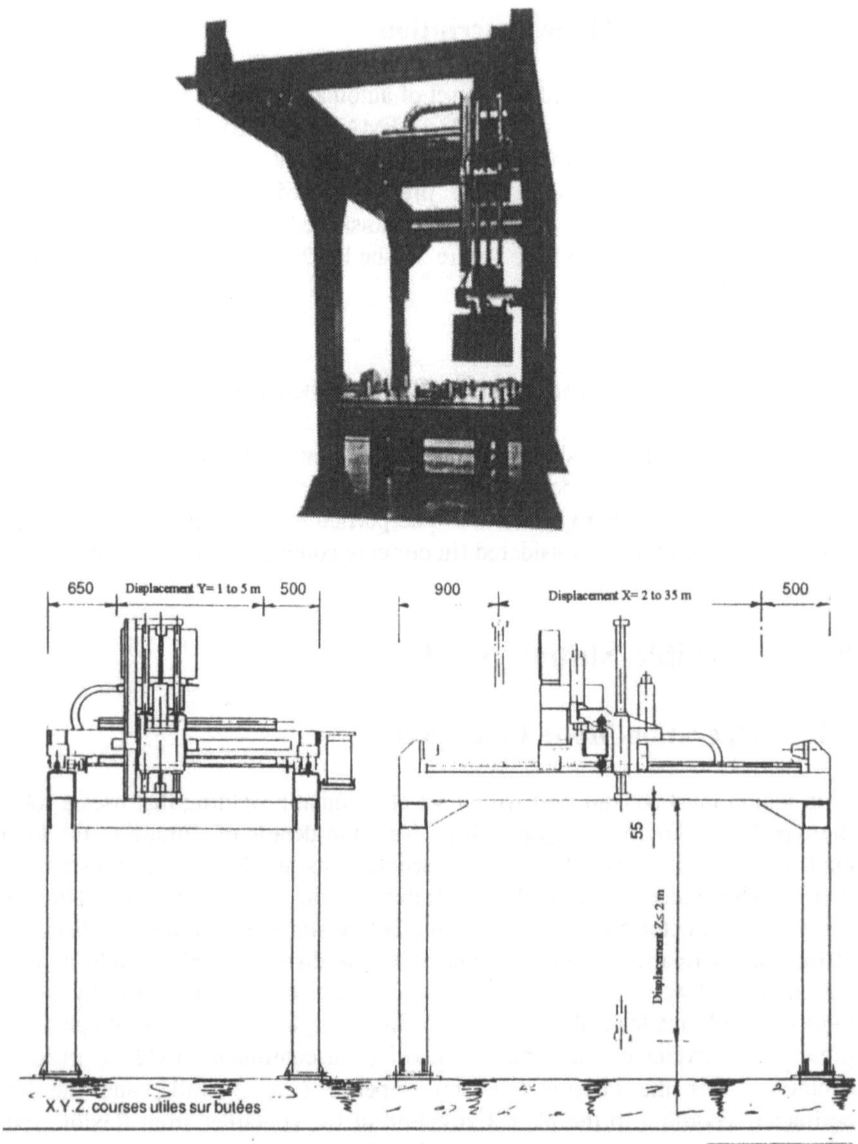

Fig. 4 General view of RP 23 Robot

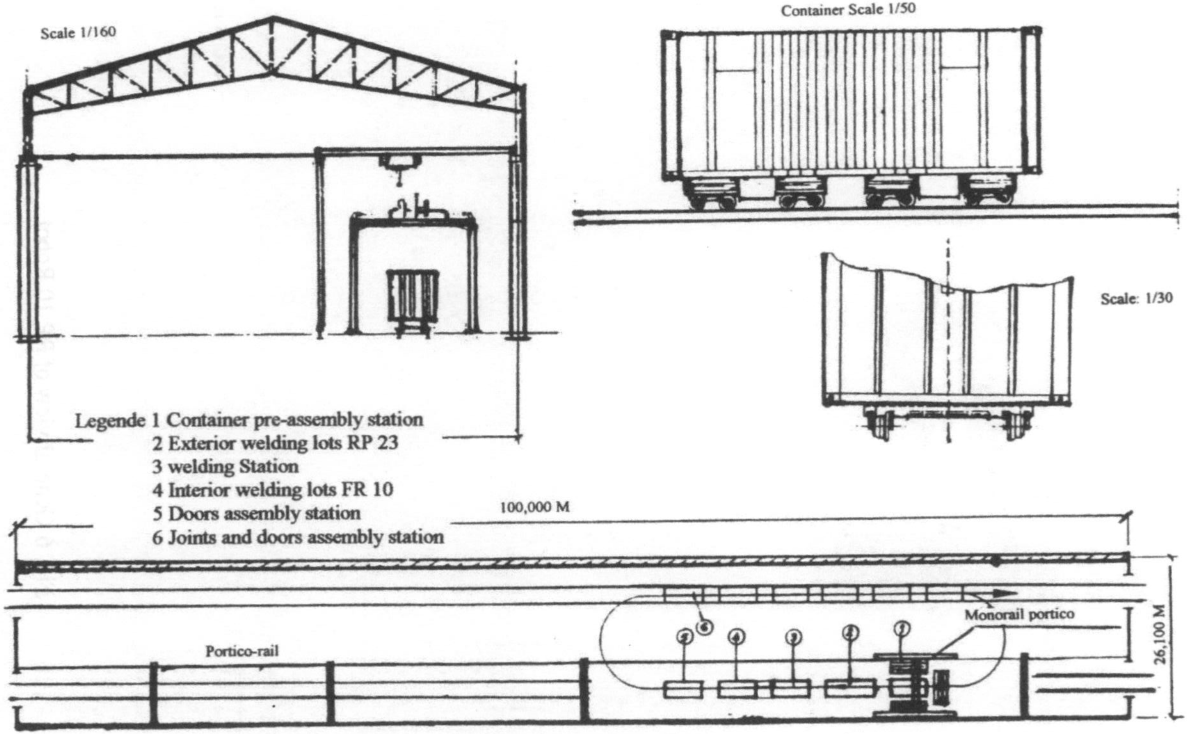

Fig. 5 General view of the container plant

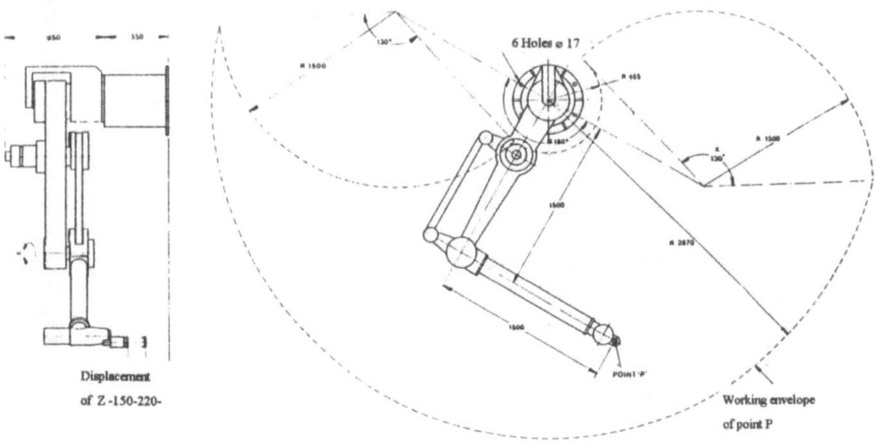

Fig. 6 General view of RP 10 Robot

6 Control System Technique Development

The control system of a flexible assembly plant must obey precise demands. Large use of a system determines whether a plant can reach the specified performance. Rapid progress in data processing techniques makes it possible to control plants using a very efficient process computer. This approach involves direct control of whole plant by a central process computer. An example of that is, when testing out subsystems, all stations must report to the central computer is attended by manifold problems such as :-

- Once the computer type has selected, it may have to be replaced if its storage capacity, reaction time, etc... prove inadequate.
- The same may happen if the plant is enlarged without allowance be made for expansion from the outset.
- When testing out subsystems, all stations must report to the central computer.
- Because of difference computers must be used depending on plant size.
- The same holds true efficient test-and-verify software.
- Some of the demands are too strict to be satisfied by the computer.

7 Conclusion

Development in sensing techniques has enabled robots to be more tolerant of poor assembly design and component variation, however, improvements, such as these are still directed at the symptom rather then the cause of the problem. Integrated design for assembly philosophies enabled us to develop products and processes more quickly and with greater confidence. Products can brought to market more rapidly, with improved performance and more customer features, yet which are cheaper and easier to manufacture. Effective design for manufacture can also in simpler and cheaper process requirements, with reduced potential for incorrect assembly and break down. These significant business advantages can, however, only be realised if senior manufacturing managers clearly understand the need of integrated design for assembly and have sufficient commitment to bring about the necessary cultural change within their organisation.

References

1. Sloberg J J, McGillem C D 1987 An Intelligent FACTORY Transport System. International Trends In Manufacturing Technology IFS LTD UK

2. Schneider F 1987 Automated Material Handling in FMS and FAS. International Trends in Manufacturing Technology IFS LTD UK

3. Hinrichs J F, Lefebvre R R 1990 Keys to the Planning and Justification of successful Robotic Arc Welding. 21 st International Symposium on Industrial Robot (ISIR), Copenhagen

4 Conde J F, Sebatian 1990 Replication Processes Between Flexible Automation and Materials Handling. 21 st International Symposium on Industrial Robot (ISIR), Copenhagen

5 NF Data Sheet 1977 Container Pieces. Edited by Afnor, France

6 Data Sheet Industrial Robots. Afma Robots Chambray-les Tours, France

27
Stereo Image Processing and Virtual Reality in an Intelligent Robot Control System

F. Tél and E. Tóth

1. Introduction

More and more robot control system requires intelligent extensions such as vision or virtual reality (VR) system in order to be capable the easy and flexible interaction with the environment or the humans. This chapter describes the stereo vision and the VR part of the intelligent control system of the Puma 560 robot and the dextrous hand developed at the Technical University of Budapest [1].

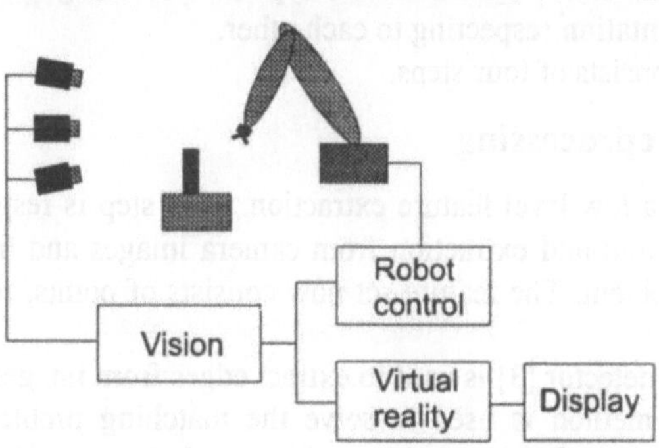

Figure 1.1: Main blocks of the intelligent robot control system

The three main blocks of the control architecture are depicted in Figure 1.1. The control system works directly on the robot. This system is responsible for handling of control signals, achieves the path planning and develops object grasping. In addition to the traditional information (joint variables) the control system uses the result of the vision system such as the type, position and orientation of the recognized objects. The VR means the interface to the human controls, it can be used to display the planned actions of the robot before the real action, determine whether there will be any collision with the environment or the another robot during the execution of the next task.

2. Stereo system

The aim of the vision part of the control system is to recognize objects from a known class in the scene and determine the relative Euclidean transformation between these objects. The transformation calculation requires that the coordinates of some features of the objects to be known in 3D coordinate system fixed to the object in the object model. The system uses uncalibrated stereo rig, hence the traditional triangulation method cannot be used. As it was shown by Faugeras [2], in this case only the projective structure of the surround world can be reconstructed, but this does not contain any metrical information. In order to obtain Euclidean solution some additional information is required, such as 3D coordinates of some points or distances between them. Different types of these constraints are introduced by Boufama *et al.* [6]. The method described herein also supposes those 3D Euclidean coordinates for some points are known, but these coordinates are object relatives, no world coordinate system is required.

Intelligent control systems usually require not only the positions of some entities in the space but also the types of them (grasp planning). Hence in the presented system an object recognition method was also developed. As will shown later, this object recognition is also used to supply the mentioned additional constraints. The recognition method is based on the predefined model database and uses probabilistic methods in order to handle the effect of the uncertainty due to the occlusions and inaccuracy in the feature detection.

The output of the stereo system is the type of the recognized objects and the position and orientation respecting to each other.

The algorithm consists of four steps.

2.1 Image preprocessing

The first step is a low level feature extraction. This step is responsible to achieve the feature detection and extraction from camera images and find the solution of the matching problem. The feature set now consists of points, lines (sections) and junctions.

The Canny edge detector [3] is used to extract edges from images.

The correlation method is used to solve the matching problem. To reduce the computational costs of the matching, the correlation is achieved only in the neighbourhood of the corner points. Every corner point is the intersection of lines extracted by Hough transformation.

To eliminate false corner points, there must be a real edge point in the neighbourhood of the calculated intersection. A junction is determined as the intersection of three or more lines and lies on real edge.

This is a heuristic method and works properly only for very simple cases. A new, more advanced low level image processing method is under development.

2.2 Reconstruction of the projective structure

The projective structure is built from point correspondences. Assuming pinhole camera model, the imaging equations can be written in linear form [4],

$$\lambda_{ij}\mathbf{m}_{ij}=\mathbf{P}_j\mathbf{M}_i \qquad (2.1)$$

where $\mathbf{m}_{ij}=[u,v,w]^T$ is the projective coordinates of the image point, $\mathbf{P}_j=[p_{ij}]$ is the unknown projection matrix with size 3×4 for the *j*th camera, $\mathbf{M}_i=[x,y,z,t]^T$ contains the unknown projective coordinates of the 3D point. Eliminating the scaling factors (λ_{ij}) and introducing a constraint in order to improve the stability of the convergence yields a nonlinear system of equations

$$\begin{aligned} X^2 &= \sum_i^n \frac{1}{\sigma_i^2}\left(x_i^2+y_i^2+z_i^2+t_i^2-1\right)^2 \\ &+ \sum_{i,j}^{m\times n}\frac{1}{\sigma_k^2}\left(u_{ij}-\frac{p_{11}^{(j)}x_i+p_{12}^{(j)}y_i+p_{13}^{(j)}z_i+p_{14}^{(j)}t_i}{p_{31}^{(j)}x_i+p_{32}^{(j)}y_i+p_{33}^{(j)}z_i+p_{34}^{(j)}t_i}\right)^2 \qquad (2.2)\\ &+ \sum_{i,j}^{m\times n}\frac{1}{\sigma_k^2}\left(v_{ij}-\frac{p_{21}^{(j)}x_i+p_{22}^{(j)}y_i+p_{23}^{(j)}z_i+p_{24}^{(j)}t_i}{p_{31}^{(j)}x_i+p_{32}^{(j)}y_i+p_{33}^{(j)}z_i+p_{34}^{(j)}t_i}\right)^2 \end{aligned}$$

in the variables $\mathbf{P}_j$ and $\mathbf{M}_i$, where *n* is the number of the 3D points and *m* is the number of views and σ_k is the standard deviation (weight) of the measured point. The method developed by Mohr *et al.* [5,6] is used to rebuild the 3D projective structure directly minimizing the equation (2.2) by Levenberg-Marquardt method [7]. This is similar to the nonlinear camera calibration process described in [4], but for this case the 3D projective coordinates ($\mathbf{M}_i$) are also unknown.

2.3 Object recognition

A 2D object recognition system is used to localize the known objects in the scene and involve the required Euclidean constraints. In order to represent the occlusions of the features and to handle the uncertainty during the recognition process, probabilistic description was choosen as done by Wells [8] and Pope [9]. The method can recognize only those objects that are predefined in a model database. The building of the database is supposed to be generated off-line. An object model in the database contains:

- Limited number of 2D views about the object. A view contains features and relations between them resulted by feature detection method. (A view here means the results of the low level processing, not the raw camera images.).
- The 3D coordinates of the features (mostly for points) in the local, object based coordinate system. These are used during the calculation of the Euclidean transformation.

For example, the database contains a 3D feature (corner point) of an object as a 2D feature (intersection two or more edges in the views, depending on the viewpoint) and the 3D Euclidean coordinates of this point in the reference frame attached to this object. The object based 3D coordinate frame can be chosen arbitrarily, but must be predefined in the model database.

The recognition process tries to find the most probable configuration representation of objects with pairing the features of the scene and the model database. The multiview case complicates the probabilistic analysis, because not only the model database but also the scene consists of more than one view about the object.

The recognition requires the description of the quality of the matching. This can be done by defining a cost function. Using the Bayesian estimation theory the quality measure can be associated with the probability $P(A|Q,T)$, where A denotes the hypothesis that the object of the model database is present in the scene, Q contains the pairings between views and T denotes viewpoint transformations. Handling these multidimensional joint probabilities is complicated, hence some simplification condition was used and the detection and the presence of the features are handled as independent events (feature independence simplification) [8,9]. This yields that the probabilities can be approximated as the product of lower dimensional distributions.

According to Pope [9], the features are represented by type t_{ij}, attribute vector (representation of independent properties of the given feature) α_{ij}, mean of the position μ_{ij}, covariance of the position Σ_{ij}. Using these, the described probability can be approximated. For example a corner point (type) can be described by number of intersecting edges in that point (attribute), the x and y coordinate of the point in the image (position) and the uncertainty of the position (can be experimentally defined, e.g. be ±n pixel if no subpixel accuracy is used or included in the response of the feature detector).

As can be seen the described method requires the calculation of the transformation between views in the scene and/or the model. This is supposed to be affine. The transformation can be described by 6 parameters $\mathbf{t}=[a_t,b_t,c_t,d_t,x_t,y_t]^T$. The pose of the features can be characterized by $\mu_k=[t_x,t_y,\beta,\gamma]^T=[t_x,t_y,\theta,s]^T$, where t_x, t_y is the position $\theta=atan(\gamma/\beta)$ is the orientation, $s=\sqrt{\beta^2+\gamma^2}$ is the scaling. Introducing β and γ yields that the effect of the transformation can be written in linear form:

$$\text{a)}\quad \mu_k' = \begin{bmatrix} t_x & t_y & 0 & 0 & 1 & 0 \\ 0 & 0 & t_x & t_y & 0 & 1 \\ \beta & \gamma & 0 & 0 & 0 & 0 \\ 0 & 0 & \beta & \gamma & 0 & 0 \end{bmatrix} \begin{bmatrix} a_t \\ b_t \\ c_t \\ d_t \\ x_t \\ y_t \end{bmatrix} \qquad \text{b)}\quad \mu_k' = \begin{bmatrix} a_t & b_t & 0 & 0 & x_t & 0 \\ c_t & d_t & 0 & 0 & 0 & y_t \\ 0 & 0 & a_t & b_t & 0 & 0 \\ 0 & 0 & c_t & d_t & 0 & 0 \end{bmatrix} \begin{bmatrix} t_x \\ t_y \\ \beta \\ \gamma \\ 1 \\ 1 \end{bmatrix} \quad (2.3)$$

To calculate the transformation and handle possible pairings, tree-search method is applied. A new feature pair is added to the sufficient node of the tree, if the features have the same type, the distance between the feature's parameters is bellow a threshold, the transformation error is small and the insertion in the tree is consistent with the actual content of the tree.

The transformation is determined by the previous level (parent) of the given node in the tree. In the first step the insertion is based only on the feature types and attributes. After every insertion phase the transformation is updated using Kalman filtering technique [10]. In this case the filter equations are much simpler because the state vector is constant respecting the time. The original equations:

$$\mathbf{x}(k+1)=\mathbf{A}(k)\mathbf{x}(k)+\mathbf{B}(k)\mathbf{u}(k)+\mathbf{w}(k) \quad P(\mathbf{w})\sim N(0,\mathbf{Q})$$
$$\mathbf{z}(k)=\mathbf{H}(k)\mathbf{x}(k)+\mathbf{v}(k) \qquad P(\mathbf{v})\sim N(0,\mathbf{R}) \tag{2.4}$$

For this case:

$\mathbf{A}(k)=\mathbf{I}$, $\mathbf{B}(k)=0$

$\mathbf{z}(k)$: position vector of the model feature (μ_k').

$\mathbf{x}(k)$: transformation to be calculated.

$\mathbf{H}(k)$: the observation matrix that can determined from the actual scene feature (μ_k) as in the equation (2.3a).

The last step of the recognition is to search for the objects that give the best representation (have the greatest probability).

If the object is recognized, the local 3D Euclidean coordinates from the model database can be attached to the features (e.g. corner points). For the recognized objects, the same features in the database and in the image can be localized, hence the 3D coordinates of the image points in the scene view can be determined respecting to the object own 3D Euclidean coordinate system.

2.4 Calculation of the Euclidean transformation

Reaching this point of the processing the coordinates of the recognized objects are expressed in two coordinate systems. The first one is the projective system that is common for all the objects. The second one is the local Euclidean frame for each object. The relative Euclidean transformation between the objects in the scene can be computed using this two types of the coordinate representation as illustrated in Figure 2.1. This calculation can be achieved in two phase.

The first phase of the displacement computation is the calculation of the transformations (collineations, ${}^{(X)}\mathbf{W}$) between the object frames and the common projective system. Using the two types of the coordinate representation of an object, the transformation can be written in the following matrix form,

$$\mathbf{W}\mathbf{M}_{P,i}-\lambda_i\mathbf{M}_{E,i}=0 \tag{2.5}$$

where $\mathbf{M}_{P,i}$ and $\mathbf{M}_{E,i}$ represent the projective and Euclidean coordinates respectively, the λ_i is the scaling factor for the ith point. Collect all of these equations into one system of equations $\mathbf{Ax}=\mathbf{b}$ (n is the number of the corresponding points)

$$\mathbf{A}=\begin{bmatrix}
\mathbf{M}_{P,1}^T & 0^T & & -M_{E,1,1} & 0 & & & \\
0^T & \mathbf{M}_{P,1}^T & & -M_{E,1,2} & 0 & & & \\
0^T & 0^T & \cdots & -M_{E,1,3} & 0 & & \ddots & \\
0^T & 0^T & & -M_{E,1,4} & 0 & & & \\
\mathbf{M}_{P,2}^T & 0^T & & 0 & -M_{E,2,1} & 0 & & \\
 & \mathbf{M}_{P,2}^T & & & -M_{E,2,2} & 0 & & \\
 & & \ddots & & & & \ddots & -M_{E,n,4} \\
0 & & & & & & 0 & 1
\end{bmatrix}$$

$$\mathbf{x} = \begin{bmatrix} w_{11}\; w_{12}\; w_{13}\; w_{14} \cdots w_{44}\; \lambda_1 \cdots \lambda_n \end{bmatrix}^T \tag{2.6}$$

$$\mathbf{b} = \begin{bmatrix} 0 \cdots 0\; 1 \end{bmatrix}^T$$

where size of **A** is (4n+1)×(n+16) and the last equation states that $\lambda_n=1$ (this is not a restriction, while all equations can be divided by $\lambda_n \neq 0$). The size of unknown vector **x** is (n+16) and size of vector **b** is (4n+1). The linear estimation of the solution can be determined in least squares (LS) sense as

$$\mathbf{x}_{n+16} = \left(\mathbf{A}^T\mathbf{A}\right)^{-1}_{n+16,n+16} \mathbf{A}^T_{n+16,4n+1}\mathbf{b}_{4n+1} \tag{2.7}$$

The second phase is the calculation of the displacement. Suppose we want to determine the transformation between objects **A** and **B**. Using the graph in the Figure 2.1 the Euclidean coordinates of the point M of object **B** in the frame of object **A** can be expressed in two ways. The first one is the direct application of the collineation ${}^{(A)}\mathbf{W}$, the second one is to transform the coordinates into the frame of object **B** using ${}^{(B)}\mathbf{W}$ then apply ${}^{(AB)}\mathbf{D}$. In equation form

$${}^{(A)}\mathbf{W}{}^{(B)}\mathbf{M}_{P,i} = {}^{(AB)}\mathbf{D}{}^{(B)}\mathbf{W}{}^{(B)}\mathbf{M}_{P,i} \tag{2.8}$$

where ${}^{(B)}\mathbf{M}_{P,i}$ are the projective coordinate of the ith point of object B. This equation can be rewritten into the form ${}^{(A)}\mathbf{V}_i = {}^{(AB)}\mathbf{D}{}^{(B)}\mathbf{V}_i$. Rescaling all of the $\mathbf{V}_i$ in order to represent Euclidean coordinates the displacement can be calculated in closed form using quaternions [11].

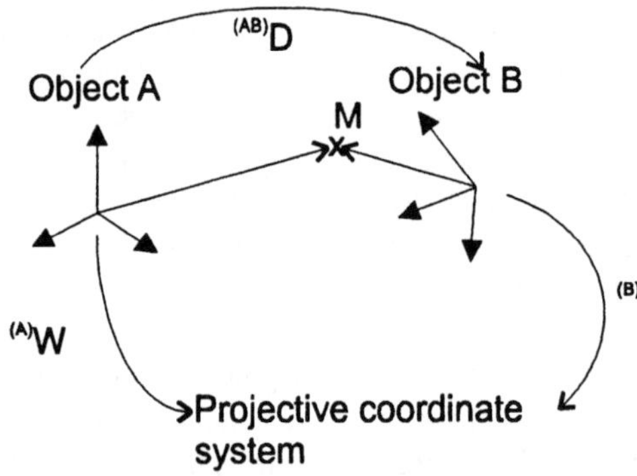

Figure 2.1: Calculation of the Euclidean transformation

In order to put all the results together, a refinement step is also developed. The relation can be written into similar form as in the equation (2.8). Expanding the resulted equation yields

$$\lambda \sum_{k=1}^{4} {}^{(A)}\mathbf{W}_{j,k}\mathbf{M}_{P,i,k} - \sum_{k=1}^{4}\sum_{l=1}^{4} {}^{(AB)}\mathbf{D}{}^{(B)}\mathbf{W}_{k,l}\mathbf{M}_{P,i,l} \quad j=1...4, \quad i=1...n \tag{2.9}$$

The unknowns are the elements of the ${}^{(AB)}\mathbf{D}$, ${}^{(A)}\mathbf{W}$, ${}^{(B)}\mathbf{W}$ and λ. Constraints must be introduced for ${}^{(AB)}\mathbf{D}$ to hold the desired form. Using the properties of the rotation matrix in the displacement, the constraints are the follows:

$$\sum_{k=1}^{3} {}^{(AB)}\mathbf{D}_{j,k}\, {}^{(AB)}\mathbf{D}_{l,k} = 0 \qquad j,l = 1...3 \text{ (orthogonality of the rows)} \tag{2.10}$$

$$\sum_{k=1}^{3} {}^{(AB)}\mathbf{D}_{k,j}\, {}^{(AB)}\mathbf{D}_{k,l} = 0 \qquad j,l = 1...3 \text{ (orthogonality of the columns)} \tag{2.11}$$

$$\sum_{k=1}^{3} {}^{(AB)}\mathbf{D}_{4,k}^{2} = 0 \text{ (first three elements of the last row are zero)} \tag{2.12}$$

$${}^{(AB)}\mathbf{D}_{4,4} - 1 = 0 \text{ (scaling is one)} \tag{2.13}$$

These systems of equations can be minimised with Levenberg-Marquardt method. The initial values of the unknowns are the results of the LS estimation and the calculation that based on quaternions.

3. Virtual reality system

The virtual reality is used to visualize the actions of the robot for the human operators and this usually has to be done in realtime. It is also a requirement to determine that the robot can do the desired action, so that it does not clash with itself and/or the surrounding items of the environment. Hence an efficient multi-level collision-detection algorithm was implemented.

There are two ways to view the robot during the visualization, it can be an overlaid image with the real camera images or can be a view from totally different viewpoint as the real cameras view the scene in the stereo system.

In order to put the virtual world together with the real environment, the calibration of the system is required. Since the stereo part does not use calibrated cameras, camera calibration process is achieved in order to display the results in the original cameras images. It means that the parameters of a camera, e.g. position, orientation, focal length, etc. are identified based on pictures taken by a real camera, and used in the visualization stage.

3.1 Collision detection

The track of a robot cannot be always fully planned in advance, because the environment may change dynamically, so there is a need for methods which can determine realtime if the robot can follow its planned path or not. On the other hand, the reference points of the path can be taught with the virtual robot without moving the actual robot.

The collision detection algorithm is hierarchical. It performs low-precision tests first, then in the further steps more and more accurate ones. The computational cost of the further steps is higher than of the previous ones, but these must be performed only on much smaller amount of data.

Since testing all objects with all other object is proportional to the square of the number of all items, we endeavoured to reduce the number of pairwise checks in the first step. Its possibilities are:

- The neighbouring segments of the robot will be reported to collide because of the resolution of the geometric model, since the polygonal representation is not accurate enough to handle a shaft rotating in a bearing. These cases must not be taken into account.

- Only the movable objects should be tested against other objects.
- There can be objects that cannot collide, e.g. two distant robots.
- The items can be grouped so they can be considered as a single object, e.g. a sophisticated gripper or dextrous hand may form a bounding volume.

The levels of the collision detection are the following.

3.1.1 Axis-parallel bounding boxes

Each object is surrounded with a box whose edges are parallel to the axes of the coodinate system which is handled as the "world" reference frame and these boxes are tested against each other. These type of boxes are depicted in the left side of Figure 3.1. These tests require comparison-type operations only. This method is not too accurate, especially when checking arbitrarily oriented long-shaped objects. The bottleneck is, however, the $O(N^2)$ computational cost (N being the number of objects). However, there is an advanced method proposed by Jonathan *et al.* [12] that can handle almost in linear time this type of test, applying incremental calculations.

3.1.2 Arbitrary bounding boxes

The local bounding boxes, which are usually much smaller than the axis-parallel ones (and besides their shapes are constant), are transformed to the world coordinate system and checked. These types of boxes are depicted in the right side of Figure 3.1.

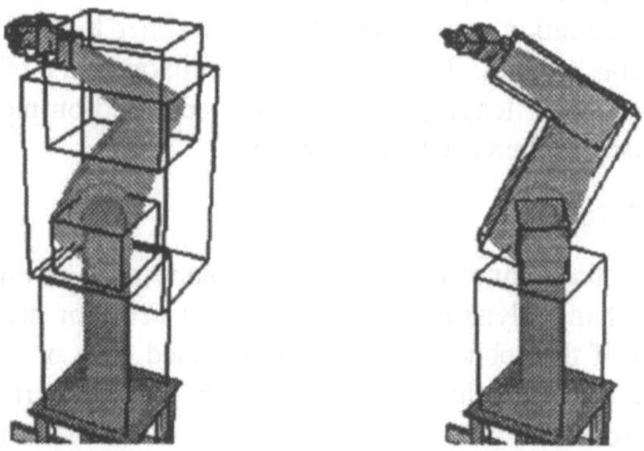

Figure 3.1: axis-parallel bounding boxes (left), arbitrary bounding boxes (right)

We tested two independent algorithms that deal with these boxes. The first one, called *three-level algorithm* is performed in three steps, of which the latter steps are executed only if the previous ones failed to give a sufficient result (similarly to the whole hierarchical collision detecting process). The second algorithm, called *axial projection*, tries to find a plane, which separates the two boxes. As our tests revealed, the axial projection method is 3...4 times faster than the three-level algorithm.

3.1.3 *Three-level test for boxes*

The steps of the algorithm can be seen in Figure 3.2:

- If all the points of the first box are on one side of a plane of the other box, they are disjoint.
- If a vertex of the first box is inside the second box, they collide.
- If there is no such vertex but one of the edges intersects a plane of the other, they collide.

Let the normalvectors of two parallel planes of the cube point toward each other. A point, substituted to a plane equation gives the signed distance from the plane. If the vertices of the first box substituted to any of the plane equations of the other gives all negative numbers, there is no collision. This test is not symmetrical, so it must be performed for both boxes.

A box contains a point if and only if it is between all three bounding plane pairs at the same time. So if a point-plane distance is positive for all six planes, the point is inside the box. The previous step produces all the distances of the possible plane-point pairs, so this stage requires only comparison-type operations.

Figure 3.2: Three possible configurations of arbitrary bounding boxes

The first two steps determine which edges intersect the bounding planes of the other box, since if the distance of point **A** from the plane is positive and the distance of **B** is negative, the line between them intersects the plane. The remaining question is that the point of intersection is inside the desired rectangle. The point of intersection can be written in the following form:

$$\bar{c} = (1-\lambda)\bar{a} + \lambda\bar{b} \quad (3.1)$$

Let the **P, Q, R** be three adjacent vertices of rectangle, then **C** can be expressed as

$$\bar{c} = \bar{q} + (\bar{p} - \bar{q})\alpha + (\bar{r} - \bar{q})\beta \quad (3.2)$$

where α and β are unambiguous. If both α and β fall to the [0,1] interval, the line intersects the rectangle.

3.1.4 *Axial projection*

The boxes projected to an arbitrary line form two intervals. If these intervals are disjoint, the line is called separating axis, and the boxes do not overlap. If the intervals overlap, additional lines must be tested. The question is that which lines should be tested. It is known that if two convex bodies are disjoint, a plane must exist, which separates the two objects, and parallel to a face of one of the objects or parallel to an edge from both objects. This means that two bodies are disjoint if exists a separating axis which perpendicular to a face or perpendicular to an edge pair. In the worst case 15 axes must be tested, 3-3 for the planes and 9 edge pair combination. If none of the axes is a separating one, the two boxes overlap [13].

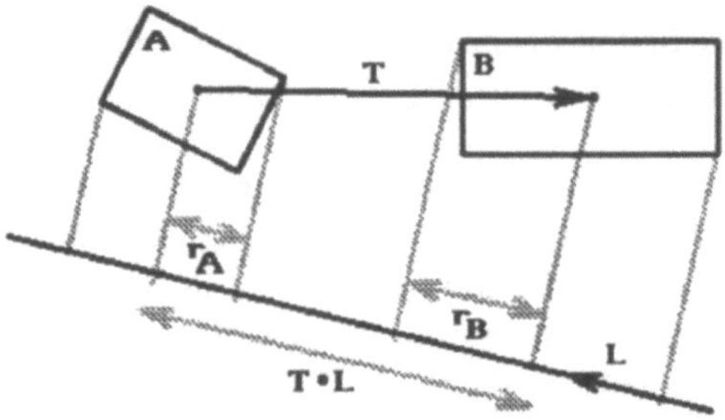

Figure 3.3: Radius test

The projected points on the axis being tested can be expressed in a form of $\bar{r}+\lambda*\bar{t}$ where $\bar{r}$ is a point on the axis and $\bar{t}$ is the direction vector. The intervals are disjoint if all λ's belonging to the first box are all smaller than all the λ's of the other. Performing this test (determining all λ's) requires several operations. Fortunately a quicker algorithm exists, which is called *radius test* [13], Figure 3.3. This projects the centres of the boxes to the axis and the radius of the projected interval (e.g. the half of its length) is determined. If the sum of the two radii is smaller than the distance between the projected centres, the boxes are disjoint. This method, using linear algebraic identities, can be brought to a form that requires the determination of the matrix, which rotates the axes of the first box to be parallel with the axes of the second. Having the matrix computed, six multiplications, five additions and five absolute value operations are needed to test one axis.

Figure 3.4: Checking the intersection of triangles

3.1.5 Arbitrarily oriented triangles

The first two bounding box methods do not give exact results, but the robot, due to the properties of the controlling system, will probably deviate from the planned track. So sometimes the exact computations are unnecessary. Otherwise, the intersections of the triangles of the model have to be checked, Figure 3.4. Testing two triangles is a two-level process:

- If all the points of the first triangle are on one side of a plane of the other, they are disjoint.
- If one of the edges intersects the other triangle, they collide. This is another bottleneck of $O(N^2)$, which can be sped up by determining first the intersection of two colliding axis-parallel bounding boxes, then the triangles from both objects which intersect this region. Finally, only these pairs must be checked against each other. If one of the objects has no intersecting triangle with the intersection, then the further tests are unnecessary.

3.2 Calibration

The usability of computer graphics heavily depends on the lifelikeness of the model and the environment. With the help of this virtual image the operator can more safely control the real robot, because the virtual robot immediately completes the given orders, so any occasional error can be avoided in time. This has special importance if there is a considerable postponement between the operator and the images taken by the real camera. This proposed e.g. in telerobotics by Bejczy [14].

3.2.1 Linear calibration based on 2D information

A point in three-space and its corresponding pair on the two-dimensional image is described by the $[x,y,z,u,v]^T$ vector. Using the pinhole model again, the projection matrix has certain constraints:

$$\|\mathbf{p}_3\|^2 = 1 \qquad \text{(Normality)} \qquad (3.3)$$

$$(\mathbf{p}_1 \times \mathbf{p}_2)\cdot(\mathbf{p}_2 \times \mathbf{p}_3) = 0 \qquad \text{(Orthogonality)} \qquad (3.4)$$

where $\mathbf{p}_i$ is the vector formed from the first three element of the ith row.

The elements of the projection matrix can be obtained as a solution of a linear equation system $\mathbf{A}\cdot\mathbf{p}=\mathbf{0}$, where $\mathbf{A}$ is a 2N×12 matrix and $\mathbf{p}$ is the vector build from the unknown camera parameters, N is the number of reference points [4]. In order to obtain unambigous result from the equation system we have to utilize the (3.3) constraint.

3.2.2 Post-orthogonalization of the projection matrix

The solution of the linear equation system does not guarantee the orthogonality, so the obtained projection matrix may be occasionally modified. There are many algorithms for orthogonalizing a matrix, but for its simplicity we used the following method [15]. For correcting the matrix $\mathbf{Q}=\left[\mathbf{p}_1{}^T\mathbf{p}_2{}^T\mathbf{p}_3{}^T\right]^T$ we compute a correction matrix $\mathbf{C}$ where $\mathbf{QC}$ is closely orthogonal. $\mathbf{C}=\left(\mathbf{Q}^T\mathbf{Q}\right)^{-1/2}$, which can be approximated as an analogy of the Taylor series of $(1+x)^{-1/2}$ around $x=0$ with the $\mathbf{X}=\mathbf{R}^T\mathbf{R}-\mathbf{I}$ substitution:

$$\mathbf{C} \cong \mathbf{I}-\frac{1}{2}\mathbf{X}+\frac{3}{8}\mathbf{X}^2-\frac{5}{16}\mathbf{X}^3+\frac{35}{128}\mathbf{X}^4-\frac{63}{256}\mathbf{X}^5. \qquad (3.5)$$

Using this, the problem can be solved with simple matrix multiplications instead of the costly operation of matrix inversion.

4. Results

We have tested the stereo method with simulated data in order to evaluate robustness under different noise conditions. As illustration let us consider 2 cubes with the size 0.4m, the transformations are translation in the range -0.3...0.7m in the direction of the axes and some rotation around the three axes. Three cameras view the scene from approximately 2.5m. Noise was added to the camera matrices in order to simulate the distortions and other effects. Gaussian noise with mean value 0 and deviation 1 is added to the 3D model database to model the inaccuracy in the recognition process. During the simulation the mean value of the pixel noise was 0, the deviation is changed between 0 and 1, so the maximum error was approximately 3-4 pixel.

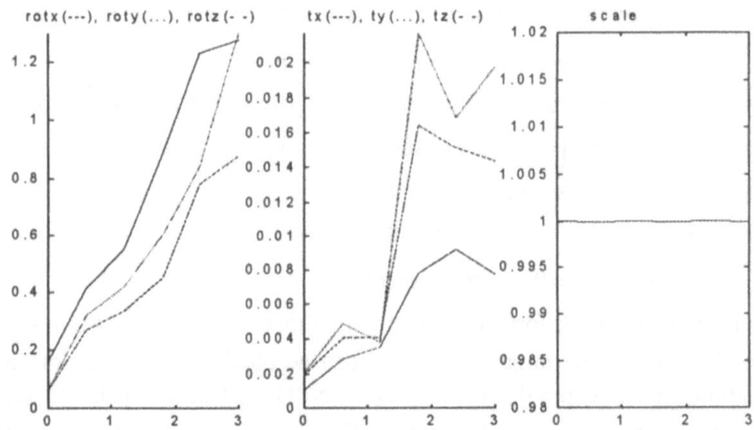

Figure 4.1: Average displacement error vs. pixel noise

The errors shown in Figure 4.1 are average values of the results of some simulations. The first part of the figure shows the errors in rotation around x, y and z axis with solid, dotted and dashed lines, respectively. The angles are given in degrees. The second part contains the translation error in the x, y, and z directions again with solid, dotted and dashed lines, respectively. The third part is the scaling, which is 1 for the Euclidean case. We found that in the worst cases the errors are near 2...4 degree in the angles and 1.5...3cm in the position. This means that the relative errors are approximately 1...1.5%, where the relative error means the ratio of the error (in position) and the overall size of the scene (approximately 2m). The scaling is very close to 1. The peaks at the pixel noise level 2 are the result of the effect of those cases, when the Levenberg-Marquadt method has not been converged properly.

From the VR point of view, our tests have shown that the better the resolution of the image used for calibration was the more orthogonalized matrix we received. If there is no possibility of getting a high-resolution image, it is profitable to use sub-pixel precision point and edge detection algorithms. In the beginning we used 256×256 images with pixel precision, and the computed matrix was far from being orthogonal. As we expected with the use of 512×512 and higher resolution images we received better results.

We have found that the quality of the result also depends on the spatial placement of the reference points. During the tests it turned out that using the points of the robot arm, as calibration pattern is not worth if there is a known box-shaped object in the scene. The reason is that the arm was narrow so the points were close to the same plane. Generally, the more place the calibration pattern occupied, the better the result was.

5. Conclusion

This chapter has shown the two "visual" part of an intelligent robot control system. The two parts are the conjugate of each other, the stereo part collects vision information about the scene, the VR generates it for visualization.

The stereo method is used to collect information about the surround world. This information is only relative one, but if one of the objects absolute pose is known (e.g. table, wall, robot) in a user defined "world" coordinate system then the absolute pose of every object in that system can be determined from the relative ones. Note that the existence of this "world" coordinate system is not required by the method.

In the current state of the stereo system, the simple feature extraction, the projective reconstructing and the calculation of the Euclidean transformation step is fully implemented and tested. The object recognition is almost fully implemented and partially tested. One of the most important goal in the future is to replace the simple, heuristic feature extractor with a new one which is capable to determine and localize wider sets of features with subpixel accuracy. This method is under development.

In the VR system the collision detection and the calibration is fully implemented and tested. The future goal is to develop methods that can handle the collisions not only for discrete steps but for the whole space which is "swept" by the robot.

6. Acknowledgement

Support for the research of stereo image processing and calibrated virtual reality for robots is provided by the Hungarian National Research programs under grant No. FKFP 0417/1997 and OTKA T 016855.

7. References

1. Lantos B, Klatsmányi P, Ludvig L, Tél F 1997 Intelligent Control System of a Robot with Dextrous Hand. In: *Proceedings of the IEEE International Conference on Intelligent Engineering Systems INES'97, Budapest*
2. Faugeras O D 1992 What can be seen in three dimensions with an uncalibrated stereo rig. In: *Proceedings of the 2nd European Conference on Computer Vision*
3. Canny J F 1986 A computational approach to edge detection. In: *IEEE Transactions on Pattern Analysis and Machine Intelligence, 8:769-798, 1986*
4. Faugeras O D 1992 *Three dimensional computer vision (A geometric viewpoint).* MIT Press
5. Boufama B, Mohr R, Veillon F 1993 *Euclidean Constraints for Uncalibrated Reconstruction*, INRIA France via Internet.
6. Mohr R, Quan L, Veillon F 1995 Relative 3D Reconstruction Using Multiple Uncalibrated Images. In: *International Journal of Robotic Research. MIT Press.*
7. Press W H, Flannery B P, Teukolsky S A, Vetterling W T 1988 *Numerical Recipes in C.* Cambridge University Press.
8. Wells W M 1993 Statistical object recognition. Ph.D. thesis, MIT
9. Pope A R 1995 Learning to recognize objects in images: acquiring and using probabilistic models of appearance Ph.D thesis The University of British Columbia
10. Welch G, Bishop G 1997 *An introduction to the Kalman Filter.* University of North Carolina at Chapel Hill, via Internet
11. Lantos B 1994 *3D Képfeldolgozási módszerek.* (Manuscript, in Hungarian) Technical University of Budapest
12. Jonathan D, Lin M. C, Manocha D, Ponamgi M. K 1995 I-COLLIDE: An Interactive and Exact Collision Detection System. *University of North Carolina, Chapel Hill*
13. Gottschalk S, Lin M C, Manocha D 1996 OBB Tree: A Hierarchical Structure for Rapid Interference Detection. *Proceedings of the 1996 SIGGRAPH Conference, University of North Carolina, Chapel Hill.*
14. Bejczy A K 1996 Calibrated virtual reality in telerobotics *California Institute of Technology, Pasadena (CA), USA,*
15. Glassner A S 1990 *Graphics Gems I.* Academic Press INC.

28
Tracking Control for Automated Bridge Cranes

O. Sawodny, H. Aschemann, S. Lahres and E.P. Hofer

1 Introduction

In order to use bridge cranes as large workspace serving robots a control system which guarantee high tracking accuracy is required. I.e., the deviation between a reference track and the actual track has to be minimized.
Until now control systems for automated cranes only provide position accuracy at the target point and damping of load oscillations for each axis [1]. But especially in case of a synchronized operation of all 3 crane axes high tracking accuracy is essential to avoid collisions.
Therefore, a control concept has been developed which achieves this goal by four main steps: a decentralized control structure, parameter dependent state space controllers, estimation and compensation of disturbances, and parameter dependent feedforward control. The decentralized control structure is based on decoupled the dynamic models of the crane axes. Thus, for controller design each axis is assumed to be independent of the dynamics of the other axes.
Here, the procedure of control design is presented exemplary for one axis. First a mathematical model of the crane axis has to be created and its parameters have to be identified by means of experiments and theoretical investigations. The analytical state space controller design leads to parameter dependent control gains. Therefore, variable load mass and length of the rope are taken into account by changing control gains.
For automatic crane systems rough operating environment as well as nonlinear dynamics are major problems. Therefore, an observer for estimating the disturbances is introduced. To improve control performance a weighted sum of reference position, velocity, acceleration, and jerk is used for feedforward control. The time dependent reference functions are generated considering the admissible maximal values of these variables to reach short travelling time.
The control structure is applied to a 5-t bridge crane. The paper closes with a detailed discussion of experimental results of the closed-loop control.

2 Model of the crane

The mechanical structure of the bridge crane is shown in figure 2.1. The crane trolley moves on the crane bridge, which is in the following named as the y-axis. The bridge itself is movable on the left and right tracks. This axis is defined as the x-axis. In case of $\varphi_l=0$ the rope coincides with the z-axis, where z denotes for the length of the rope.
The crane has been equipped with torque controlled motors and sensors to measure the bridge position on the left and right track x_{B1} and x_{B2}, the trolley position y_K and the angle φ_l between rope and trolley. The angle measurement provides the projected angle into the zx-plain φ_{lx} and yz-plain φ_{ly}. For crane control a decentralized control concept is used. Each axis has its own controller which is working independent of the other axes. In the following the modeling of the y-axis is presented.

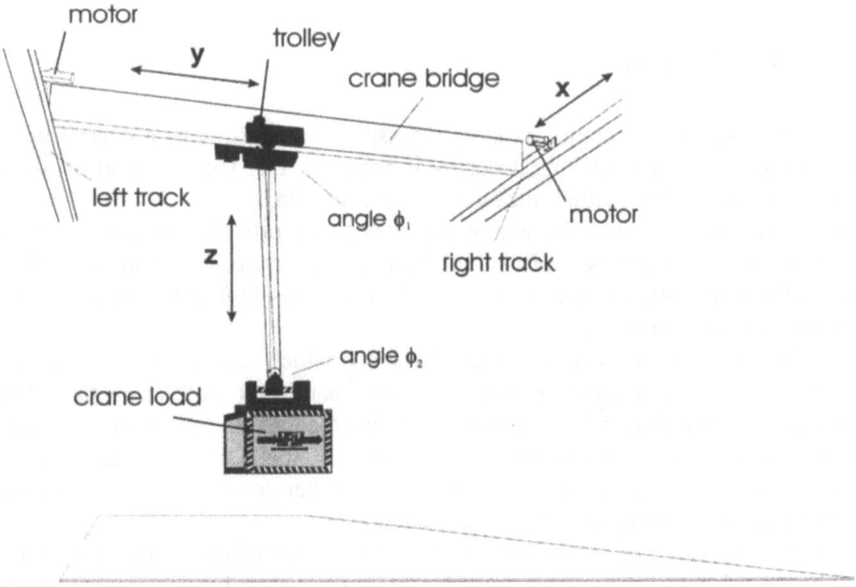

Fig. 2.1: Crane structure.

First the equations of motion for the crane in the yz-plain are formulated. A accelerating force

$$F_{By} = m_L g \cdot \sin \varphi_{1y} \tag{1}$$

acts on the crane load mass m_L, where g is the gravity acceleration. So the equation of motion for the load is given by

$$m_L \cdot \ddot{\varphi}_{1y} = -\frac{m_L \cdot g}{z} \cdot \sin \varphi_{1y} \tag{2}$$

The reacting force in the rope also affects the movement of the crane trolley. The equation of motion for the crane trolley with mass m_K is then

$$m_K \cdot \ddot{y}_K = F_y - F_{RCy} - b_K \cdot \dot{y}_K - m_L \cdot g \cdot \sin\varphi_{1y} \tag{3}$$

where Fy is the motor force, F_{RCy} the Coulomb friction force, and b_K the constant for the velocity proportional friction. (3) is linearized, so $\sin\varphi \approx \varphi$. Because the control variable is the position of the load, the new variable

$$y_L = y_K + z \cdot \sin\varphi_{1y} \tag{4}$$

is introduced. Therefore, the resulting equations describing the dynamics of the y-axis can be obtained as

$$\begin{aligned} m_L \cdot \ddot{y}_L &= \frac{m_L \cdot g}{z} \cdot (y_K - y_L) \\ m_K \cdot \ddot{y}_K &= F_y - F_{RCy} - b_K \cdot \dot{y}_K - \frac{m_L \cdot g}{z} \cdot (y_K - y_L) \end{aligned} \tag{5}$$

The equations (5) are transformed into state space representation form. The nonlinear Coulomb friction cannot be considered in the linear representation of the model. The state space model is then

$$\underline{\dot{x}}_y = \underline{A}_y \underline{x}_y + \underline{b}_y \underline{u}_y$$

$$\underline{x}_y = \begin{bmatrix} y_L \\ \dot{y}_L \\ y_K \\ \dot{y}_K \\ F_y \end{bmatrix} \quad \underline{A}_y = \begin{bmatrix} 0 & 1 & 0 & 0 & 0 \\ -\frac{c}{m_L} & 0 & \frac{c}{m_L} & 0 & 0 \\ 0 & 0 & 0 & 1 & 0 \\ \frac{c}{m_K} & 0 & -\frac{c}{m_K} & -\frac{b_K}{m_K} & \frac{1}{m_K} \\ 0 & 0 & 0 & 0 & -\frac{1}{T} \end{bmatrix} \quad \underline{b}_y = \begin{bmatrix} 0 \\ 0 \\ 0 \\ 0 \\ \frac{1}{T} \end{bmatrix} \tag{6}$$

with $\frac{m_L \cdot g}{z} = c$.

The dynamics of the motor is described by a first order delay element. All parameters except the time constant T of the motor and the velocity proportional friction constant b_K are given from technical data of the crane. The motor model is identified by a frequency response analysis. The velocity proportional friction constant is evaluated by a friction measurement according to [2]. As shown in figure 2.2, one gets the characteristic curve of the friction force depending on the trolley velocity. In the state space model the proportional part is taken into account by b_K, while the nonlinear part is neglected in control design because of observer based disturbance rejection.

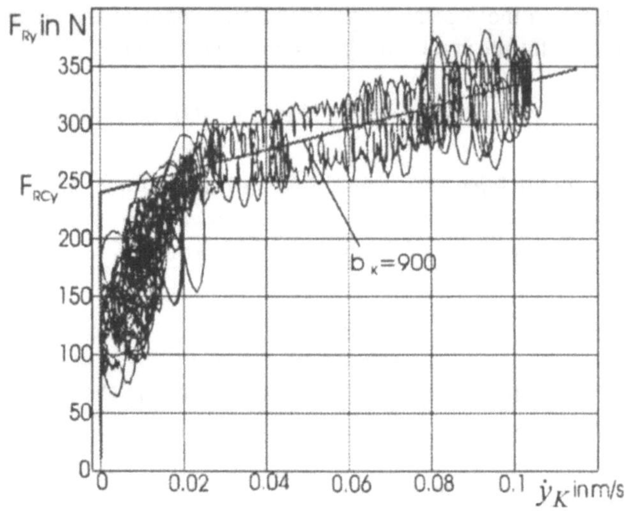

Fig. 2.2: Friction force depending on the trolley velocity.

3 Control structure

To improve performance of crane control with the goal of minimizing deviation between reference and actual track of the load three structural actions are taken (Fig. 3.1) [3].

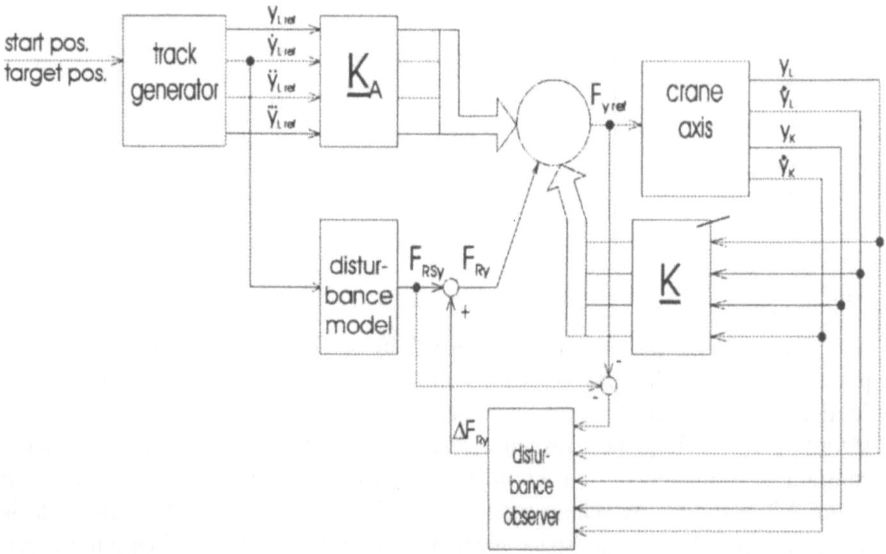

Fig. 3.1: Control structure.

Because of changing load masses and rope lengths the bridge crane is a parameter varying system. This has to be taken into account at the control design by use of a parameter depending state space controller. To reduce the influence of disturbances, like wind effects by loading containers on a ship or the coulomb friction an observer for the disturbances is introduced. All upcoming disturbances are reduced on a force, which negative value is given on the process input. To support the disturbance estimation by the observer a disturbance model evaluates the necessary force to compensate the known parts of the coulomb friction in advance. To improve the tracking accuracy the time functions of reference position, velocity, acceleration, and jerk are generated considering the maximum of these variables as well as they reach in minimized time their maximum. These time functions are weighted in the sense of a feedforward control to guarantee a high tracking accuracy.

3.1 Parameter dependent state space controller

Based on the state space model of the crane axis an analytical parameter dependent controller can be designed taking variable parameters like rope length and load mass into account. Compared to a robust design using constant controller gains this approach yields a important improvement of control performance. Figure 3.2 shows an exemplary movement of the y-axis of the crane from position *0* m to position *2 m*. In case of a load mass of m_L=*350 kg* and a rope length of z= *2 m* the control behavior is except to the remaining load swinging at the target point, caused by Coulomb friction, quite good.

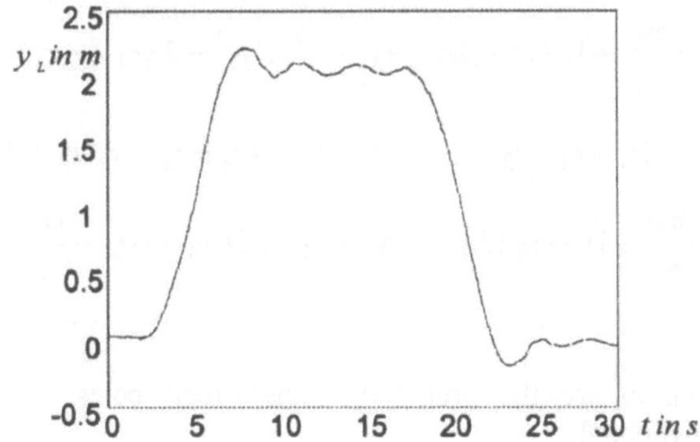

Fig. 3.2: Crane movement with a robust controller. (m_L=*350 kg*, z= *2m*)

Now the robust controller is used for a load mass of *15 kg* and the same rope length. As shown in figure 3.3, an overshoot of approximately *20 cm* at the target point occurs. This experiment shows the necessity of controller gain adaptation, which is achieved here by analytical calculation depending on the varying system parameters.

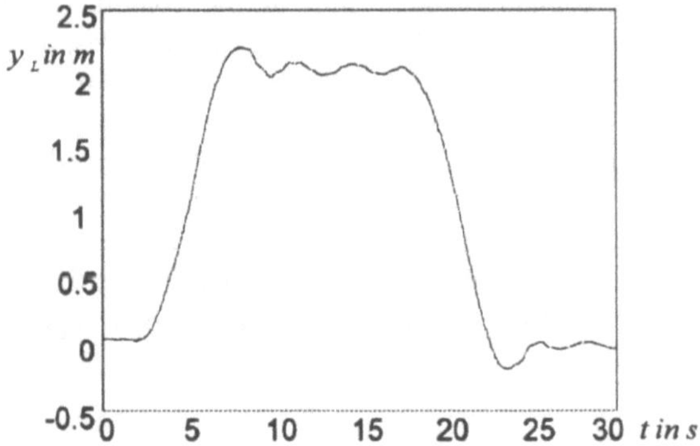

Fig. 3.3: Same experiment, as in fig. 3.2, but with a load mass m_L=*15 kg*.

Comparing the characteristic polynomial based on the proposed closed-loop poles and the determinant of $\underline{A}_y-\underline{b}_y\underline{k}_y^T$ a system of linear equations for the control gains depending on the system parameters can be derived. By this means the controller gains become

$$k_4 = -b_K - (r_4 + r_3 + 2r_1)m_K \tag{7}$$

$$k_3 = -c(\frac{m_K}{m_L} + 1) + m_K(2r_1r_3m_K + r_1^2 + q_1^2 + 2r_1r_4 + r_3r_4)$$

$$k_2 = m_K(2r_1 + r_4 + r_3 - \frac{m_L}{c}(r_1^2r_3 + q_1^2(r_4 + r_3) + r_1^2r_4 + 2r_1r_3r_4))$$

$$k_1 = c(\frac{m_K}{m_L} + 1) - m_K(2r_1r_3 + r_1^2 + q_1^2 + 2r_1r_4 + r_3r_4 - \frac{m_L}{c}r_3r_4(q_1^2 + r_1^2))$$

where r_1, q_1, r_3, r_4 are the proposed closed-loop poles according to the characteristic polynomial

$$p(s) = (s - r_1 + iq_1)(s - r_1 - iq_1)(s - r_3)(s - r_4)\,. \tag{8}$$

Therefore, the controller gains are changed depending on the actual values for the mass of the load and the rope length. Due to the slowly change of these parameters this does not cause any instabilities.

3.2 Estimation of disturbances

Disturbances have a great influence on the efficiency of an automatic crane system. E.g. wind decreases the handling output of a container crane. Especially, Coulomb friction has a dominant effect on the dynamic behavior and leads to a suboptimal behavior of the crane control and reduces reachable velocity and accelerations. So the estimation of disturbances becomes important.
For disturbance estimation a reduced disturbance observer is designed [4]. Therefore, the disturbance force F_{Ry} is introduced as an additional state variable. The disturbance is assumed as section wise constant, according to

$$\dot{F}_{Ry} = 0 \tag{9}$$

So $\underline{x}_y, \underline{A}_y, \underline{b}_y$ are modified to $\underline{x}_{yS}, \underline{A}_{yS}, \underline{b}_{yS}$ considering F_{Ry}

$$\underline{x}_{yS} = \begin{bmatrix} F_{Ry} \\ y_L \\ \dot{y}_L \\ y_K \\ \dot{y}_K \\ F_y \end{bmatrix}, \quad \underline{A}_{yS} = \begin{bmatrix} 0 & 0 & 0 & 0 & 0 & 0 \\ 0 & 0 & 1 & 0 & 0 & 0 \\ 0 & -\frac{c}{m_L} & 0 & \frac{c}{m_L} & 0 & 0 \\ 0 & 0 & 0 & 0 & 1 & 0 \\ \frac{-1}{m_K} & -\frac{c}{m_K} & 0 & \frac{c}{m_K} & -\frac{b}{m_K} & \frac{1}{m_K} \\ 0 & 0 & 0 & 0 & 0 & -\frac{1}{T} \end{bmatrix}, \quad \underline{b}_{yS} = \begin{bmatrix} 0 \\ 0 \\ 0 \\ 0 \\ 0 \\ \frac{1}{T} \end{bmatrix} \tag{10}$$

The reduced observer estimates the assumed disturbance force F_{Ry}. Depending on the proposed observer pole λ_0 the observer gain is then

$$l_4 = \lambda_0 \cdot m_K . \tag{11}$$

Therefore, the disturbance observer can be represented by the following differential equation

$$\dot{F}_{Ry} = \frac{l_4}{m_K} \cdot F_{Ry} + l_4 \cdot \left[\ddot{y}_K - \frac{c \cdot y_L}{m_K} + \frac{c \cdot y_K}{m_K} + \frac{b \cdot \dot{y}_K}{m_K} - \frac{1 \cdot F_y}{m_K} \right] . \tag{12}$$

The negative estimated value for the disturbance F_{Ry} can then be used to compensate for nonlinear friction effects. To support the disturbance estimation the known part (fig. 2.2) of Coulomb friction is reproduced in the disturbance model. Instead of differentiation of the measured y_K the reference velocity of the load y_{Lref} is used. Output of the disturbance model is the friction force F_{RSy}. The coulomb friction is described by the equation

$$F_{RSy} = F_{RCy} \tanh(\frac{\dot{y}_K}{\varepsilon}) \qquad (13)$$

where ε is determined by fitting the resulting curve into the friction measurement (fig. 2.2). Because compensation has a relevant effect on stability of the resulting system it has to be proved again.

3.3 Track generation and limited control inputs

For use of the crane as a robot a track generator transform the information of start and target point in the workspace into the time functions for position, velocity, acceleration, and jerk of the three axes. Hereby, the track generator has to consider the synchronizing of the axes and the maximum reachable velocity, acceleration and jerk. As a track planning method the method of constant jerk or the polynomial method [5] can be used. For example, the generated time functions for a straight line movement from the starting point to the target point are shown in figure 3.4.

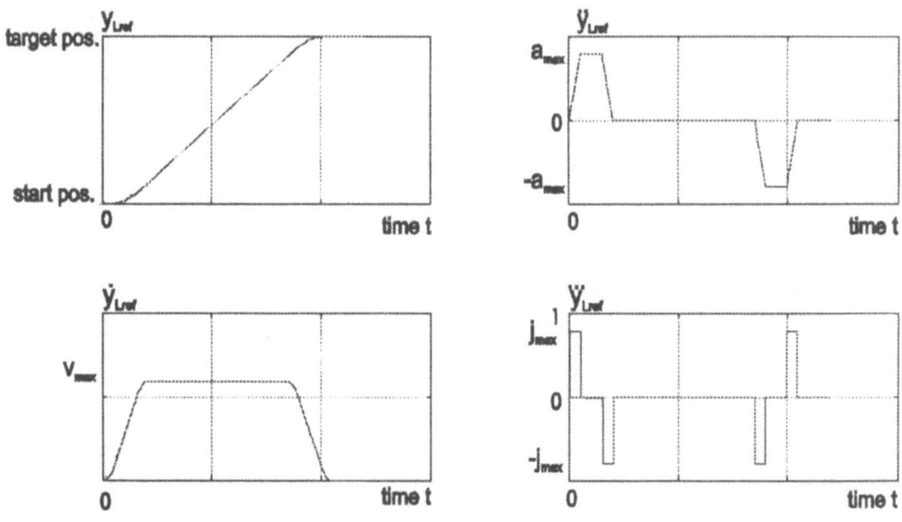

Fig. 3.4: Generated time functions for position, velocity, acceleration and jerk for the y-axis.

The track generation considers the dynamic limitations of the process as well as to reach these maximum values as fast as possible. So at first the jerk becomes its maximum value until the acceleration reach the allowed maximum. The acceleration now keep the maximum until the velocity reach the maximum. Due to this method a very fast run of the automatic crane is achieved.

To improve the dynamic behavior a weigthed sum of reference position, velocity, acceleration and jerk is used for feedforward control (fig. 3.5).

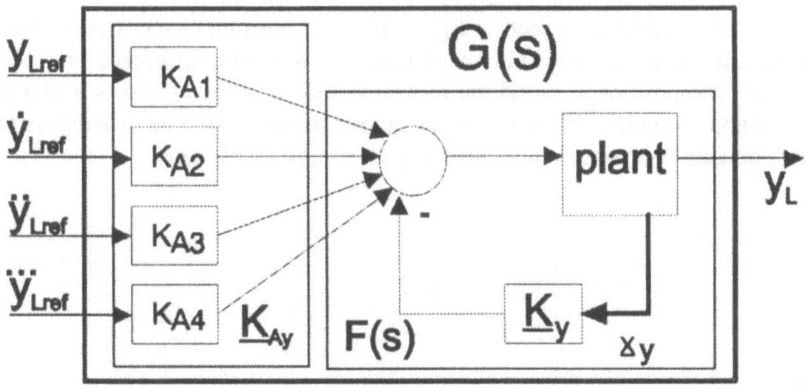

Fig. 3.5: Structure of the feedforward control for position, velocity, acceleration and jerk.

To evaluate the weighting factors for the feedforward control the closed-loop transfer function *F(s)* of the crane axis with the parameter dependent controller is required. Then, the transfer function from reference load position y_{Lref} to measured load position y_L is then

$$\frac{y_{Lref}}{y_L} = G(s) = F(s) \cdot (K_{A1} + K_{A2} \cdot s + K_{A3} \cdot s^2 + K_{A4} \cdot s^3) \tag{14}$$

Depending on the weighting factors K_{Ai} the following equation is given

$$\frac{y_{Lref}}{y_L} = \frac{..b_2(K_{A1}, K_{A2}, K_{A3}) \cdot s^2 + b_1(K_{A1}, K_{A2}) \cdot s + b_0(K_{A1})}{...a_2 \cdot s^2 + a_1 \cdot s + a_0}. \tag{15}$$

Steady-state accuracy concerning constant, linear, quadratic and cubic reference functions of time is achieved, when the following conditions are fulfilled

$$\frac{b_0}{a_0} = 1 \; ; \frac{b_1}{a_1} = 1 ; \frac{b_2}{a_2} = 1 ; \frac{b_3}{a_3} = 1 \tag{16}$$

This system of linear equations can be solved with respect to the weighting factors $K_{A1} .. K_{A4}$.

4 Experimental results

The described control concept was applied to a 5-t-bridge crane. The first experiment (fig. 4.1) shows a crane movement from position $y = 0$ to $y = -2m$ with a load mass of *350 kg*. As seen in fig 4.1 the load position changes very smoothly and reaches the target point after approximately *8 s*. The difference between the reference load position x_{Lref} and the measured load position x_L shows a maximum value of approximately *10 cm* during the acceleration phase. The differences at the target point are less then *1 cm* without any remaining oscillations.

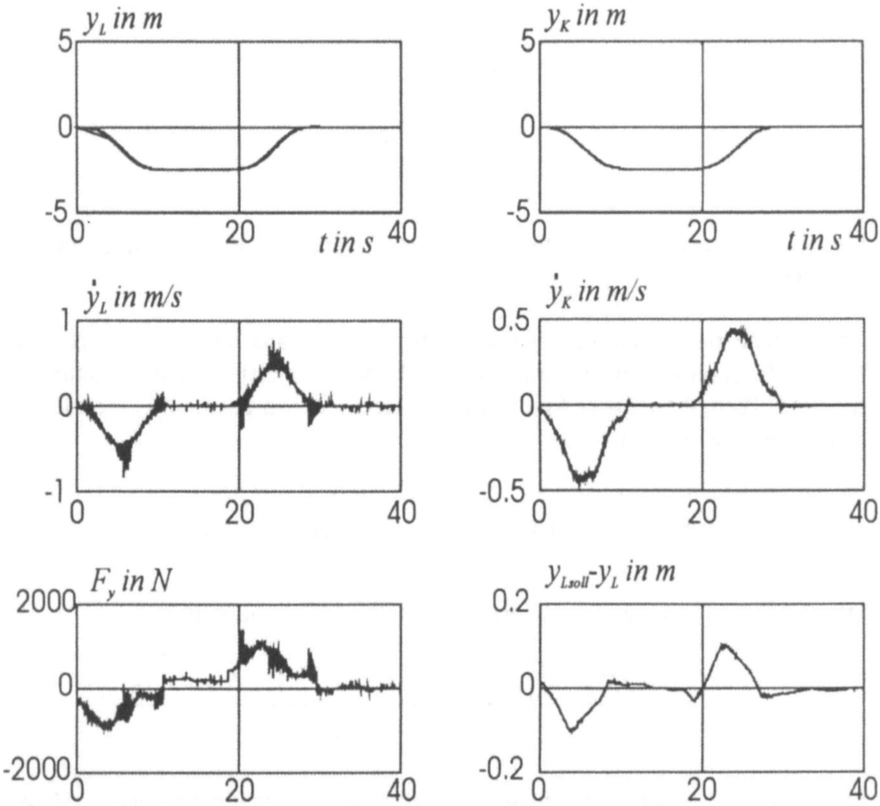

Fig. 4.1.: Exemplary movement with a rope length of z=*6 m* and a load mass of m_L=*350 kg*.

To visualize the performance of the control in case of changing rope lengths and load masses the second experiment is with a load mass of m_L=*1250 kg* and a rope length of z=*4 m*. The motion of the load is very smooth and soft. The maximum differences between the reference load position and the measured position reach in case of a *1,25 t*-load only a maximum of *13 cm*.

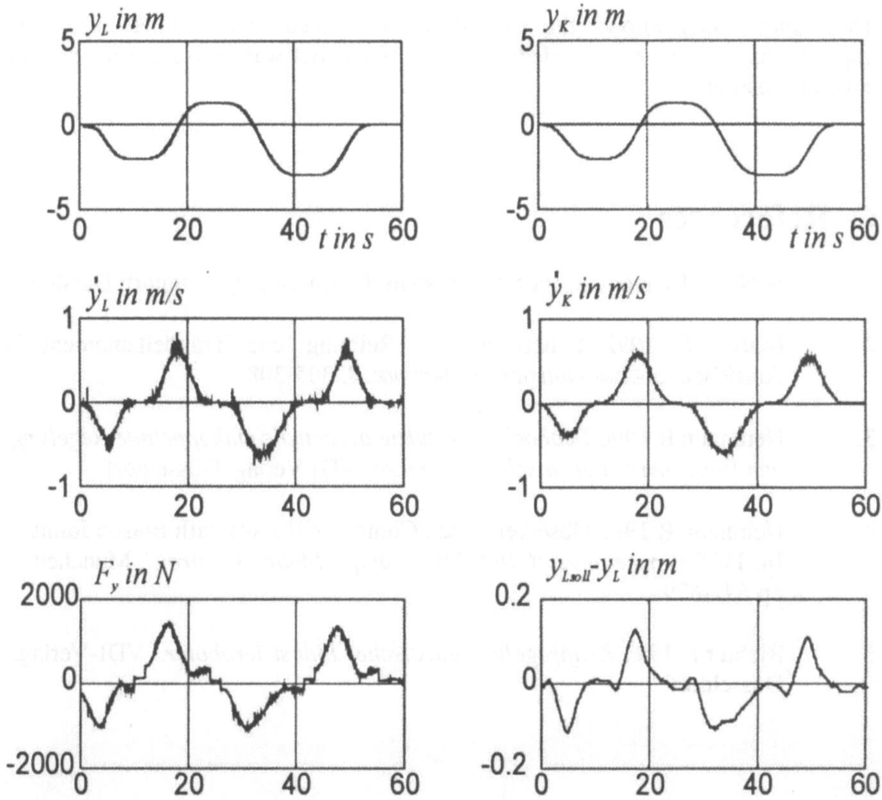

Fig. 4.2: Exemplary movement with a rope length of *z=4 m* and a load mass of $m_L=1250\ kg$.

5 Conclusion

The presented control concept for an automated bridge crane performs an exact positioning of the load, complete damping of the load oscillations and a high tracking accuracy. This is achieved by use of parameter dependent state space feedback controller which considers the varying parameters of load mass and rope length. A disturbance observer reduces the deviation between reference and actual track and compensates nonlinear Coulomb friction effects. Additionally, a parallel model of Coulomb friction supports the disturbance observer. The track deviation can be reduced applying a weighted sum of reference position, velocity, acceleration, and jerk to the control input as feedforward control. The weighting factors are determined for each crane axis by considering the closed-loop transfer function.

This functionality allows the use of an automated crane for a lot of new applications, like the rope guided operation of a pallet warehouse or the automatic concrete casting.

6 References

1. N N 1994 Automatisches Kransystem. Research project report, Duisburg

2. Maron C 1992 Schätzung von Reibung und Trägheitsmoment bei Antrieben. *at-Automatisierungstechnik* 8: 305-308

3. Neumann R 1996 *Beobachtergestützte dezentrale entkoppelnde Regelung von Robotern mit elastischen Gelenken.* VDI Verlag, Düsseldorf

4. Neumann R 1995 Observer Based Control of Robots with Elastic Joints. In: 1995 *Proceedings of IFAC Workshop "Motion Control"*, München, pp 672-679

5. Richter K 1991 *Kraftregelung elastischer Industrieroboter.*. VDI-Verlag, Düsseldorf

29

Robust Feedforward Control Design for Variation of Resonant Frequency and its Application to Crane Systems

T. Miyoshi and K. Terashima

1. Introduction

In high-speed mechanical transfer systems, inappropriate command inputs can generate the vibration of plant, and make the control performance worse. It is therefore important to study command inputs, namely feedforward control, in order to almost eliminate the residual vibration.

The authors have already reported a feedforward control without residual vibration for time-varying vibration mechanism such as a crane system hoisting up and down[1]. The analytical solution of its feedforward control inputs is represented by finite-power series of time. However, from the practical point of view, guaranteeing the robustness in feedforward control will be essential because feedback-loop may not always be implemented.

So far, researches of feedforward guaranteeing the robustness have been developed by Meckl and Kinceler[2], Yamaura and Ono[3] and etc. But Meckle evaluates the performance index using only a few points in the ranges of frequencies around the natural frequency of the system; they don't consider the whole range of variety of resonant frequency. In the case of LQG method by Yamaura, the problem is remained on the choice of weighting function.

This paper presents an *Algorithm* useful for any single-degree of the freedom system which has the perturbation of resonant frequency. Proposed robust control input is represented by higher-power series of time than that of the analytical solution designed for nominal model. The control input function is solved by satisfying the boundary condition at the initial time and the final time, and by minimizing the performance index. Performance index given in this paper evaluates the square of amplitude of the residual vibration in a phase plane, and it is corresponded to total energy of all the range of the perturbation of resonant frequency.

The usefulness of robust feedforward control proposed in this paper is demonstrated by experiments of an overhead traveling crane system for the various constant length of rope.

2. Derivation of Nominal Control Input

In this paper, we consider the traveling crane system as shown in Figure 1.

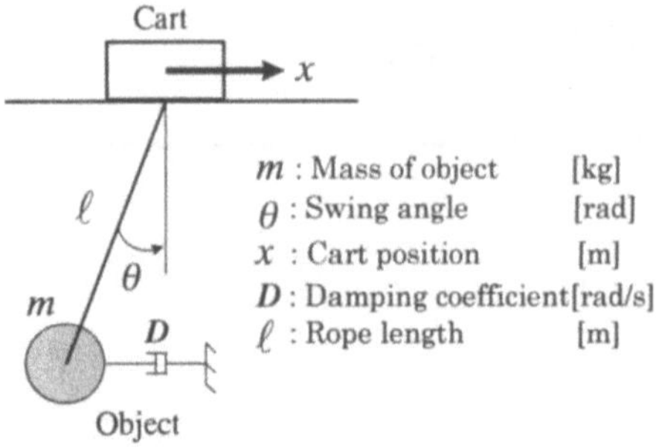

Figure 1: Model of crane system

A state equation is written as

$$\dot{\boldsymbol{x}} = A\boldsymbol{x} + Bu \tag{1}$$

where

$$A = \begin{bmatrix} -\frac{D}{m\ell^2} & 0 & -\frac{g}{\ell} & 0 \\ 0 & 0 & 0 & 0 \\ 1 & 0 & 0 & 0 \\ 0 & 1 & 0 & 0 \end{bmatrix}, B = col.\{\frac{1}{\ell}, 1, 0, 0\}, \quad \boldsymbol{x} = col.\{\dot{\theta}, \dot{x}, \theta, x\}, \quad u = \ddot{x}.$$

Usually, the vibration of plant is excited when a cart is moved by inappropriate command inputs. However we require the control input such as does not excite residual vibration after acceleration interval.

Hence we consider a boundary condition without residual vibration at the initial and the final time of acceleration

$$\left.\begin{array}{l}\begin{pmatrix}\dot{\theta}(0)\\ \dot{x}(0)\\ \theta(0)\\ x(0)\end{pmatrix} = \begin{pmatrix}0\\0\\0\\0\end{pmatrix}\\ \begin{pmatrix}\dot{\theta}(T)\\ \dot{x}(T)\\ \theta(T)\\ x(T)\end{pmatrix} = \begin{pmatrix}0\\V\\0\\0.5VT\end{pmatrix}\end{array}\right\} \tag{2}$$

,where T : final time[s] of acceleration intervals, V : final velocity[m/s] of the cart at time T, $x(T) = 0.5VT$ means that a cart moves over time T[s] in the mean velocity of 0.5V[m/s].

The control input can be obtained without residual vibration by solving the Eq.(1) under the condition of Eq.(2).

Now, we shall consider a time-invariant linear system $\dot{\boldsymbol{x}} = A\boldsymbol{x} + Bu$, which is a single input and the order n of state variables. If the system is controllable and matrix A has no zero eigenvalues, then the system can be transited by control inputs represented by $(2n-1)$-th power series of time from the arbitrary states $\boldsymbol{x}(0)$ at $t=0$ to the arbitrary states $\boldsymbol{x}(t_1)$ at $t=t_1$(See Theorem 1 in Appendix). And if the system is controllable and all eigenvalues of A is zero, then the system can be transited by $(n-1)$-th power series of time from arbitrary $\boldsymbol{x}(0)$ to arbitrary $\boldsymbol{x}(t_1)$(See Theorem 2 in Appendix). And if the system has both zero and non-zero eigenvalues, the system can be transited by $(2n_\lambda + n_0 - 1)$-th power series of time from arbitrary $\boldsymbol{x}(0)$ to arbitrary $\boldsymbol{x}(t_1)$, where n_λ is the number of non-zero eigenvalues and n_0 is the number of zero eigenvalues.

The crane system has two non-zero eigenvalues and double zero eigenvalues, i.e. a pendulum dynamics causes non-zero eigenvalues and a cart dynamics causes zero eigenvalues. Therefore the control input which eliminates residual vibration is the 5th-power series of time.

Particularly, when we can neglect the dumping, the control input is given as a simple analytical solution of 4th-degree (i.e. the coefficient of 5th-degree is zero) by

$$u(t) = \frac{30V}{gT^3}\{\ell(2-12\tau+12\tau^2) + gT^2(\tau^2-2\tau^3+\tau^4)\} \tag{3}$$

$$\tau = \frac{t}{T} \qquad (0 \le t \le T, 0 \le \tau \le 1)$$

where τ means a normalized time variable.

We call the above control input "nominal control input" because it guarantees to eliminate residual vibration only for nominal plant.

3. Derivation of Robust Control Input

From the practical point of view, the designed control input is expected such as does not generate residual vibration even if the resonant frequency of plant changes. We call this control input "robust control input" because it has robustness against the perturbation of resonant frequency.

In the crane system, a nominal control input is a power series of the 5th-degrees as shown above. In this section we consider a robust control input as a $(5+n)$-th power series. It has n-unknown coefficients and we have to solve these unknown coefficients by optimization techniques. We call this freedom as n-th degree of freedom.

A dynamic equation of crane systems which have the resonant frequency varied from nominal resonant one is represented by

$$\frac{1}{\omega_\delta^2}\ddot{\theta}_\delta + \theta_\delta = \frac{1}{g}u \tag{4}$$

where $\omega_\delta = \sqrt{\frac{g}{\ell+\delta}}$: resonant angular frequency[rad/s], δ : perturbation of rope length[m], θ_δ : swing angle[rad] at the states varied from nonimal ones.

At first, we substitute control input $u(t)$ of $(5+n)$-th power series to Eq.(4), and we get the analytical solution which includes n unknown coefficients, by solving Eq.(4) subject to initial condition $\theta_\delta(0) = 0, \dot{\theta}_\delta(0) = 0$.

Secondly, we solve n unknown coefficients by minimizing the performance index J given by

$$J = \int_{\omega_0}^{\omega_1} (\theta^2(T,\omega) + \frac{1}{\omega^2}\dot{\theta}^2(T,\omega))d\omega \tag{5}$$

where ω_0 and ω_1 are respectively lower and upper bound of the perturbation of resonant frequency, $\theta(T,\omega)$ is the swing angle at the end of acceleration for the system with the resonant frequency ω, and $\theta^2(T,\omega) + \frac{1}{\omega^2}\dot{\theta}^2(T,\omega)$ is the square norm of swing angle in a phase plane at the final time T of acceleration.

As shown in Figure 2, if we drive a nominal system using the nominal control input, the orbital starts from origin and returns to origin at the end of acceleration(solid). But if the system is un-nominal state (i.e. the rope length is different from a nominal length), then the orbital can't return to origin and draws a circle as shown in dotted line. Norm of $||P||^2$ plays a role of a direct index of residual vibration, and therefore a control input minimizing the performance index which integrates $||P||^2$ within all the range of varied resonant frequency reduces the residual vibration comprehensively.

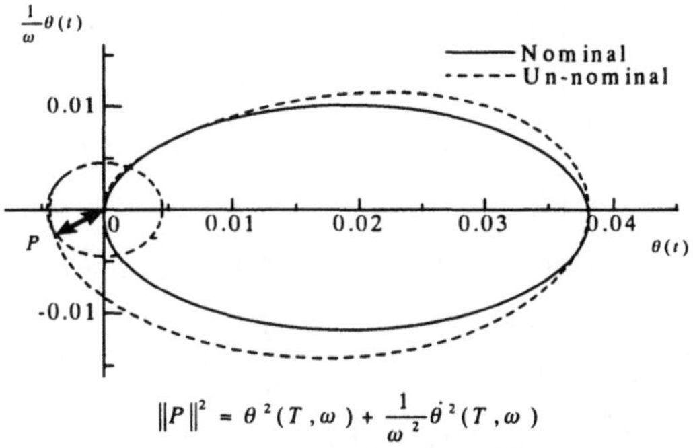

Figure 2: Phase plane of swing angle

4. Relationship between Robustness and Frequency Characteristics

Bhat and Miu[4] have shown that a necessary and sufficient condition for zero residual vibration at the end of acceleration is that the Laplace transform of the bounded control input has zero component at natural frequency of the system.

But the discussion is limitted to a specific frequency; it doesn't discuss the frequency characteristics of the control input which minimizes the residual vibration on the whole range of the perturbation of the resonant frequency. So, let us clear the relationship between minimizing the residual vibration and the frequency characteristics of the control input for single-degree of freedom systems.

Consider the following system,

$$\ddot{\theta} + \omega^2\theta = u \tag{6}$$

A response due to the control input u is written as

$$O(t,\omega) = \int_0^t e^{j\omega(t-\tau)}u(\tau)d\tau \tag{7}$$

in the complex phase plane. Where real axis represents $\dot{\theta}/\omega$ and image axis represents θ. Then a performance index J can be rewritten as

$$\begin{aligned} J &= \int_{\omega_0}^{\omega_1} (\theta^2(T,\omega) + \frac{1}{\omega^2}\dot{\theta}^2(T,\omega))d\omega \\ &= \int_{\omega_0}^{\omega_1} (Im[O(T,\omega)]^2 + Re[O(T,\omega)]^2)d\omega \\ &= \int_{\omega_0}^{\omega_1} O(T,\omega)O^*(T,\omega)d\omega \end{aligned}$$

$$= \int_{\omega_0}^{\omega_1} \left[\int_0^T e^{j\omega(T-\tau)} u(\tau) d\tau \int_0^T e^{-j\omega(T-\tau)} u(\tau) d\tau \right] d\omega$$

$$= \int_{\omega_0}^{\omega_1} \left[\int_0^T e^{-j\omega\tau} u(\tau) d\tau \int_0^T e^{j\omega\tau} u(\tau) d\tau \right] d\omega. \qquad (8)$$

Since $u(t)$ is zero after the end of acceleration ($t \geq T$),

$$J = \int_{\omega_0}^{\omega_1} \left[\int_0^\infty e^{-j\omega\tau} u(\tau) d\tau \int_0^\infty e^{j\omega\tau} u(\tau) d\tau \right] d\omega$$

$$= \int_{\omega_0}^{\omega_1} U(j\omega) U^*(j\omega) d\omega \qquad (9)$$

where $U(s)$ is Laplace Transform of $u(t)$.

This implies that the optimal robust control input which minimizes the integration of residual vibration has the least energy within the perturbation range of frequency domain.

5. Experimental Results

We demonstrate the usefulness of proposed control input by the experiments of an overhead traveling crane system using the different control input and the rope length.

Let nominal rope length be $\ell = 0.5$[m], acceleration time $T = 2$[s], final velocity $\dot{x}(T) = 0.5$[m/s], cart position at the end of acceleration $x(T) = 0.5$[m] and the domain of integration of performance index $\omega_0 = 4.42719$[rad/s]($\ell = 0.5$[m])$\sim$ $\omega_1 = 3.41565$[rad/s]($\ell = 0.84$[m]). Then the nominal control input by Eq.(3) and the robust control input which has one-degree of freedom ($n = 1$) for the nominal control inputs are written respectively as

Nominal control inputs:

$$\ddot{x}(t) = 0.1913266 - 0.5739798t + 2.161992t^2 - 1.875t^3 + 0.46875t^4 \qquad (10)$$

Robust control inputs:

$$\ddot{x}(t) = 0.74644 - 6.1251t + 20.0922t^2 - 29.7232t^3 + 23.29782t^4 - 9.520224t^5 + 1.586704t^6. \qquad (11)$$

Figures 3 and 4 illustrate the acceleration of the cart (i.e. control input) and frequency characteristics.

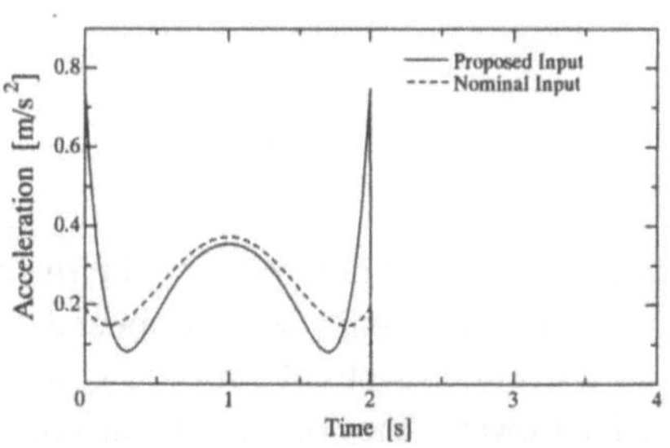

Figure3: Acceleration of the cart

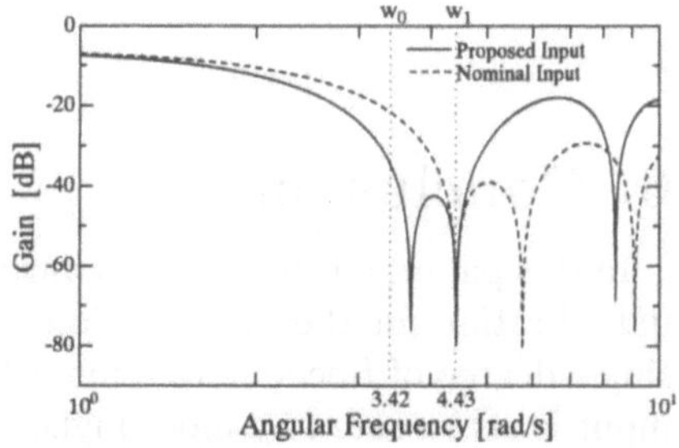

Figure4: Frequency characteristics

Note that robust control input has very big acceleration at initial and final times. Although we can reduce the acceleration by solving the optimal problem subject to "acceleration zero" as the initial and the final condition, the shape of acceleration remains similar because it has very big jark at initial and final times. Since the proposed robust control input has less energy than nominal control input in most frequencies within the domain of integration, it is expected that the robust control input reduces the residual vibration more than nominal control input.

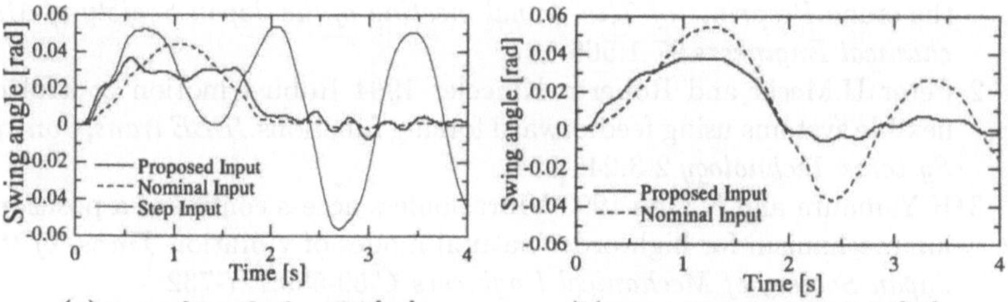

(a) rope length $\ell = 0.5$[m] (b) rope length $\ell = 0.84$[m]

Figure 5: Residual vibration of crane system

Figure 5(a) illustrates experimental results in the case of $\ell = 0.5$[m] of nominal rope length. As a comparison, we also show the step response which satisfies the boundary condition for the cart. While a step command can not reduce the residual vibration at all, both nominal control input and robust control input reduce the residual vibration. It seems that robust control input slightly excites a higher vibration mode. However, Figure 5(b) illustrates that proposed robust control input reduces the residual vibration more effectively than nominal control input in un-nominal rope length of $\ell = 0.84$[m].

6. Conclusion

In this paper, a robust feedforward control input which reduces the residual vibration for the perturbation of resonant frequency was proposed for single degree of freedom system such as a crane system. Proposed control input is represented by more higher order power series of time than nominal control input which reduces the residual vibration only for the nominal plant. A robust control input was solved by minimizing the performance index J which means to minimize the energy of frequency component on perturbation domain. Finally, the usefulness of the proposed robust feedforward control input was demonstrated by a series of experimental results of the crane system with respect to the perturbation of the rope length.

References

1 T.Miyoshi and K.Terashima 1997 Study of the vibrationless reference input for the time-variant vibration mechanism and its application to the crane.*Preprints of 74th-Anual meeting of the Japan Society of Mechanical Engineers* 97-1:566-567

2 Peter H.Meckl and Roberto Kinceler 1994 Robust motion control of flexible systems using feedforward forcing functions.*IEEE trans,Control Systems Technology* 2-3:245-254

3 H.Yamaura and K.Ono 1993 Vibrationless access control of a positioning mechanism for high-order natural modes of vibration.*Trans. of the Japan Society of Mechanical Engineers C* 59-559:727-732

4 Bhat.S.T.and Miu.D.K.1990 Precise point-to-point positioning control of flexible structures.*Trans.ASME,J.Dyn.Syst.Meas.Control* 112-4:667-674

Appendix

Theorem 1 *Let a time-invariant linear system $\dot{x} = Ax + Bu$ be the single input and order n. If the system is controllable and eigenvalues of A have no zero, then the system can be transited by the control input represented by $(2n-1)$-th power series of time from the arbitrary states $x(0)$ at $t = 0$ to the arbitrary states $x(t_1)$ at $t = t_1$.*

[Proof]
A state vector can be represented by

$$\begin{aligned} \boldsymbol{x}_1 &= \int_0^{t_1} e^{A(t_1-\tau)}Bu(\tau)d\tau + e^{At_1}\boldsymbol{x}_0 \\ &= e^{At_1}\int_0^{t_1} e^{-A\tau}Bu(\tau)d\tau + e^{At_1}\boldsymbol{x}_0. \end{aligned} \tag{12}$$

The above equation becomes by parts-integration as the followings;

$$\begin{aligned} e^{-At_1}\boldsymbol{x}_1 - \boldsymbol{x}_0 &= \left[(-A)^{-1}e^{-A\tau}Bu(\tau)\right]_0^{t_1} - \int_0^{t_1}(-A)^{-1}e^{-A\tau}B\dot{u}(\tau)d\tau \\ &= \left[(-A)^{-1}e^{-A\tau}Bu(\tau)\right]_0^{t_1} - \left[(A^{-2})e^{-A\tau}B\dot{u}(\tau)\right]_0^{t_1} \\ &\quad +...+(-1)^{2n-1}\left[(-A)^{-2n}e^{-A\tau}Bu^{(2n-1)}(\tau)\right]_0^{t_1} \\ &= -\left[\sum_{i=1}^{2n} A^{-i}e^{-A\tau}Bu^{(i-1)}(\tau)\right]_0^{t_1} \end{aligned} \tag{13}$$

Note that $u^{(2n)}(\tau)$ is zero when $u(\tau)$ is $(2n-1)$-th order power series of time.

Hence we have Eq.(14) by multipling the both sides of Eq.(13) in terms of A^{2n}.

$$A^{2n}\left[e^{-A\tau}\boldsymbol{x}(\tau)\right]_0^{t_1} = -\left[e^{-A\tau}\sum_{i=1}^{2n}A^{2n-i}Bu^{(i-1)}(\tau)\right]_0^{t_1} \tag{14}$$

Since

$$\begin{aligned} \sum_{i=1}^{2n} A^{2n-i}Bu^{(i-1)}(\tau) &= (B, AB, A^2B, ..., A^{2n-1}B)\begin{pmatrix} u^{(2n-1)}(\tau) \\ \vdots \\ u^{(0)}(\tau) \end{pmatrix} \\ &= (B, AB, ..., A^{n-1}B, A^nB, A^nAB, ..., A^nA^{n-1}B)\cdot \\ &\quad \begin{pmatrix} u^{(2n-1)}(\tau) \\ \vdots \\ u^{(0)}(\tau) \end{pmatrix} \\ &= (V_c, A^nV_c)\begin{pmatrix} u^{(2n-1)}(\tau) \\ u^{(2n-2)}(\tau) \\ \vdots \\ u^{(0)}(\tau) \end{pmatrix}, \end{aligned}$$

Eq.(14) can be rewritten as

$$A^{2n}\left[e^{-A\tau}\boldsymbol{x}(\tau)\right]_0^{t_1} = -\left[e^{-A\tau}(V_c, A^n V_c)\begin{pmatrix} u^{(2n-1)}(\tau) \\ u^{(2n-2)}(\tau) \\ \vdots \\ u^{(0)}(\tau) \end{pmatrix}\right]_0^{t_1}. \tag{15}$$

V_c is a matrix representing observability, and it is full rank by assumption.Then, $(V_c, A^n V_c)$ is raw full rank $(rank = n)$.

When arbitrary $\boldsymbol{x}(t_1)$ and $\boldsymbol{x}(0)$ satisfy Eq.(15), it holds that

$$\begin{pmatrix} A^{2n}\boldsymbol{x}(t_1) \\ A^{2n}\boldsymbol{x}(0) \end{pmatrix} = -\begin{pmatrix} V_c, A^n V_c & 0 \\ 0 & V_c, A^n V_c \end{pmatrix}\begin{pmatrix} u^{(2n-1)}(t_1) \\ u^{(2n-2)}(t_1) \\ \vdots \\ u^{(1)}(t_1) \\ u^{(0)}(t_1) \\ u^{(2n-1)}(0) \\ u^{(2n-2)}(0) \\ \vdots \\ u^{(1)}(0) \\ u^{(0)}(0) \end{pmatrix}. \tag{16}$$

Since $u(\tau)$ is $(2n-1)$-th power series of time represented by

$$u(\tau) = u_0 + u_1\tau + u_2\tau^2 + \ldots + u_{2n-1}\tau^{2n-1},$$

then we have

$$\begin{pmatrix} u^{(2n-1)}(\tau) \\ u^{(2n-2)}(\tau) \\ \vdots \\ u^{(2)}(\tau) \\ u^{(1)}(\tau) \\ u^{(0)}(\tau) \end{pmatrix} = \begin{pmatrix} (2n-1)! & 0 & \dots & 0 & 0 \\ (2n-1)(2n-2)...2\tau & (2n-2)! & \dots & 0 & 0 \\ \vdots & \vdots & \vdots & \vdots & \vdots \\ (2n-1)(2n-2)\tau^{2n-3} & (2n-2)(2n-3)\tau^{2n-4} & \dots & 0 & 0 \\ (2n-1)\tau^{2n-2} & (2n-2)\tau^{2n-3} & \dots & 1 & 0 \\ \tau^{2n-1} & \tau^{2n-2} & \dots & \tau & 1 \end{pmatrix} \begin{pmatrix} u_{2n-1} \\ u_{2n-2} \\ \vdots \\ u_2 \\ u_1 \\ u_0 \end{pmatrix}$$

$$= G_{tmp}(\tau)\boldsymbol{u}$$

Where $\boldsymbol{u}$ is a vector with unknown coefficients, $G_{tmp}(\tau)$ is always full rank ($rank = 2n$) for any τ since $G_{tmp}(\tau)$ is a lower-trianguler matrix. Therefore Eq.(16) can be rewritten as

$$\begin{aligned} \begin{pmatrix} A^{2n}\boldsymbol{x}(t_1) \\ A^{2n}\boldsymbol{x}(0) \end{pmatrix} &= -\begin{pmatrix} V_c, A^n V_c & 0 \\ 0 & V_c, A^n V_c \end{pmatrix} \begin{pmatrix} G_{tmp}(t_1)\boldsymbol{u} \\ G_{tmp}(0)\boldsymbol{u} \end{pmatrix} \\ &= -\begin{pmatrix} V_c, A^n V_c & 0 \\ 0 & V_c, A^n V_c \end{pmatrix} \begin{pmatrix} G_{tmp}(t_1) \\ G_{tmp}(0) \end{pmatrix} \boldsymbol{u} \end{aligned} \tag{17}$$

Note that the first block matrix of right hand side in Eq.(17) is $2n \times 4n$ and raw full rank ($rank = 2n$), and the second block vector of right hand side in Eq.(17) is $4n \times 2n$ and column full rank ($rank = 2n$).

Coefficients of power series of control inputs which make the state transit from x_0 to x_1 can be reduced to

$$\boldsymbol{u} = -\left\{ \begin{pmatrix} V_c, A^n V_c & 0 \\ 0 & V_c, A^n V_c \end{pmatrix} \begin{pmatrix} G_{tmp}(t_1) \\ G_{tmp}(0) \end{pmatrix} \right\}^{-1} \begin{pmatrix} A^{2n}\boldsymbol{x}(t_1) \\ A^{2n}\boldsymbol{x}(0) \end{pmatrix}.$$

This concludes the proof. □

Theorem 2 *Let a time-invariant linear system $\dot{\boldsymbol{x}} = A\boldsymbol{x} + Bu$ be a single input and order n. If the system is controllable and all the eigenvalues of A are zero, then a system can be transited by control input represented by $(n-1)$-th power series of time from the arbitrary states $\boldsymbol{x}(0)$ at $t = 0$ to the arbitrary states $\boldsymbol{x}(t_1)$ at $t = t_1$*

[Proof]

When all eigenvalues are zero, Eq.(12) can be written by parts-integration as

$$
\begin{aligned}
e^{-At_1}\boldsymbol{x}_1 - \boldsymbol{x}_0 &= \left[e^{A\tau}BU_{(1)}(\tau)\right]_0^{t_1} - \int_0^{t_1} -e^{-A\tau}ABU_{(1)}(\tau)d\tau \\
&= \left[e^{A\tau}BU_{(1)}(\tau)\right]_0^{t_1} + \left[e^{-A\tau}ABU_{(2)}\right]_0^{t_1} + \ldots \\
&\quad + \left[e^{-A\tau}A^{n-1}BU_{(n)}\right] + \int_0^{t_1} e^{-A\tau}A^nBU_{(n+1)}(\tau)d\tau, \quad (18)
\end{aligned}
$$

where $U_{(n)}(\tau)$ means n-th integral of $u(\tau)$. Since A^n is zero by assumption (It is clear by Caylay-Hamilton theorem), Eq.(18) can be written as

$$
\left[e^{-A\tau}\boldsymbol{x}(\tau)\right]_0^{t_1} = \left[e^{-A\tau}\sum_{i=1}^{n} A^{i-1}BU_{(i)}(\tau)\right]_0^{t_1} \tag{19}
$$

Since

$$
\begin{aligned}
\sum_{i=1}^{n} A^{i-1}BU_{(i-1)}(\tau) &= (B, AB, A^2B, \ldots, A^{i-1}B)\begin{pmatrix} U_{(1)}(\tau) \\ U_{(2)}(\tau) \\ \vdots \\ U_{(n)}(\tau) \end{pmatrix} \\
&= V_c \begin{pmatrix} U_{(1)}(\tau) \\ U_{(2)}(\tau) \\ \vdots \\ U_{(n)}(\tau) \end{pmatrix},
\end{aligned}
$$

Eq.(19) can be rewritten as

$$
\left[e^{-A\tau}\boldsymbol{x}(\tau)\right]_0^{t_1} = \left[e^{-A\tau}V_c \begin{pmatrix} U_{(1)}(\tau) \\ U_{(2)}(\tau) \\ \vdots \\ U_{(n)}(\tau) \end{pmatrix}\right]_0^{t_1}, \tag{20}
$$

where V_c is a matrix representing observability and it is full rank by assumption.

When arbitrary $\boldsymbol{x}(t_1)$ and $\boldsymbol{x}(0)$ satisfy Eq.(20), it holds

$$\begin{pmatrix} \boldsymbol{x}(t_1) \\ \boldsymbol{x}(0) \end{pmatrix} = \begin{pmatrix} V_c & 0 \\ 0 & V_c \end{pmatrix} \begin{pmatrix} U_{(1)}(t_1) \\ U_{(2)}(t_1) \\ \vdots \\ U_{(n-1)}(t_1) \\ U_{(n)}(t_1) \\ U_{(1)}(0) \\ U_{(2)}(0) \\ \vdots \\ U_{(n-1)}(0) \\ U_{(n)}(0) \end{pmatrix}. \tag{21}$$

Since $u(\tau)$ is $(n-1)$th power series of time represented by

$$u(\tau) = u_0 + u_1\tau + u_2\tau^2 + \ldots + u_{n-1}\tau^{n-1},$$

then we have

$$\begin{pmatrix} U_{(1)}(t_1) \\ U_{(2)}(t_1) \\ \vdots \\ U_{(n-1)}(t_1) \\ U_{(n)}(t_1) \\ U_{(1)}(0) \\ U_{(2)}(0) \\ \vdots \\ U_{(n)}(0) \end{pmatrix} = \left(\begin{array}{ccccc|cccc} \frac{1}{n}\tau^n & \frac{1}{n-1}\tau^{n-1} & \ldots & \tau & 1 & 0 & \ldots & \ldots & 0 \\ \frac{1}{n(n+1)}\tau^{n+1} & \frac{1}{(n-1)n}\tau^n & \ldots & \ldots & \tau & 1 & 0 & \ldots & 0 \\ \vdots & \vdots & & & & \ldots & \ldots & \ldots & \vdots \\ \vdots & \vdots & & & & & \tau & 1 & 0 \\ \frac{1}{n(n+1)\ldots(2n-1)}\tau^{2n-1} & \ldots & \ldots & \ldots & \ldots & \ldots & \ldots & \tau & 1 \\ 0 & \ldots & \ldots & \ldots & 0 & 1 & 0 & \ldots & 0 \\ 0 & & & & & 0 & 1 & 0 & \vdots \\ \vdots & & & & & & \ldots & \ldots & \vdots \\ 0 & \ldots & \ldots & \ldots & \ldots & \ldots & \ldots & 0 & 1 \end{array}\right) \begin{pmatrix} u_{n-1} \\ u_{n-2} \\ \vdots \\ u_1 \\ u_0 \\ c_{n-1} \\ c_{n-2} \\ \vdots \\ c_0 \end{pmatrix} = U_{tmp}\boldsymbol{U},$$

where $\boldsymbol{U}$ is a vector of unknown coefficients including $\{c_{n-1}, c_{n-2}, ..., c_0\}$ which is the integration constants. Although their integration constants need in the course of solving the coefficients of control input, they are not related to the coefficients of $u(\tau)$ directly. Then Eq.(21) can be rewritten as

$$\begin{pmatrix} \boldsymbol{x}(t_1) \\ \boldsymbol{x}(0) \end{pmatrix} = \begin{pmatrix} V_c & 0 \\ 0 & V_c \end{pmatrix} U_{tmp} \ \boldsymbol{U}.$$

The coefficients of power series of control input which makes the state transit from x_0 to x_1 can be reduced to

$$\boldsymbol{U} = \left\{ \begin{pmatrix} V_c & 0 \\ 0 & V_c \end{pmatrix} U_{tmp} \right\}^{-1} \begin{pmatrix} \boldsymbol{x}(t_1) \\ \boldsymbol{x}(0) \end{pmatrix}. \tag{22}$$

This completes the proof. □

30

Path Planning and Navigation of Overhead Traveling Crane with Three-Dimensional Transport Based on a Diffusion Equation Strategy

K. Terashima and M. Suzuki

1 Introduction

In order to heighten the effectiveness and flexibility of factory automation, the optimization of automatic transportation systems is vital. The overhead travelling crane, with its high flexibility and potential for moving freely in three-dimensional space, has been a staple of most such transportation systems. Although there has been a substantial body of research into these systems, no prior studies have addressed the present topic; namely, the development of an autonomous mobile overhead travelling crane of complex transfer trajectory with and without hoisting or lowering motions on a straight and curved transfer path system [1]~[5]. In multi-item, small-sized production systems, the start and goal points of crane transfer are frequently various. The first step in realizing an autonomous mobile crane system is to establish the motion path planning as such to compute a precisely specified path between the start and the goal based on a given geometrical map, and to compute the motion control with respect to both the tracking control and the sway-suppression control for off-line application, although on-line motion planning and execution in changing indoor environments will be a succeeding problem.

In addressing this issue, in our previous research [3]~[5] we proposed a method for establishing an autonomous mobile crane system with a capacity for obstacle recognition and path-planning for off-line application, as well as for suppression of sway during the transfer of the object on a complex transfer trajectory. In the visual system, an environment map was generated by the Slid-Ray Projection method [4],[5]. In the planning system, the obtained environment map was used to derive an optimal transfer path such that a minimum energy or a minimum distance path using a Branch and Bound method [4],[5]. In the control system, a fixed-pole variable controller was used to accomplish motion dynamical control [3]. However, an explosion of necessary calculations is occurred with the increase in the number of obstacles in the planed path when using the Branch and bound method, and the obtained path had to be revised in the smooth curve at each obstacle edge, because a path is connected between a point and a point in its method. Since the steep curve transfer generates the swing of the crane's transfer objects, we need an easy method to achieve a smooth and safety path. In recent years, many studies for path planning based on the potential field have been done focusing on mobile robots [6],[7] with two-dimensional transport; however, studies have not been carried out for path planning in regard to three-dimensional transport on an overhead travelling crane.

This paper presents a method for map-based path planning for an overhead travelling crane attempting smooth collision-free paths by using a potential approach based on three-dimensional diffusion equations, and the obtained path is applied to a navigation control system in order to realize both obstacle avoidance and swinging-suppression for transfer objects.

2 Strategy of Path Planning

2.1 Strategy outline

First, a crane's (transfer object's) collision-free path between a start and a mission-dependent goal point is generated by off-line simulation of a diffusion process, and by evaluating the gradient of the computed concentration distribution functions. Secondly, for both position and swinging-suppression control, command inputs (reference trajectory given at every time-step) for the path obtained in potential fields must be given in order to implement a feedback control system. The magnitude of the command inputs is decided by considering the crane's maximum velocity and maximum acceleration, while a moving direction is calculated in terms of a proposed vector-composite method. For the command inputs obtained, a time-varying controller using a fixed-pole approach is applied to obtain the position control and swinging-suppression control. Especially, in excessive curved paths, a holding technique is proposed using a steering angle at a subgoal that allows a crane to stop with maximum deceleration along the velocity direction at a given time. If *a steering angle* is larger than a certain angle, command inputs are held for a certain time at the same position in order to compensate for the tracking delay which may generate a collision. Definition of *a steering angle* will be submitted in section 3.2. Finally, the usefulness of the proposed method is demonstrated by simulations and experiments.

2.2 Diffusion process

Assuming that the motion of a real crane can be reduced to a problem in finding a path in known environments, the following physical analogy can be employed for path-planning purposes. Goal point G of the crane's collision-free motion path is considered to be the location of a virtual source like a scent or a perfume. While the concentration at point G is kept constant, the substance diffuses steadily into the surrounding space [6].

The diffusion process is modeled by an unsteady or dynamic diffusion equation as described by Fick's second law:

$$\frac{\partial C}{\partial t} = D\left(\frac{\partial^2 C}{\partial x^2} + \frac{\partial^2 C}{\partial y^2} + \frac{\partial^2 C}{\partial z^2}\right) \tag{1}$$

By application of standard finite difference methods, the following time and space discretized model of Eq.(1) can be obtained for a grid point or node:

$$\begin{aligned} C_{t+1,i,j,k} &= \mathrm{r}\,(C_{t,i+1,j,k} + C_{t,i,j+1,k} + C_{t,i,j,k+1} \\ &\quad + C_{t,i-1,j,k} + C_{t,i,j-1,k} + C_{t,i,j,k-1}) \\ &\quad + (1-6\mathrm{r})\,C_{t,i,j,k} \end{aligned} \tag{2}$$

Boundary condition:

$$C_{t,i,j,k} = \begin{cases} 1 & \text{(Goal point)} \\ 0 & \text{(Wall, Obstacles)} \end{cases} \tag{3}$$

Initial condition:

$$C_{0,i,j,k} = \ 0 \ \text{(except for Goal)} \tag{4}$$

where $\mathrm{r} \equiv \mathrm{D}\,\Delta t/(\Delta h)^2, i \in x, j \in y, k \in z, C_{t,i,j,k}$; concentration of time t and space (i,j,k), Δt; sampling time, Δh; grid size (0.05m), D; diffusion coefficient, $r; 0 < \mathrm{r} \leq 1/6$ (stable region).

If complete and instantaneous absorption of any substance reaching the obstacle points or natural and artificial space boundaries is assumed, the concentration values will always remain equal to zero at the corresponding point sets. As a result of the diffusion process, a concentration distribution develops over time and space. In its equilibrium state, the distribution shows a monotonically decreasing concentration along a path between the goal point G, i.e. the point of maximum potential concentration, and an arbitrary crane starting point S inside the substance field space. Parameter r is determined by D, Δt and Δh, which is related with the stability and oscillation of solution [6],[7]. In this paper, the value of r=1/12 is determined through the simulation analysis, which is not determined in the physical sense. When the value of r is larger more than 1/10, the obtained path becomes oscillatory, and the value of r is selected as r=1/12. Then, when diffusion coefficient D in gas is about $10^{-4}\mathrm{m}^2/\mathrm{s}$, time step Δt becomes 2.08 sec.

Figure 1.(a) and (b) show the obstacles in a work space of 2.0x1.0x1.0m and 3-dimensional potential concentration in (X,Y)-field for each height of Z, while Figure 1.(c) represents obstacles. In the simulation, a scent travels in calculation steps of 75 from a goal to a start. Because the steady potential field was obtained in about 5 times of 75, the potential field obtained in 375-steps was used for the path planning in this paper. We note that a path avoids obstacles and walls with the increase of calculation steps.

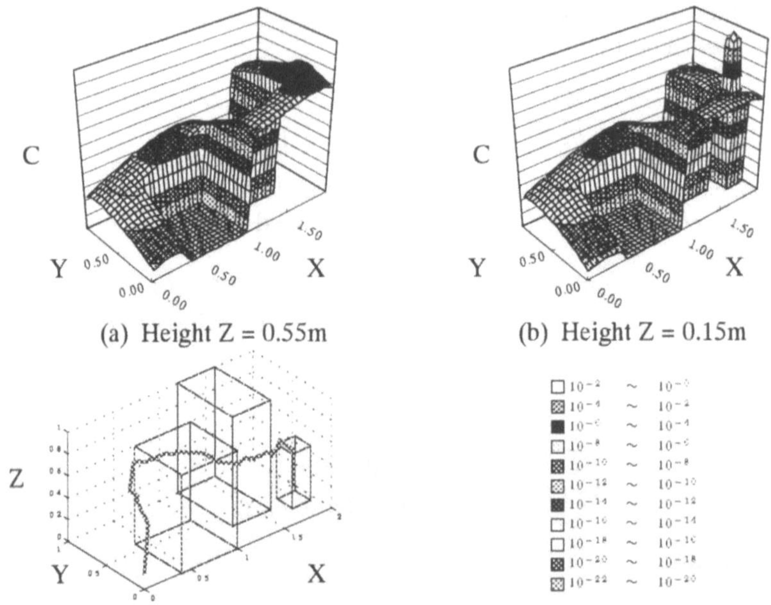

(a) Height Z = 0.55m (b) Height Z = 0.15m

(c) Obstacle and path

Figure 1: Potential field and path

2.3 Gradient computation

Based on the above-mentioned properties of the distribution function, the steepest ascent path, following the gradient from an arbitrary starting point to the goal point, can be computed. The unique path between S and G can be found by following a steepest ascent strategy. For the purpose, we give the following two preliminaries.

Preliminary 1

(*Extension to 3-dimensional case of Ascent algorithm 1 in [6]*):

(1) Pick the starting point $\mathbf{r}_s$;

(2) In the set of immediate neighbors (M=26) of $\mathbf{r}_s$, select the grid point $\mathbf{r}_p$, with maximum concentration $C_{\infty;r}$

(3) Continue with the same procedure starting from $\mathbf{r}_p$ until $\mathbf{r}_g$ is reached

The path is given by the sequence of grid points ($\mathbf{r}_s$, $\mathbf{r}_{p_1}$, $\mathbf{r}_{p_2}$, ..., $\mathbf{r}_{p_n} = \mathbf{r}_g$). Figure 2.(c) is a path by the algorithm of Preliminary 1 based on the

3-dimensional diffusion concentration for every Z, where (a) and (b) are one example. This path proves to be comparatively rough with a tendency for the crane to touch obstacle corners.

In a 2-dimensional case, smoother paths can be generated by application of second order gradient interpolation during the steepest ascent search and by permitting the selection of appropriate path points $\mathbf{r}_{p_i}$ outside the set of discrete grid points r which is described as Ascent algorithm 2 by G.Schmidt [4].

Preliminary 2

(*Ascent algorithm 2 in [6] for 2-dimensional space systems*):

(1) Pick the starting point $\mathbf{r}_s$. Evaluate the gradient direction at $\mathbf{r}_s$ based on the second order polynomial interpolation of $C_{\infty;r}$ over $\mathbf{r}_s$ and its immediate neighbors (M=8).

(2) Select $\mathbf{r}_{p_1} = \mathbf{r}_s + \eta \,\mathrm{grad}\, C_{\infty;r}$.

(3) Evaluate gradient at the grid point next to $\mathbf{r}_{p_1}$ as mentioned above.

(4) Select $\mathbf{r}_{p_2} = \mathbf{r}_{p_1} + \eta \,\mathrm{grad}\, C_{\infty;r}$.

(5) Continue with the same procedure starting from $\mathbf{r}_{p_2}$ until $\mathbf{r}_g$ is reached.

However, in the 3-dimensional case, it is impossible to use a second-order gradient interpolation. And, it is inexact to use the approximation of (3) in Preliminary 2 when the grid size is large. Further, the 3-dimensional example requires a trial-and-error technique for the appropriate choice of parameter η. Instead of parameter η, it will be possible to give the magnitude for the direction of movement in a more systematic way by considering the allowable maximum acceleration and velocity constrained by the hardware.

We then propose a vector-composite method to determine the smooth path, instead of second-order interpolation as shown in Figure 2.

Preliminary 3

(*A vector-composite method to determine the direction of movement*):

(1) Draw the potential concentration of 26 grid points(M=26) around the present position.

(2) Calculate the gradient between the present position and each grid point.

(3) Determine the direction of movement in the next step by the vector-composite of each vector by the following equation:

$$U_{now} = \sum_{i=\frac{x}{\Delta h}-1}^{\frac{x}{\Delta h}+1} \sum_{j=\frac{y}{\Delta h}-1}^{\frac{y}{\Delta h}+1} \sum_{k=\frac{z}{\Delta h}-1}^{\frac{z}{\Delta h}+1} \left\{ \left| \frac{C_{(i,j,k)} - C_{P_{\mathrm{now}}}}{P_{(i\Delta h, j\Delta h, k\Delta h)} - P_{\mathrm{now}}} \right| \frac{P_{(i\Delta h, j\Delta h, k\Delta h)} - P_{\mathrm{now}}}{\left| P_{(i\Delta h, j\Delta h, k\Delta h)} - P_{\mathrm{now}} \right|} \right\} \tag{5}$$

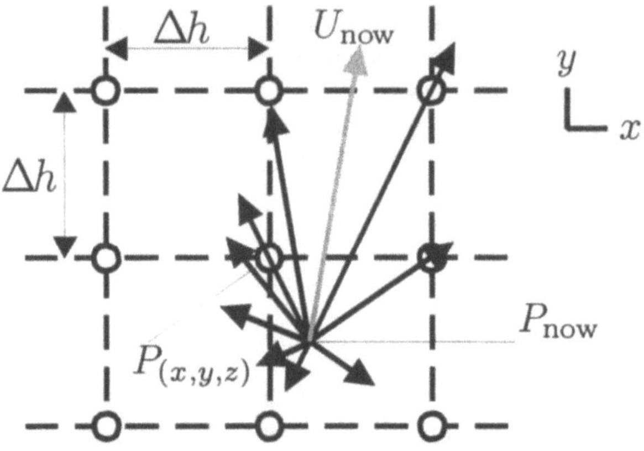

Figure 2: Proposed vector-composite method

In this paper, the number of grids is 16000, because a transfer space is 2.0x1.0x1.0m and grid interval is 0.05m. We sometimes need to interpolate between grids, because it is not favorable to use the approximation of (3) in preliminary 2 in case that grid interval is long. So, since a cube consisting of 8 vertices is small in the whole configurated space, the concentration value inside a cube may be represented by a linear interpolation.

By preparing a composite-vector method and a linear interpolation method of the concentration between grids, we can get a reference trajectory given every time-step by considering the limits of acceleration and velocity due to the motor capacity of the crane. For simplicity, we explain the concept using the example of a 2-dimensional case, and then extending to 3-dimensional case using the obvious example of a cube. The square part in Figure 3 shows the allowable region of velocity in the next step when the maximum acceleration has been used for the present velocity. P_{now} denotes the present position of the crane, V_{now} presents velocity, U_{now} the proposed direction calculated from the potential field and V_{next} reference velocity at the next sampling. However, because velocity imposes limits on the maximum speed, the actual allowable region of velocity becomes a shadow part in square one. Figure 4 shows a method to generate reference velocity at next sampling time.

Preliminary 4 (*A method to determine magnitude of movement*):
Two kinds of cases are considered. Case (a) is such that the direction of U_{now} is included in the allowable region of velocity at the next sampling time. Then a farthest point of intersection from the present position is selected among the intersections with the planes, where points of intersection between U_{now} and each plane of a cube are calculated by using a straight line equation of U_{now} and 6 plane equations for a cube. Case (b) is such the case that the direction of U_{now} is not included in the allowable region of velocity at the next sampling

time. A perpendicular line is then drawn from the tip of V_{now} to the line of U_{now}, and a nearest point of intersection between its perpendicular line and a plane of the cube is calculated. Then, V_{next} is determined as shown in case (b) of Figure 4. Hence, by tracing these points, we can get reference trajectories every time-step. In this paper, the width of time-step is 0.01 sec.

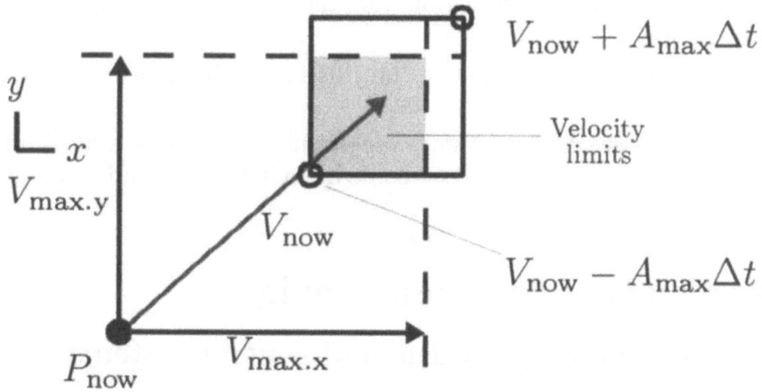

Figure 3: Allowable region of velocity at next step under limits of acceleration and velocity

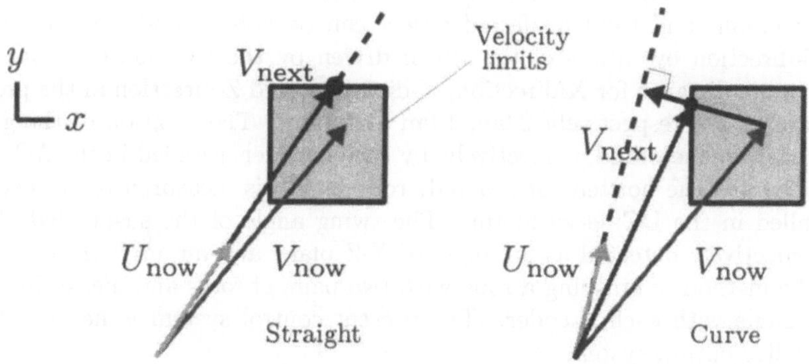

Figure 4: Determination of reference velocity at next step

Now, we summarize the procedure to get a reference trajectory or command input:

<u>**Main Result 1**</u>

(*Procedure of obtaining a reference trajectory in a 3-dimensional case*):

(1) Determine Start point S and Goal point G, and generate the potential field.

(2) Determine the direction to proceed at the next step from the present position in terms of a vector-composite method gathering each gradient between potential concentration of the present position and a potential position in its surrounding 26 grids.

(3) Move a crane only one sampling time, where the direction U_{now} is determined by a vector-composite method described in Preliminary 3, and the magnitude along its direction is calculated by using the allowable maximum values of acceleration and velocity in Preliminary 4.

(4) Calculate the potential concentration at the present position obtained in terms of a linear interpolation from the potential concentration in 8 grids of a cube.

(5) Repeat the procedure (1) - (4) until the crane has reached its goal point.

We remark that, in rare cases, Main Result 1 may drop a local minimum, and that this deficiency can be overcome by switching Preliminary 1 and Main Result 1.

3 Application to Crane Navigation

3.1Experimental apparatus and control systems

Figure 5 shows the apparatus of an experimental overhead travelling crane built in the laboratory. A girder and a cart are independently driven in X-direction as travel motion and Y-direction as transverse motion by each AC servo motor, and the transferred object can be hoisted and lowered toward in Z-direction by means of the winch driven by the DC servo motor. The maximum distance for X-direction, Y-direction, and Z-direction in the present apparatus are respectively 2.0m, 1.0m and 1.0m. The position of the girder and cart are measured, respectively, by each encoder installed in the AC servo motors, and the hoisted (or lowered) rope length is measured by an encoder installed in the DC servo motor. The swing angle of the suspended object is respectively detected as an angle of X-Z plane and an angle in Y-Z plane by the method of grasping a rope with two pairs of forks and measuring each fork angle with each encoder. The present control system is hence a three input-five output system.

The model of an overhead travelling crane representing 3-D in its travelling, traverse, and hoisting motions has already been built by the authors using the Newton and Euler method. A fixed-pole variable gain controller was used for the LPV (Linear Parametric Variable system), which adjusts the change of rope length. Pole assignments were determined by considering both the tracking properties to the reference trajectory (command inputs) and the suppression of sway of the transfer-object, where the controller for X-, Y-, and Z-direction was independently designed. The environment map was generated by the Slid-Ray Projection method. The details are omitted due to the space limitation of this paper and since these subjects have been discussed in authors previous papers.

Figure 5: Laboratory experimental apparatus

3.2 A holding method using a steering angle and control results

In off-line simulation, based on the given environmental map, a reference trajectory, i.e. command inputs, is calculated following the algorithm of Main Result 1. For the command inputs given by reference trajectory, a fixed-pole variable controller is applied to both the tracking control and the suppression of the transfer objects sway.

Now, in the crane control, a transfer delay for the command input occurs due to the consideration of both the tracking control and the anti-sway control, because both are trade-off relations. In the control design, a compromise between both has been conducted by the appropriate adjustments of control gain using computer simulations. However, in the curve transfer, this problem is serious, and hence we sometimes encounter collisions with obstacles due to the transfer delay for command input, as shown in Figure 6. If controller gain is given as a large value following the reference trajectory, a large sway in transfer-objects will be generated. If control gain is given as a small value to suppress this sway, it will take a longer transfer time. Further, it is very difficult to beforehand determine the command inputs, considering the suppression of sway in the present three-dimensional complex trajectory. Yet it is possible, in the simple case of straigh transfer, to find an optimum control input and an optimum trajectory by using a nonlinear dynamical optimization technique. So, in a curve transfer, we use other adjustments in addition to controller's gain. Namely, in order to avoid collisions with obstacles in a curve trajectory, we propose *a holding method*, using *a steering angle* at a subgoal to improve the tracking property for the reference values of a crane. Let us define y_t as the present position of a crane at time t; r_t as reference position at time t; v_t, the velocity at reference position at time t; L_{nf}, the minimum distance $L_{nf} = {v_t}^2/(2A_{max})$ which can be stopped at the maximum deceleration, i.e., A_{max} when the reference position of a crane moves along the direction of the present velocity v_t at time t. We define *a steering angle* as follows:

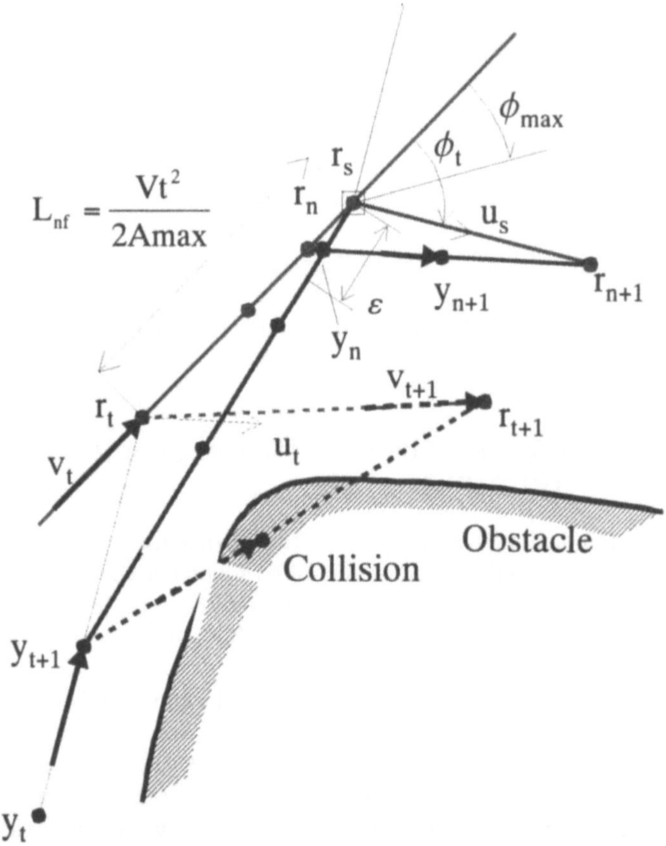

Figure 6: Concept of a holding method to avoid obstacle in curve transfer

<u>**Definition 1**</u> (*Steering angle*):

Steering angle ϕ_t at a subgoal, meaning the stopping-position is defined by the angle between the direction of the velocity at the present position of r_t and the direction of the velocity calculated by a vector-composite method at r_s, where r_s is a position in which a crane can be stopped in the future with the distance of L_{nf} by using the maximum deceleration along the direction of

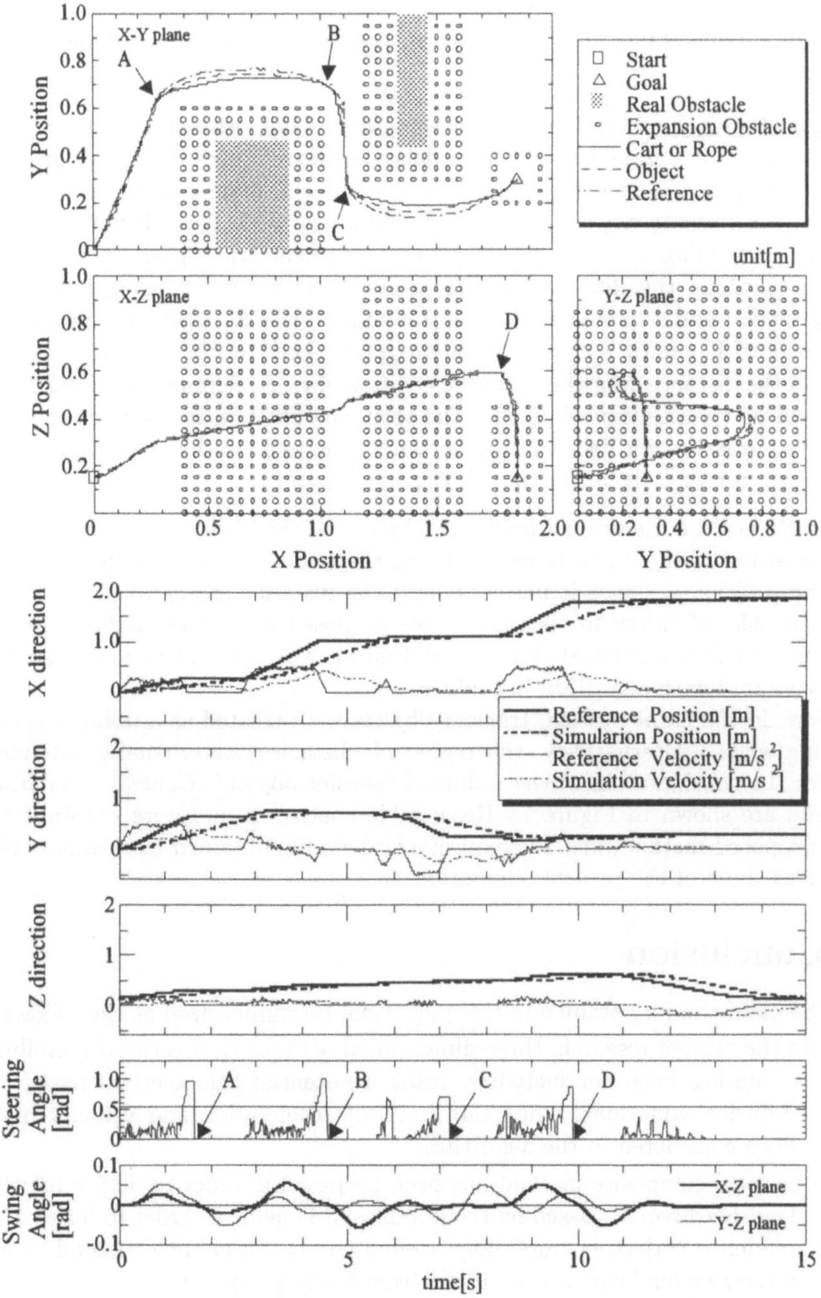

Figure 7: Simulation results with a holding method using a steering angle

the present velocity v_t, as seen in Figure 6.

Now, *a holding method* using *a steering angle* at a subgoal is summarized as follows:

Main Result 2 (*A holding method using a steering angle at a subgoal*):
Assume that a crane has a transfer delay for the reference trajectory because of the suppression of swing of the transfer objects, mainly in curve transfer. Then, if *a steering angle* ϕ_t has the relation of $|\phi_t| > \phi_{\max}$ as shown in Figure 6, the succeeding reference position for the present crane position y_t should be given along the direction of the present velocity v_t by using maximum deceleration and stopping at the position of $r_s(= r_t + L_{nf})$, where we need the distance of L_{nf} until a crane can be stopped. And until a crane approaches within the distance ε for the reference position of r_s, a reference position should be held at the position of r_s, and the reference position is restarted when a crane approaches within ε for r_s as shown in Figure 6. In this paper, $\phi_{\max}$ and ε are respectively given as $\phi_{\max}$=0.55rad and ε=0.07m.

We do not use the present holding technique at the position of r_t. For, when the position of the crane is very near r_t, the crane cannot stop immediately at the position r_t, since it needs at least the distance of L_{nf} to stop. Then, for the sake of safety in such cases, we proposed *a holding method* using *a steering angle* at a subgoal. We remark that an Algorithm of Main Result 2 is incorporated with algorithm in Main Result 1.

Now, in the crane system, transfer objects were treated as a point without having scale. Alternatively, the region of obstacles was virtually extended under the consideration of the radius of transfer objects. Control simulation results are shown in Figure 7. Reasonable control results were obtained by the proposed method, and the experimental results obtained were almost the same as those of the present simulation.

4 Conclusion

The main results obtained in this paper can be summarized as the follows:

(1) In the present research, three-dimensional path for the overhead travelling crane has been conducted by using a potential field method based on diffusion equations, when the limits of acceleration and velocity have been considered in the algorithm.

(2) A vector-composite method has been proposed in order to determine the transfer direction based on the potential 3-D field. In order to follow the reference trajectory, especially around corners, a holding method using a steering angle at a subgoal has been newly proposed.

(3) The usefulness of the method described above has been demonstrated through control simulations and laboratory experiments focusing on obstacle avoidance and the reduction of vibration.

REFERENCES

1. Sakawa Y, Shindo Y 1982 Optimal control of container crane. *Automatica* 18:257-266
2. Hamalainen J J, Martlinen A, Baharora L, Virkkunen J 1995 Optimal path planning for trolly crane. fast and smooth transfer of load. *IEE Proc. - Control Theory Appl.* 142:51-57
3. Kaneshige A, Terashima K, et al 1997 Motion control of an overhead travelling crane with hoisting motion and curve trajectory. *Trans. of the Japan Society of Mechanical Engineers(C)* 63:921-928
4. Kaneshige A, Terashima K, et al 1997 Autonomous mobile crane system considering obstacle recognition and optimal path planning. *Trans. of the Japan Society of Mechanical Engineers(C)* 64:487-494
5. Kaneshige A, Suzuki M, Yin Lei, Terashima K 1997 Autonomous mobile crane system with optimal path planning and environmental recognition for obstacle avoidance. *Proc. of the 2nd Asian Control Conference, July 22-25, Seoul, Korea* I:597-600
6. Schmidt G, Azarm K 1993 Mobile robot path planning and execution based on a diffusion equation strategy. *Advanced Robotics* 7:479-490
7. Azarm K, Schmidt G 1994 Integrated mobile robot motion planning and execution in changing indoor environments. *Proc. of IROS'94, Munich, Germany* :298-395

REFERENCES

[illegible]

PART V

SPECIFIC APPLICATIONS

31

Teleworkplace in Laser Material Processing

W. Schmidt, H.-G. Grundmann, G. Olthoff and R. Borger

1 Introduction

Five years ago teleworking was an subject required applications. Meanwhile the situation has changed completely. There are reasons for that, e.g.the globalisation of markets, and in generally, the enhanced international status of work.

Furthermore, in the early nineties components have been created for an effective manufacturing process with regard to data processing, telecommunication, picture transmission, and modification of computerized numerical controls. The progressive trend leads to an increasing use of multimedial technology in manufacturing process forcibly.

To combine that for material processing it would be possible to reduce cost-extensive parts in the field of engineering. Concerning this laser material processing is a manufacturing method that includes both short cycle times on the one hand and high cost operations on the other hand in industrial applications. Therefore, laser technology can be able to use teleworkplace strategies in an excellent way.

Nowadays the expectations awakened concerning the production engineering and economical efficiency of the multimedial technologies have not been fulfilled to the greatest extent. Comparative assessments in relation to the conventional technologies exist.

A lot of companies have been investigated with regard to use of teleworkplace technologies. The result shows that advantages could not fully scoop out up to now. Most of the conventional process sequences are based on a discontinued flow of information. In our opinion there are a lot of reasons for that: often the interfaces are not exactly adapted between the several modules, furthermore, the industrial users underestimate the cost-performance ratio of such a solution.
The task is to activate new focal points of application for continual low cost tele-

working projects. Companies in mechanical engineering and especial mechanical engineering offering to use such technologies. Therefore, we have started to create a continued project consiting of a modular concept that is capable of development dependent on the special interests of small and medium sized companies.

The object of the contribution is to present the production engineering state of development of these technologies with a special example of application. The project shows first results of our department of mechanical engineering in cooperation with small and medium sized businesses. By use the manufacture can be completely controlled and monitored by teleworkplace of engineering centre. By the way, low cost teleworking become the focus of attention worldwide. Following the project infrastructure is shown in a short way.

2 Modular structure of the conception

Computer aided planning modules are important components in the flexible and cost oriented manufacturing. The objective is to point out effective possibilities for process planning and quotation costing to small and medium sized businesses.

A comprehensive system of constructional and technological fundamentals of laser beam machining meets these interests (fig.1) /1/.

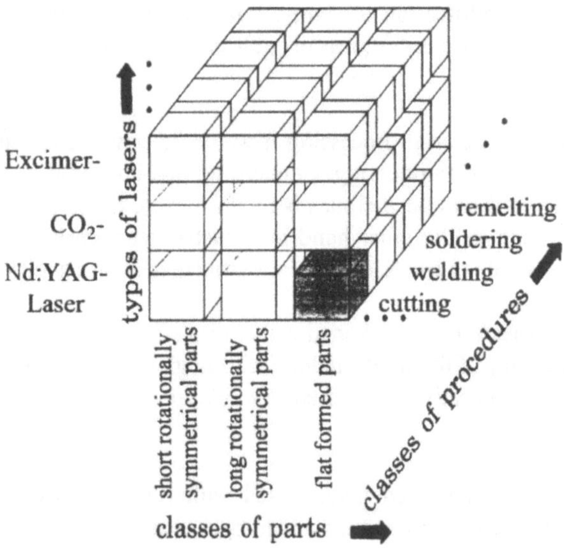

Fig.1: Modular structure of the conception

The system registers the dependences between classes of parts, classes of procedures and types of lasers.

There were two reasons for the creation of the system:

- The results of published papers were not precise enough and not helpful for decisions in practice.
- A quickly determined expertise concerning technical, economical and qualitative results is needed.

The system is based on special classifiers for both constructional data and technological data. The structure of classifiers can be varied in a simple way with regard to special market conditions or interests of businesses.

An example for a database is shown in fig.2. The data base includes all informations for laser hardening of a locking disk.

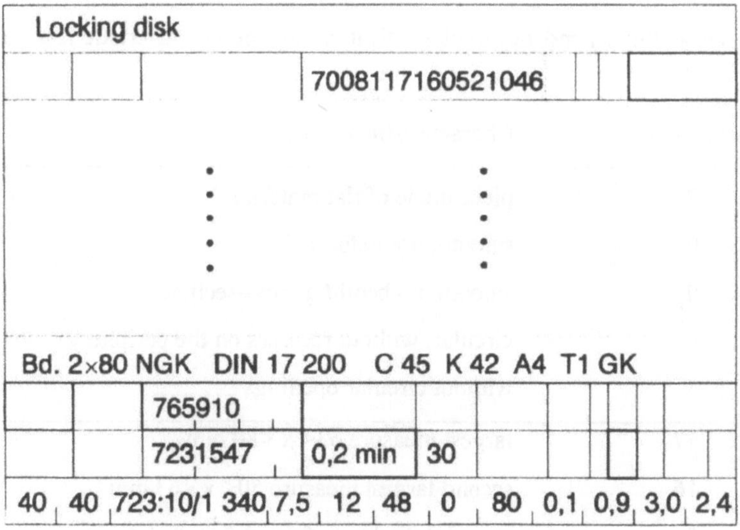

Fig.2: Example for a special data base

The CAD/CAM-systems are equipped with data management systems which use the feature-based description of parts and processes. The search for identifying the right class and the connections between parts and processes is carried out by an alphanumerical dialog.

For this purpose, following it will be demonstrated the basic functions of the system by another example of laser cutting of a modified logo of Athens`university consisting of stainless steel (fig.3).

According to the methodology of the complex system, the logo is numerically coded. The classifier is stored in the computer.

Fig.3: Example of workpiece for purpose of demonstration

Fig.4 shows the dependence of clas-sification no. and characteristic feature for the logo.

Classification no.	Characteristic feature
7	piece made of flat material
0	smooth, not deformed
0	smooth, no bending cross-section
6	circular, without recesses on the periphery
3	without circular openings
17	largest measure 63< x <80 mm
16	second largest measure 50< x <63 mm
06	material thickness from 0.5 up to 1.6 mm
1	stainless steel, not quenched and not tempered
0	not annealed
0	protected against corrosion
4	surface finish quality 6 µm < R_z <10 µm
9	quality of the fit IT 10...11

Fig.4: Excerpt of constructional classifier

The coding of process is done in the same way by use of technological classifier. The classifier is able to specify the technological data as follows:

Process	State	Condition	Equipment
843	17	39	865910.

Thus, the connection is created between constructional and technological data for laser cutting in the same way as shown in fig.2 for laser hardening.The computer-based evaluation of the data fields is possible.

Furthermore, additional parameters are presented for material specification, set-up parameters for machine tool, setting parameters for the laser, and cutting time. Additional, in the data base are stored available materials, data of several workcentres, statistical cost accounted workpieces, and given quotations.

In such a way it would be possible to evaluate a request of market in a shortest time and to hand over a quotation costing for the potential customer. That is the content of <u>step 1</u> and <u>step 2</u>. In summary, these steps include the digital precalculation of machining with help of repetitive workpieces or similar workpieces, furthermore, the digital offer for service provider of market.

By the way, such a comprehensive system of constructional and technological fundamentals of laser beam machining delivers further important findings, such as the formulation of technological laws and rules. The verbal description of such a rule for laser hardening of a sharp-edged contour is shown in fig.3. In this case laser hardening is selected to show the widespread performance of the system.

<u>Rule no. 023</u>

- indirect irradiation

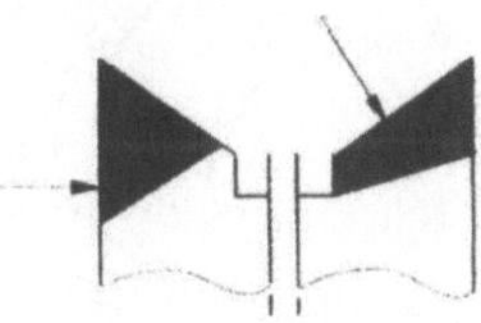

characteristics:

larger dephts and widths of hardening
reduced danger of melting and overheating

cause:

more homogeneous effect of integral energy density with utilization of an imperfect heat conduction specific with regard to form and function

- direct irradiation

characteristics:

reduced hardening dephts and/or increased

danger of melting and overheating

cause:

imperfect heat dissipation
increased integral energy density at the edge of the workpiece

conclusions:

⇒ edges, corners are to be irradiated indirectly (better heating effect, simplified positioning, less reworking)

⇒ edges with chamfers or radiusses are more suitable

Fig.5: Example of a technological rule for laser hardening

Further steps run at teleworkplace as follows:

Step 3:
The file generation function of logo is done with a special software supporting file dxf.

Excerpt from file dxf

```
0
SECTION
2
HEADER
9
$ACADVER
1
AC1012
9
$DWGCODEPAGE
3
ansi_1252
9
$INSBASE
10
0.0
20
0.0
30
0.0
9
$EXTMIN
10
0.6201998948
20
0.577705856
30
0.0
9
$EXTMAX
10
23.9337620794
20
36.1243345739
30
0.0
9
$LIMMIN
10
0.0
20
0.0
9
$LIMMAX
10
12.0
20
9.0
9
$ORTHOMODE
70
0
9
REGENMODE
70
1
9
$FILLMODE
```

Fig.6: Drawing of logo and result of digitalization

Step 4:
It contains the data transformation of file dxf into file cnc of machine tool with the help of technological set-up parameters similarly locking disk.

Excerpt from file nc

```
F1000
G1 Z5
G1 G90 X0 Y0
(———————— Start contour 1 ————)
G00 X5.954 Y2.678
(———————— Laser on ————————)
G1 Z1
M8
M3
G1 Z0
S300
(———————— Contour cut ————)
G01 X5.954 Y4.737
G01 X5.954 Y9.979
G03 X6.943 Y9.919 I7.896 J33.732
G03 X8.097 Y14.155 I6.869 J12.214
G03 X5.762 Y13.822 I7.159 J12.38
G01 X5.862 Y16.319
G01 X2.413 Y16.319
G01 X2.876 Y22.943
G01 X0.62 Y22.943
G01 X0.62 Y25.719
G01 X2.895 Y25.719
G02 X11.602 Y34.325 I11.564 J25.656
G02 X19.732 Y32.102 I7.486 J3.3
G01 X19.732 Y29.518
G03 X11.508 Y31.852 I3.891 J-10.653
```

Technological parameters

Material		**: stainless steel**
Thickness	**mm**	**: 0.5**
Voltage	**V**	**: 400**
Puls Frequency	**ms**	**: 0.34**
Frequency	**Hz**	**: 477**
Compensation value of radius	**mm**	**: 0.15**
"Burn a hole in" distance	**mm**	**: 1**
Working plane	**mm**	**: 1**
"Burn a hole in" time	**s**	**: 0.3**
Rapid feed	**m/min**	**: 5**
Cutting feed rate	**m/min**	**: 1**
Cycle time	**s**	**: 1.9**
Cutting length	**m**	**: 3.54**

Fig.7: Nc machining program with special technological parameters

Step 5:
A special data simulation program is able to locate and to remove any errors in-time before the cnc program gets access to machine tool. Such errors could be e.g. not continued contours at workpiece drawing caused by cad software, and partially overlapped lines at technical drawing.

Step 6:
External data file transmission mode is working via Internet or other special networks such as Intranet or local area networks to send cnc files from teleworkplace to machine tool /2/.

Step 7:
The step is one of the most important steps regarding the whole project. It represents both the control of manufacture at machine tool and the online- multimedia quality monitoring at teleworkplace of engineering centre.

Based on online picture transmission the digital in-process quality monitoring is increasing in modern industrial manufacturing, e.g. laser material processing, high speed cutting (hsc) at the field of turning, milling and drilling a.s.o. In many cases it is difficult for the engineer to evaluate and make use of the advantages of in-process quality monitoring for problematic parts, and sensible technological parameters. Special process variants help to influence the unwanted forming of burs or flashes in the cutting process. A further example is to control the material inhomogenities with explosive material ejection. So there are many reasons for changes in the surface topography of workpieces. Therefore we are preparing a special TV camera recording system. New potentials for use arise where conventional methods fail. Such potentials could be closed loop controlled micro-processing.

Some examples of monitoring of lasermachined contours of workpieces are shown in fig.8 and fig.9. In fig.8 at the top (left) the bore with a diameter of 1.1 mm is extremely corresponding with the drawing data. The influence of defects of shape are eliminated by precise measurement of the smallest and the greatest dimension of bore /3; 4/.

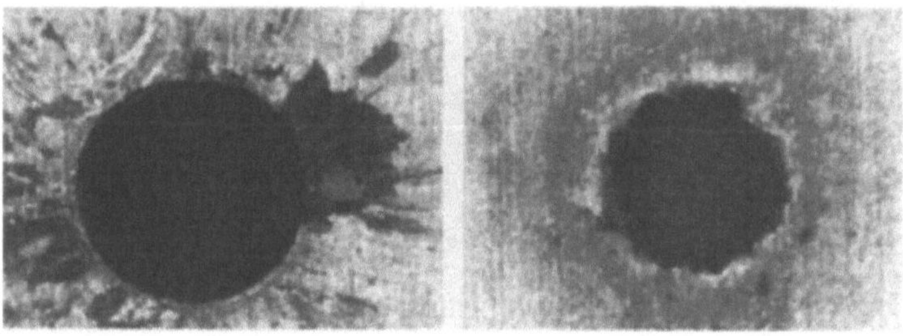

Fig.8: Example of qualitative evaluation of bore diameters

By the way, at the bottom (right) the comparison of contour shows dimensional and shape deviations. These variations of the machined part are brought about by different causes. One of the principal cause is the focus position of laserbeam with regard to workpiece surface. In special event the focus is positioned at the bottom of workpiece. Otherwise, there is a computer stored technical rule describing that in such a case the focus position should be only located about 30 % into the workpiece. That means the data change module must be activated to set the new value of working distance between workpiece surface and output of laser nozzle. The focus position which is worked in this way for the laser machine indicates that -under the specific operational conditions laid down- the deviation lies within the manufacturing tolerance.

The excerpt of workpiece (see fig.9) includes the monitoring of a standard according to test condition. The comparison of the nominal dimension of workpiece drawing versus the actual dimension of lasercut devotes no difference. That means the set-up parameters of machining process on the one hand and the operational scatter of the laser machine tool on the other hand are created in an excellent way.

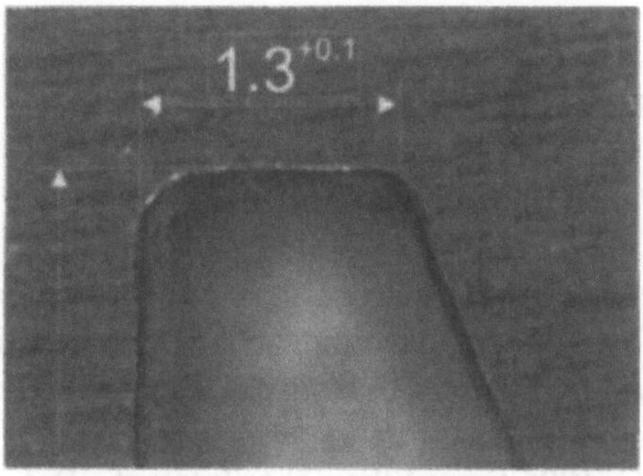

Fig.9: Test example of quality assurance by means of teleworking

Step 8:
The module in this step is a data change module installed for changing data in above mentioned steps. It supports a customer-oriented cooperation. Our experience shows that the customers often wish a changing of workpiece data before the cutting process starts immediately. In other case if the result of setting parameters does not show the required workpiece quality or cutting time there is no problem to change the set-up parameters at teleworkplace and to send these ones to machine tool once more. In this way the machine worker will also be a teleworker in future.

3 Summary

Based on a new type of cooperative value added industrial material processing needs international thinking engineers and students having special experience at the field of teleworking /5/.

By use of teleworkplace strategies laser machining systems can be online controlled and monitored continuously and worldwide. The demonstrated system is established on the market. The areas of application have been determined for diverse laser-beam technologies.

The concept of project can be used for other methods, e.g. water jet cutting, and high speed cutting at milling machines. Future extensions of use of multimedia technologies in medium sized businesses will be determined primarily by relativizing the price-performance ratio. In a.m. project, marketable proportions have already been reached today. In a short time, some of our partners have brought the method in use, e.g. in relation to rapid prototyping procedures as well as simultaneous engineering /6/.

4 Reference list

Book Chapters or Papers in Proceedings

1. Schmidt W, 1992 A comprehensive system of constructional and technological fundamentals of laser beam machining. *3rd FOCAL SPOT Seminar of United Technologies Industrial Lasers, Hartford, CT, presented at Orlando, FL.*

2. Grundmann, H-G, 1995 From Ethernet to Internet. *Dpt. of Electrical Power Technology, University of Paderborn, Germany*

3. Schmidt, W, Olthoff, G, Grundmann, H-G, 1997 Teleworking technologies in engineering applications. *Symposium at University of Paderborn, Soest*

Journal Articles

4. Eversheim, W, 1996 Telecooperation gains advantages with regard to competitiveness. *Management magazine, no.5, BWI-ETH Zurich*

Books

5. Reichwald, R, 1998 Telecooperation. *Springer Verlag, Berlin Heidelberg*

Expositions

6. Schmidt W, Olthoff G, Börger R, 1998 VISION LASER Precision and flexibility. *Hannover Fair, Hall 18 Groundfloor/Stand A 20*

32
Batch and Adaptive Instantaneous Blind Source Separation Algorithms Applied to Pyroelectric Sensor Response in Real Time

M. Haritopoulos, Y. Naudet, R. Huez and A. Billat

1 Introduction

This work is a part of an experimental device which aim is to determine the profile of humidity inside a biopolymer sample by a photothermal method. This one consists in measuring the variation of the temperature of the sample excited by the beam of a laser diode. Then, we have to realize a digital synchronous filter between these two signals. The Bode diagrams obtained allow us to find the required humidity profile. The detection of temperature is realized by a pyroelectric sensor, which is a disk of pyroelectric material covered by electrodes. This type of sensor being also piezoelectric, the received signal is corrupted by noise of acoustic, mechanical, and thermal origin.

We created a double-sensor that we associated to a method of blind source separation (BSS) to de-noise the exciting signal [1]. This double-sensor is obtained by the separation of the inferior electrode in two unequal surfaces. This is achieved by tracing a demetalized thin circle which center is the same as that of the disk. The BSS consists in restoring several source signals from the sensors responses. In our case we have two sources, perceived on the sensors in a different ratio. The first source is the signal resulting from the laser diode, and the other is the noise.

In a first approach, we used batch algorithms of BSS on our signals. During the usual working of the experimental set-up, this method proves to be very efficient.

The improvement being done by an on-line digital synchronous filter, it lead us to use BSS algorithms also working on-line. This will allow a restitution of source signals despite the mixing matrix changes. These changes consist here in an accidental displacement of the sensor, a shift of the beam focusing due to working conditions, and in the fact that the sensors have a different perception of the noise.

In the following of this chapter, we show the results of the application of the adaptive algorithms to simulated and experimental signals. Finally, we will compare the experimental results with those obtained by the application of a 'sliding' batch algorithm.

2 Blind source separation concept, algorithms

2.1 Notions of BSS

A BSS problem is stated classically in the following way. Consider x(t) the $m \times 1$ vector, resulting from the response of *m* sensors receiving the contributions of *n* sources that are supposed statistically independent and s(t) the $n \times 1$ vector of these source signals. Then, the observations x(t) are of the form :

$$\mathrm{x}(t) = A\ \mathrm{s}(t), \tag{1}$$

where *A* is a $m \times n$ matrix called *mixing matrix*. This equation is characteristic of a mixture known as instantaneous, which is the simplest form.

The aim is thus to restitute the sources s(t) while assuming no a-priori information to be available about them. Thereafter, it is considered that one has at least as many observations than of sources ($m \geq n$). In order to be coherent with our experimental device, let *m* and *n* be equal to two. Finally, any algorithm source separation yields to the restitution of a vector of independent components, which is the estimated of the sources. One can use an independance criterion provided by higher order statistics and based on the cancellation of the crosscumulants [2].

2.2 The SOBI algorithm

The SOBI (Second Order Blind Identification) algorithm [3] is based on the decomposition in eigenelements of the cross-correlation matrix between the observation vector spatially whitened taken at different moments, that is to say :

$$\underline{R}(\tau) = UR_S(\tau)U^H, \tag{2}$$

with $R_S(\tau)$ and $\underline{R}(\tau)$, the covariance matrices of the source signals and of the whitened observations and U a unitary matrix of transformation. As U is a unitary matrix and $R_S(\tau)$ a diagonal one, the covariance matrices are diagonalized under the same unitary transformation. The identification thus consists of a joint diagonalization of the whitened observations.

2.3 The Hérault and Jutten neural network

The HJ algorithm [4] consists of a not supervised one layer neural network with recurrent inhibited connections. The output expression is the following :

$$y_i = x_i - \sum_{j=1}^{n} w_{ji} y_j \quad i \neq j . \tag{3}$$

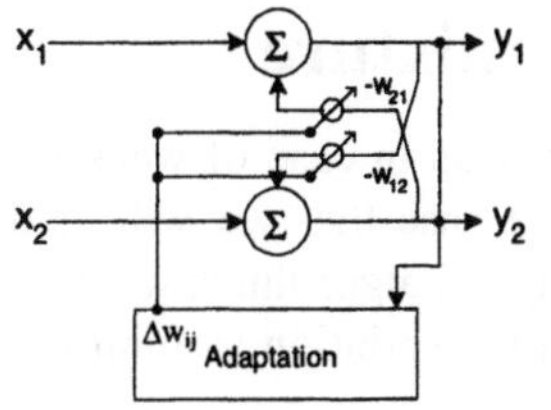

Fig. 1 : *HJ recursive network.*

If the network is stable, one has finally :

$$y_{k+1} = (I+W)^{-1}x_k = (I+W)^{-1}As_k \qquad (4)$$

The weights w_{ij} are adapted according to a 'stochastic Robbins-Monro' type rule, the functional to be minimized being a product of non-linear functions of the outputs :

$$w_{ij} = w_{ij} + \mu\, f(y_i - \langle y_i\rangle)g(y_j - \langle y_j\rangle),\ i \neq j, \qquad (5)$$

where f and g are odd and different non-linear functions and <.> is the average operator. The initial version of the algorithm uses a constant or decreasing adaptation step μ ; this last possibility being inadequate in non-stationary context.

2.4 The EASI algorithm

The EASI (Equivariant Adaptive Separation via Independence) algorithm family, such as they are introduced by Cardoso and *al.* in [5], is based on the serial updating of the $n \times m$ separation matrix B such as :

$$y(t) = B(t)\, x(t), \qquad (6)$$

where y(t) is the vector of the estimated sources. The update of B is done according to the natural (or relative) gradient descent rule :

$$B_{k+1} = B_k - \mu\, \nabla\phi(B_k)\, B_k, \qquad (7)$$

where $\phi(B)$ is an objective function of B whose partial derivatives with respect to the elements of B exist and μ is a small positive and constant scalar.

For n non-linear functions $\{g_1, \ldots g_n\}$ arbitrarily chosen, one defines :

$$g(y) = [g_1(y_1), \ldots, g_n(y_n)]^T \qquad (8)$$

$$Hg(y) = yy^T - I + g(y)y^T - yg(y)^T. \qquad (9)$$

With these notations, the EASI algorithm associated with g is written as follows :

$$B_{k+1} = B_k - \mu\, Hg(y_k)\, B_k \qquad (10)$$

and in its normalized form :

$$B_{k+1} = B_k - \mu\left[\frac{y_k y_k^T - I}{1+\mu y_k^T y_k} + \frac{g(y_k)y_k^T - y_k g(y_k)^T}{1+\mu\left|y_k^T g(y_k)\right|}\right] B_k\,. \qquad (11)$$

This last version of EASI preserves the unitary performance of the algorithm and its stability without introducing any constraint on the separating matrix itself.

Let us note finally, that EASI benefits the equivariance property which for a particular estimator $\hat{A}$ of A results in :

$$\hat{A}(MX_n) = M\hat{A}(X_n),\ \forall M \text{ an invertible matrix}, \qquad (12)$$

where X_n is the $m \times 1$ data vector ($X_n = AS_n$ with $S_n = [s_1, \ldots, s_n]$).

3 Adaptation in a non-stationary mixture

Any separation adaptive algorithm allows a certain follow-up in case of weak non-stationarities of signals or of the mixture ; it is this last case that we will study. However, by using a constant step, necessary with the original algorithms, we do not obtain the expected results. The pursuit is possible when the variation is minimal; it is not assured in case of abrupt significant change.

A first interesting approach is to detect any change in the mixing matrix ; what permits to take advantages of a decreasing step (better speed and precision), which is reset as soon as such a variation arises. Such a detection can be carried out by observing the form of the estimated source cumulants, which will be calculated by an adaptive way. We will use a recursive non-biased version of the estimation technique of the fourth order cumulant tensor proposed by P.O. Amblard and J.M. Brossier in [6] and easily extensible to the calculation of all crosscumulants until this order. This estimator takes the following recursive form :

$$\hat{cum}_4(k)=(1-\alpha)\hat{cum}_4(k-1)+\alpha\left(x(k)^4-3x(k)^2x(k-1)^2\right), \tag{13}$$

where α is a constant << 1.

3.1 HJ : self-adaptation by a cumulant function

A sufficient independence condition of two random variables is the cancellation of crosscumulants to all orders. Because this solution is not a realistic one, we will be limited to fourth order statistics only. In the context of an adaptive separation method, one can affirm that the more we approach independent solutions, the more crosscumulants will tend towards zero. We thus propose to modify the adaptation step by a function of cumulants of the network outputs. Thus, the algorithm is suited to a follow-up in non-stationary mixture, and more especially when some abrupt changes occur at the level of the mixing matrix of the system. In a general way, the adaptation step proposed here takes the following form :

$$\mu_{ij}(k)=\mu_0 E\left\{\frac{K(y(k))}{\eta(y(k))}\right\}, \quad \forall i \neq j, \tag{14}$$

where k is the discrete time, and μ_0 a constant parameter fixing the initial value of the step. It is admitted that it should be adjusted in order to obtain the best possible separation from the beginning. It is besides always the case in actual situation where successive tests permit to find the best compromise between precision and speed. $K(y)$ is a direct function of the output crosscumulants ; $\eta(y)$ is a term of normalization. This method allows a better adaptation, each step which is associated with a coefficient to be estimated (weights of the network), being changed according to a certain gap to the independence. According to the simulation results obtained with different K and η, we finally propose the following form :

$$\mu_{ij}(k)=\mu_0 E\left\{\frac{\sqrt{Cum_{31}(y,k)^2+Cum_{13}(y,k)^2}}{MAX_k\{Cum_{31}(y,k)\}}\right\}, \forall i \neq j, \tag{15}$$

The numerator K can be extended to its most rigorous form: $K(y,k)=\sqrt{Cum_{31}(y,k)^2+Cum_{22}(y,k)^2+Cum_{13}(y,k)^2}$, but this does not improve the performances significantly. The parameter of normalization $MAX\{Cum(y,k)\}$ is calculated in an adaptive way. It must permit to prevent that the added cumulant term makes explode the adaptation step. So, to ensure a correct value of the step, one can put $\eta(y)=MAX\{K(y)\}$, which limits to one the function values but imposes the same evolution of the step for each coefficient, involving some correct average performances but less accurate.

This simple modified step that we will call self-adaptive permits, as the simulations confirm it, to improve the algorithm reaction speed to a change of mixture. Let us note, however, that the added function brings a certain number of additional computations, which we will have to take into account. This method, although not being optimised (it is only based on an observation) requires less computations than those proposed by Thi [7] which use the derivatives of non-linear functions or cumulants.

3.2 EASI coupled with a fourth order cumulants criterion

The adaptation step μ such as it is proposed in the initial algorithm is a positive constant or a decreasing scalar following an α/t law. In both cases, the algorithm is unable to adapt itself if the observations are non stationary mixtures. To deal with this case, we propose here to introduce a criterion based on the evolution of the crosscumulants 1-3 and 3-1 slope. Thus, in every update step of the separating matrix B, we calculate the fourth order crosscumulant estimates of the outputs by using the recursive algorithm of Amblard [6]. Then we follow the slope evolution of that latter.

The test of slope is just a derivative filter looking for the strong values of the filtered signal and constitute an approach of level rupture detection in a signal. The recursive version of this filter [8] is written here in the following way :

$$f_{t+1}(c)=f_t(c)+\left(\frac{c_{t+n}-2c_t+c_{t-n}}{n^2}\right),\quad t\geq n+1 \tag{16}$$

where $2n$ indicates the length of the sliding window on which we calculate the averages. Its initialisation is given by :

$$f_i(c)=\frac{m_i^1-m_i^2}{n},\quad i\geq n,$$
$$\text{with } m_i^1=\frac{1}{n}\sum_{j=1}^{n}c_{i+j} \text{ et } m_i^2=\frac{1}{n}\sum_{j=0}^{n-1}c_{i-j}. \tag{17}$$

Lastly, it is decided if there is a rupture of level by testing the absolute values of $f_t(c)$, where c is the set $\{Cum_{31}(y), Cum_{13}(y)\}$. If one of these values is superior to a threshold of reference λ, then it is considered that the detection of the transient is successfull and the adaptation step μ is reset.

It is obvious that this adaptive method has several disadvantages. First, a judicious choice of the threshold λ as well as of the length of the sliding window is needed in order to achieve a compromise between false alarms and miss. Moreover, one introduces higher order statistics that do not work explicitly in EASI.

Nevertheless, this test combined with a decreasing adaptation step permits us to detect a change in the mixing matrix, even if this one is progressive or very weak, although in this case, a constant training rate offers in actual fact the same efficiency as an adaptive step.

4 Simulations

The simulations carried out take into account a Gaussian noise of variance equals to one and a sinusoid to which we added a fast square component in order to simulate the distortion due to the passage of the pure sinusoid in our experimental device. The sources are mixed in the following way :

- according to the matrix A1 = [1 , 0.8 ; 0.7 , 1] until the sample 1500,
- then according to A2 = [1 ,0.4 ; 0.3 , 1] until the sample 3000.

The figure below shows the source signals, the mixtures (signals of the sensors), and the estimates of the sources after separation around the change of matrix.

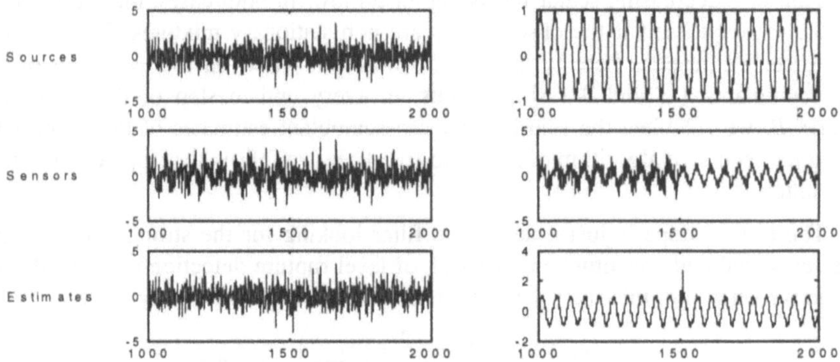

Fig. 2 : *experimental signals.*

After a correct adjustment of the parameters, the two self-adaptive algorithms restore the sources. The reaction to the change of matrix is fast, about 20 to 500 samples according to the used technique (HJ or EASI).

4.1 HJ : self-adaptation contribution

In order to illustrate the working of the self-adaptive step, the figure (3) shows the evolution of the absolute values of the three crosscumulants and the *K* function. The peak is highly visible on Cum_{31} and is reflected on the shape of the steps, which are a normalized and smoothed version of the influence of the cumulants.

The graphs of convergence (Fig. 4) show the attraction areas obtained by a constant and a self-adaptive step with the same initial value. The envelopes are smaller in the second case, which shows a gain in precision. Notice that, whereas

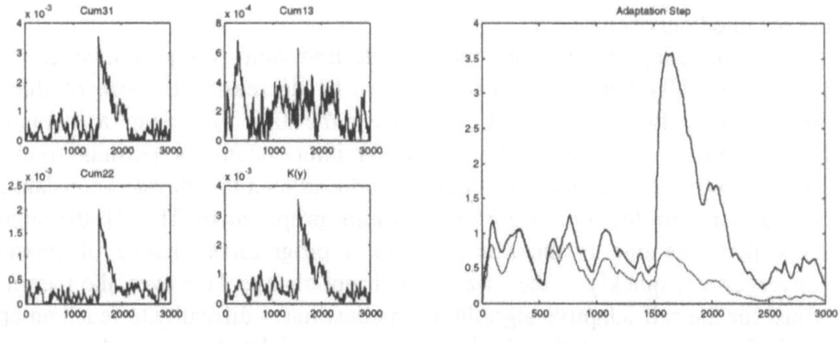

Fig. 3 : *crosscumulants of the outputs and K function (left), adaptation step evolution (right).*

they converge both towards the point of the first matrix A1, the self-adaptation is faster and more accurate in the case of the second point. Other results obtained, in particular with two changes of matrix strengthen this assumption.

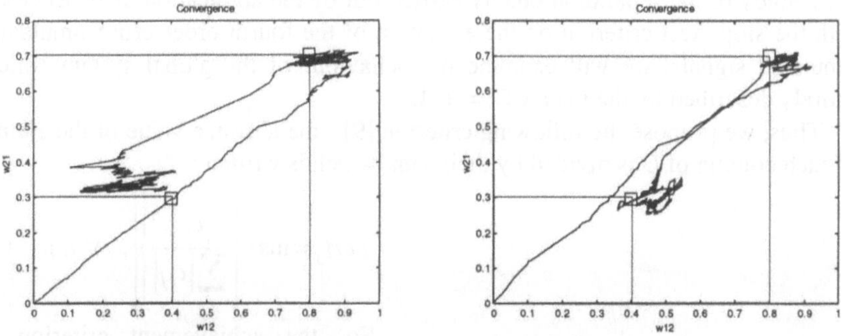

Fig. 4 : *weight convergence, constant step left, self-adaptation right.*

Finally, the achievement can be calculated for each sample according to a quadratic error criterion of the form :

$$Perfs(k) = 1 - \frac{\sum_{i,j=1,2;\, i \neq j} (a_{ij}(k) - w_{ij}(k))^2}{MAX_k \left\{ \sum_{i,j=1,2;\, i \neq j} (a_{ij}(k) - w_{ij}(k))^2 \right\}} \quad (18)$$

where a_{ij} et w_{ij} are respectively antidiagonals coefficients of the mixing matrix A and of the estimated weights matrix W. The figure beside shows the evolution of this criterion for a self-adaptive step (full line), and a constant step (dotted line). The ordinate equal to one corresponds to a good average separation. One will notice the speed of convergence of the self-adaptation after the change of the coefficients of matrix A. The quality of separation is very good as

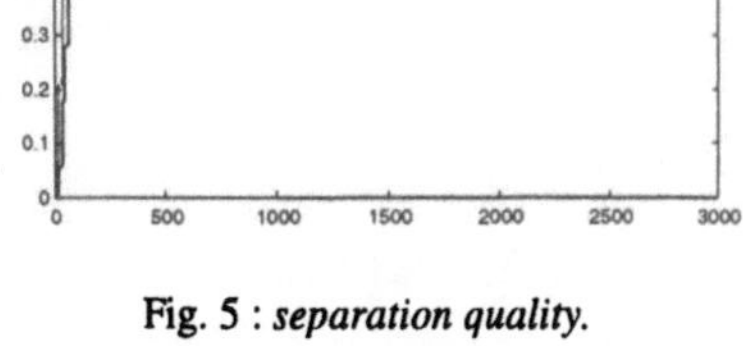

Fig. 5 : *separation quality.*

the obtained curves prove it.

To conclude, we can say that the desired aim is reached, since the self-adaptation really brings a gain of precision by decreasing the size of fluctuations around the convergence point. Moreover, the fact of having a changing step proportional to a certain extent to the mixture statistics permits to avoid an unhooking of the algorithm that could be induced by a too elevated constant step. In comparison with the more rigorous algorithm proposed by Thi [7], the simulations show that this one is more adapted to a progressive change of mixture and converges less quickly in the case of an abrupt change. However, the parameters to adjust for the self-adaptive algorithm can constitute a difficulty to reach an optimum result. Let us also note that the irregular shape of the step can induce somewhat a noisy restitution, though it is often negligible.

4.2 EASI : performances

For a study of the separation quality carried out by the application of EASI coupled with the slope test criterion on the evolution of the fourth order crosscumulants on simulated signals, we will examine the behaviour of the global system which is entirely described by the matrix $C_t = B_t A$.

Thus, we propose the following criterion [9] : the absolute value of the elements of each column of C is divided by their sum, which is written :

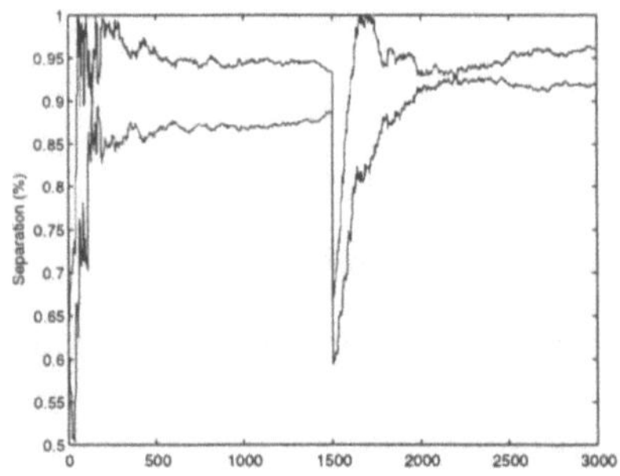

Fig. 6 : *performance criterion for EASI.*

$$perf_j = \max\left\{\frac{|c_{ij}|}{\sum_j |c_{ij}|}\right\}_i, \ \forall\, i, j. \quad (19)$$

So, the achievement criterion *perf* takes his values into the interval [0.5, 1] where the lower bound corresponds to a separation of 0%, and the superior bound to a perfect separation (100%). On the figure beside, one can follow the evolution of the maximal values of every column of C, calculated according to the previous criterion. Thus, we note a fast convergence near the starting point as well as a good readjustment after the change of the mixing matrix,

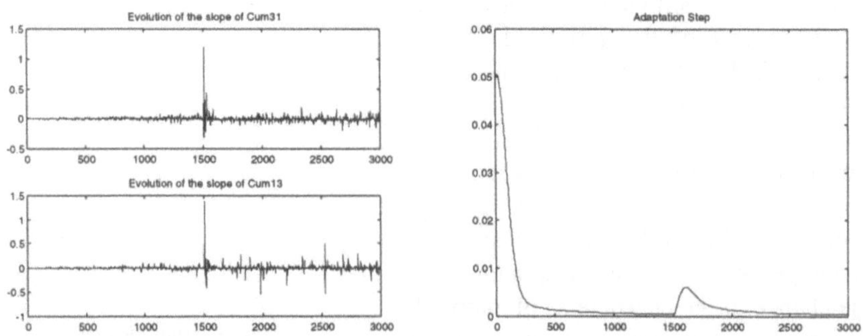

Fig. 7 : *evolution of the cumulant slope (left), and of the adaptation steps (right).*

with a global separation rate of approximately 90%.

The separating matrix B_0 has been randomly initialized, while the adaptation step evolves according to the law : $\mu_{k+1} = \varepsilon + \mu_k\ ((k / N) / 2 + 1)$, where ε, k and N are respectively a very small positive constant, an iteration index, and the number of samples. The length of the sliding window on which we calculate the slope is fixed to 6 and the detection threshold value λ to 0.5.

The slope test locates the rupture from 1 to 20 steps after it occurred effectively. As for the HJ algorithm, the rupture is clear on the curve of the cumulant evolution and also on that of their slopes (fig. 7). The decreasing adaptation step is reset as soon as the slope test detects the transient signal, we can note it on the figure, what allows the algorithm to escape quickly from the local minimas and to re-converge relatively quickly towards another value.

The decreasing step can possibly be replaced by a constant and small one without having a considerable impact on the algorithm performances. Moreover, according to the experimental studies, the performances of EASI are appreciably the same in the case of a small change of the mixing matrix with a constant step or with the self-adaptation. We finally notice great fluctuations after the algorithm has begun to run and after the readjustment (Fig. 7). This does not disturb the contribution in precision of this adaptive version. It will be seen thereafter, that results are promising in an actual case of more progressive change of the mixing matrix.

5 Results

The results presented here concern the de-noising of the exciting beam in the

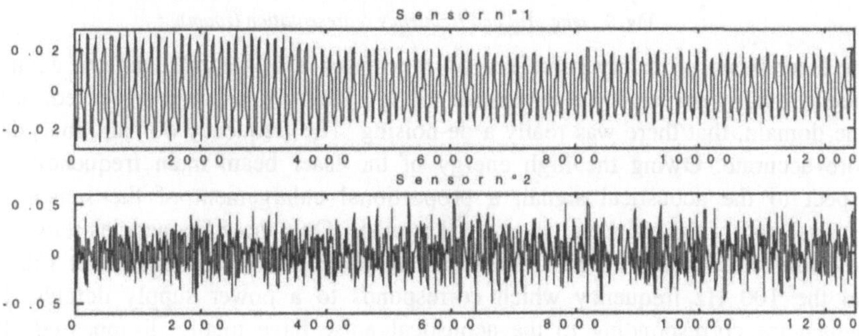

Fig. 8 : experimental signals.

presence of an acoustic noise (90 Hz ≤ f ≤ 500 Hz) when a defocusing of the laser beam is occuring. The exciting beam includes on the one hand, the exciting frequency (in this case : 80 Hz) and on the other hand, other frequencies (harmonic of the 80 Hz, and a frequency of 100 Hz resulting from the power supply) inherent to its generation. The source mixing matrix goes from :

$$C = \begin{bmatrix} 1 & 0.1 \\ 0.2 & 1 \end{bmatrix} * S \text{ to } C = \begin{bmatrix} 1 & 0.1 \\ 0.9 & 1 \end{bmatrix} * S$$

with $C = \begin{bmatrix} \text{Peripheral sensor} \\ \text{Central sensor} \end{bmatrix}$ and $S = \begin{bmatrix} \text{Acoustic signal} \\ \text{Laser beam} \end{bmatrix}$.

The defocusing time is included between 0,3 (sample 3000) and 0,55 seconds (sample 5500). The figure (8) shows the sensor response.

We apply the three algorithms (SOBI, HJ, EASI) on these signals. The batch algorithm is used as reference to measure the separation quality before and after the change of the mixing matrix. The other studied point concerns the achievement of the convergence speed after the change of the mixing matrix.

5.1 Precision

The separation quality is computed in such an area where the algorithms have converged. The figure (9) shows the signals and their estimates, respectively, in the time and frequency domains (psd). They are obtained by the application of the SOBI algorithm.

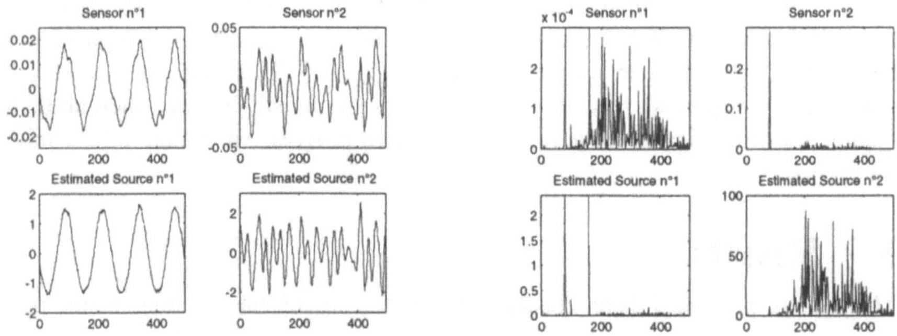

Fig. 9 : *temporal and frequency representation (zoom).*

The lazer beam is mainly perceived on the sensor n°1 (central), and the acoustic noise is mainly perceived on the sensor n°2 (peripheral). It can be observed, in the time domain, that there was really a de-noising. The frequency domain analysis is more accurate. Owing the high energy of the laser beam main frequency with respect to the acoustical signal, a proportional enlargement of the sensor n°1 response and of the estimated source n°1 is made. Only three frequencies remain on the estimated source n°1 figure: the 80 Hz main frequency, its harmonic at 160 Hz, and the 100 Hz frequency which corresponds to a power supply default. The frequencies corresponding to the acoustical noise have totally disappeared. The signals obtained by the EASI and HJ algorithm are approximately the same.

We normalize the signals of the laser beam estimated source and those of the central sensor in the frequency domain with respect to the main peak psd (80 Hz). Then, by subtracting the actual estimated source (the three frequencies 80, 100, and 160 Hz) from the normalized signals, we obtain residuals, the energy of which is computed. The ratio of residuals is computed: "noise before separation / noise after separation".

	SOBI	HJ	EASI
Ratio (dB)	11	9	8

The lower accuracy of the adaptive algorithms compared to the batch one finds its origin in the fluctuations around the solution. These fluctuations result from the way in which they were adapted. They are nevertheless relatively accurate : we have a ratio of 8 dB in the case of the least accurate algorithm, which is a good ratio because the best restitution provided by the batch algorithm gives an 11 dB value.

5.2 Convergence speed

Since the use of a 'sliding' batch algorithm introduces a loss of information equals to the length of a batch (2000 samples), it is necessary to insure that the convergence of the adaptive algorithms is quick enough so as to avoid this loss. The convergence speed is therefore analysed by studying the evolution curves, on one hand, of the weights for the HJ network, and on the other hand, of the normalized coefficients of the demixing matrix for EASI.

5.2.1 HJ algorithm

We notice on the convergence curve (Fig. 10, left), a dispersion in the first

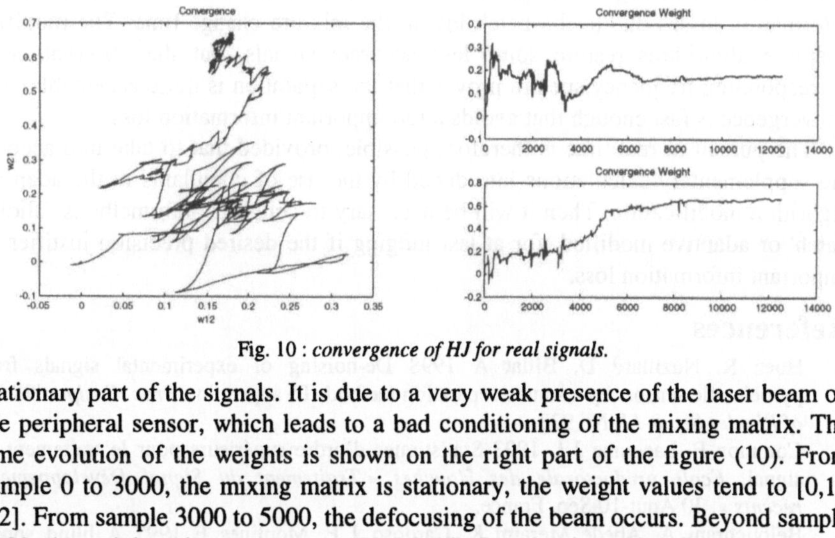

Fig. 10 : *convergence of HJ for real signals.*

stationary part of the signals. It is due to a very weak presence of the laser beam on the peripheral sensor, which leads to a bad conditioning of the mixing matrix. The time evolution of the weights is shown on the right part of the figure (10). From sample 0 to 3000, the mixing matrix is stationary, the weight values tend to [0,15 0,2]. From sample 3000 to 5000, the defocusing of the beam occurs. Beyond sample 5000, the weight values converge to [0,5 0,65]. Despite these high fluctuations, it is estimated that 500 samples or less can be sufficient for convergence.

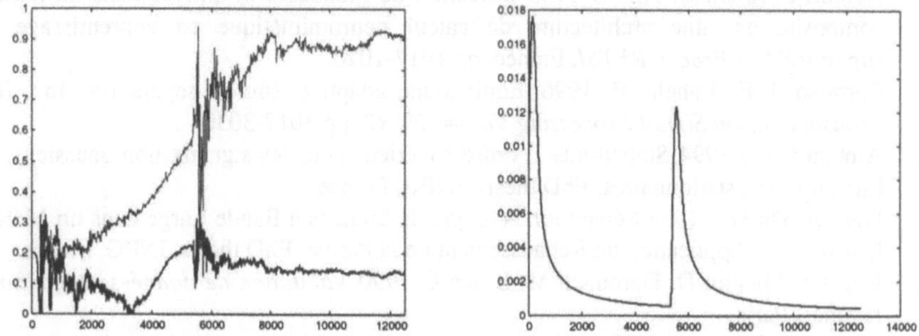

Fig. 11 : *evolution of B antidiagonal coefficients (left), and of the adaptation step (right).*

5.2.2 *EASI algorithm*

The EASI algorithm, associated with the slope test of crosscumulants and applied to the actual signals gives quite good results (Fig. 11). Despite the transient part of the curve which runs from sample 3000 to 5500, the slope test succeeds in detecting the change of the mixing matrix at k=5300. The results are noised but a convergence (in average) can be achieved on 500 samples after defocusing. The reached convergence value is [0,1 0,2] at the sample 500 and [0,15 0,8] at the sample 7000.

6 Conclusion

In this chapter, we adapted three algorithms of source separation to a non-stationary context where the mixing matrix coefficients vary. Their application on simulated signals where the changes are very fast proves the interest of a self-adaptation training step, the results being very satisfactory. In the case of signals obtained experimentally, the performances are degraded by the presence of an additive sensor noise. The batch algorithm gives some very accurate results in spite of the information losses due to the batch lost at the mixture change time. The modified adaptive algorithms restore some less accurate signals, but the observation of corresponding frequency spectra proves that the separation is quite acceptable. The convergence is fast enough that avoids a too important information loss.

The pursuit in real time is therefore possible, provided that to take into account the supplementary calculations introduced by the use of cumulants in the adaptive algorithm modification. Then it will be necessary to consider both methods, 'sliding batch' or adaptive modified, for at last judging if the desired precision justifies an important information loss.

References

1. Huez R, Nuzillard D, Billat A 1998 De-noising of experimental signals from pyroelectric sensors by a source separation method, To appear in : *Proc. Eusipco'98 Isle of Rhodes Sep 8-11*, No 933.
2. Common P, Lacoume J L 1993 Statistiques d'ordre supérieurs pour le traitement du signal, *Ecole prédoctorale des Houches « Traitement du Signal Développements récents »*, 30 Aout-10 Sep. France.
3. Belouchrani A, Abede Meraim K, Cardoso J F, Moulines E 1997 A blind source separation technique using second-ordre statistics. In : *IEEE Transactions on Signal Processing Vol 45, No 2*, pp 434-443.
4. Herault J, Jutten C, Ans B 1985 Détection de grandeurs primitives dans un message composite par une architecture de calcul neuromimétique en apprentissage non supervisé. In : *Proc. GRETSI*, France, pp 1017-1020.
5. Cardoso J F, Laheld B 1996 Equivariant adaptive source separation. In : *IEEE Transactions on Signal Processing Vol 44, No 12*, pp 3017-3030.
6. Amblard P O 1994 Statistiques d'ordre supérieur pour les signaux non gaussiens, non linéaires, non stationnaires. PhD thesis, INPG, France.
7. Nguyen Thi H L 1993 Séparation Aveugle de Sources à Bande Large dans un Mélange Convolutif : Application au Rehaussement de la Parole. PhD thesis, INPG, France.
8. Ragot J, Maquin D, Darouach M, Bloch G 1990 *Validation de données et diagnostic.* Hermes, Paris
9. Barros A K, Ohnishi N 1997 Pre-filtering non-stationary signals to improve blind source separation, *13th International Conference on Digital Signal Processing*, Greece.

33

Build-Time Estimation Tools for Rapid Prototyping Stereolithography Systems

J. Giannatsis, V. Dedoussis and L. Laios

1 Introduction

A few years ago a new group of technologies appeared in the growing field of Advanced Manufacturing systems, the so called Rapid Prototyping & Manufacturing (RP & M) technologies or better Layer Manufacturing (LM) technologies. All these technologies share the same basic operational concept which is to build parts from various materials (polymers, metals, paper etc.) in a layer-by-layer fashion, almost directly from CAD data and without employing any kind of manufacturing tools. The above features, combined with the high degree of automation employed in these systems and the small amount of time required for production, made RP & M systems an attractive solution for a great deal of applications in the automotive, aerospace and consumer products industry.

Stereolithography, a process which builds objects by successive solidification of polymer resin layers, was the first to open the road for the wide adoption of RP & M technologies at the beginning of the 90's. Among the several research issues, which emerged with the gradual expansion of the number of installed Stereolithography units, was the trade-off problem between fabrication quality and fabrication cost. Cost for Stereolithography (and for most of the RP & M technologies) is directly related to the amount of time required for the complete fabrication of parts. Therefore, the prediction of the time required for the completion of certain "jobs" or "builds" (batches of products fabricated simultaneously) is of critical importance.

The estimation of "build-time" is useful for the decision making processes which precede fabrication, such as optimum placement and orientation of parts inside the build space, selection of build parameters for the fabrication and obviously job pricing and scheduling. Especially for job scheduling, build-time estimation is very important, since Stereolithography systems are relatively highly automated (no human supervision is usually required) and the fabricated parts are usually of unique nature (prototypes or models).

In order to predict build-time accurately, several parameters concerning the system, the material, the build style and the geometry of the part should be taken into account. The effects of the first three groups of parameters is fairly accurately and in a straightforward manner evaluated by the operator of the system. The effect of certain geometric properties of the part, however, is not straightforward and requires some computer processing. Geometry information of the part is usually provided either in the form of STL file or in the form of slice(s) file. The first type is the *de facto* communication standard between CAD and RP systems and employs tessellated triangles in order to describe the outer surfaces of the part. Slice files, on the other hand, include basically the 2D description of all the slices-layers composing the part. Several build-time estimation tools for Stereolithography have been proposed in the last few years, utilising these two types of geometric information [1], [2], [3], [4].

The purpose of this chapter is to investigate the degree of accuracy of build-time estimators based on different types of geometry representation. Two software tools, one using STL file geometry information and another one using slice file information, have been developed after an extensive study of the EOS-Stereos Desktop S Stereolithography system installed at the University of Piraeus. The study covered its basic operational subsystems, namely the laser scanning and the resin recoating mechanisms. The results of this study, as well as the basic geometric concepts underlying the two approaches and their software implementation are presented in detail. Build-time estimations reported indicate that the tools proposed in this chapter are sufficiently accurate and quite reliable.

2 Theoretical Basis

"Build-time" is defined as *the amount of time that is required by the Stereolithography system to fabricate a single or a group of parts*. Similar to other RP technologies the fabrication is achieved by gradually adding layers of solid material, polymerised photosensitive resin in our case. Build-time, therefore, of a part is expressed as:

$$Build\,Time = \sum_{i=1}^{n} TLayer(i) \tag{1}$$

where *TLayer(i)* is the time required for the addition of the ith layer and *n* is the total number of layers.

The addition of a single layer consists of the following two steps:

1. *Recoating*, during which a thin layer of uncured resin is spread across the platform or above the previously solidified layers.
2. *Scanning*, in which the laser beam scans the appropriate area on the surface of the liquid resin, thereby solidifying the uncured resin layer and adding it to the previously fabricated layers.

Thus, the time required for the fabrication of a single, say the ith, layer-slice is evaluated as the sum of the recoating and scanning time:

$$TLayer(i) = TRecoat(i) + TScan(i) \tag{2}$$

Both recoating and scanning time can also be analysed into smaller periods of time. Recoating time for a single layer can be calculated by adding the time required for the lowering of the platform, the time required for the recoater (a sort of blade used for spreading the resin) movement and the resting (or delay) periods. Resting periods are needed after the recoater and platform movement, so that leveling of the resin surface is achieved (pre-scan delay), and after the end of the scanning phase, so that the polymerisation of the newly formed layer is completed (post-scan delay). Thus:

$$TRecoat(i) = TPlatform(i) + TRecoater(i) + TPRe(i) + TPOst(i) \tag{3}$$

Recoater and platform time are expressed as:

$$TRecoater(i) = \frac{Rd(i)}{Rv(i)} \tag{4}$$

and

$$TPlatform(i) = \frac{Pd(i)}{Pv(i)} \tag{5}$$

where *Rd* and *Pd* denote the distance covered by the recoater and platform movement respectively, while *Rv* and *Pv* denote their corresponding velocities.

Scanning consists of two procedures, contouring (or border scanning) and hatching (see Fig. 1). During contouring the scanning control mechanism directs the laser beam to "draw" the borders (contours) of the slice area. These borders are in the form of polylines (set of vertex points, connected contiguously by straight line segments) and enclose the slice area. Layer hatching is carried out in order to solidify the area inside the borders. In conventional hatching styles this involves the "drawing" of closely spaced parallel vectors (line segments defined by a pair of points) in the X- and Y-wise sense, with X and Y being the two axes defining the platform plane [5]. Scanning time, therefore, reads:

$$TScan(i) = TContour(i) + THatch(i) \tag{6}$$

Contouring and hatching time can be calculated, if the distance that the laser beam travels and the velocity at which it moves are known. The corresponding relations are:

$$TContour(i) = \frac{Cl(i)}{Cv(i)} \tag{7}$$

and

$$THatch(i) = \frac{Hl(i)}{Hv(i)} \tag{8}$$

where *Cl* and *Hl* denote the total length of all vectors drawn during hatching and contouring (hatching and border length) respectively and *Cv* and *Hv* are the corresponding scanning velocities.

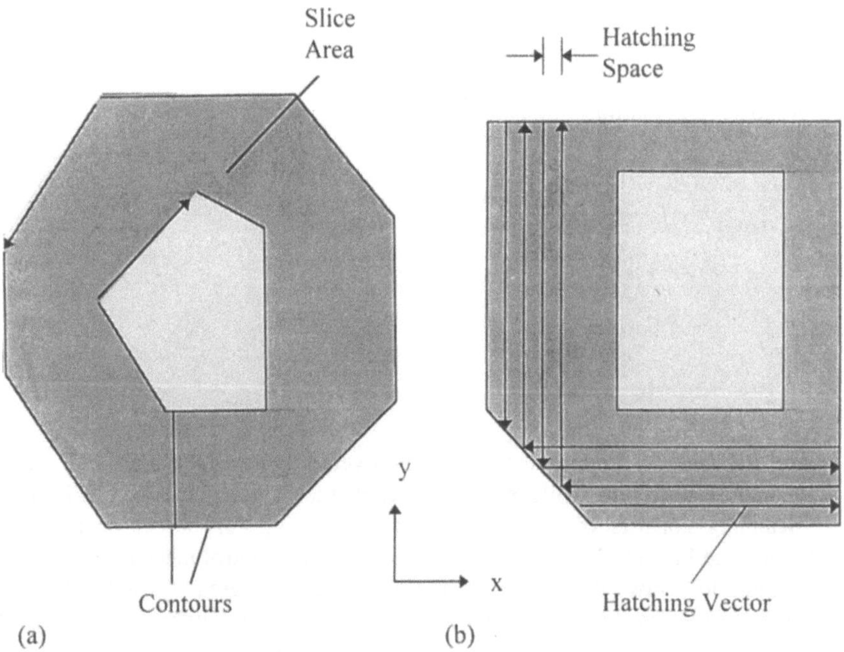

Figure 1. The two scanning procedures: (a) Contouring, (b) Hatching.

Contouring and hatching velocities are expressed in terms of the layer curing depth (how deep the layer solidifies), a parameter which is specified by the operator of the system, the material properties and the laser characteristics. The corresponding expressions employed by EOS-Stereos systems are:

$$Cv(i) = \sqrt{\frac{2}{\pi}}\frac{P_L}{W_0\, E_c\, e^{(Cd(i)\,/\,Dp)}} \tag{9}$$

$$Hv(i) = \frac{n\,P_L}{hs\,E_c\,e^{(Cd(i)\,/\,Dp)}} \tag{10}$$

where P_L denotes the laser power, W_0 is the laser beam half width, Cd is the curing depth, hs is the hatching space (distance between neighbour scanning vectors), n is the number of times the slice area is scanned, E_c is the critical energy, and Dp is the penetration depth. The last two parameters are basic resin properties. Equation (9) has been first proposed by Jacobs [5]. Equation (10) is used in the scanning control software of EOS and has been experimentally verified, following the procedure (*Windowpanes*) also proposed by Jacobs [5].

Equations (2)-(10) indicate that the layer fabrication time depends on system parameters (P_L, W_0), resin properties (Dp, Ec), build style parameters (Rd, Rv, Pd, Pv, Cd, hs, resting times) and slice geometrical properties, namely the contour length and the slice area. The values for most of the above parameters are directly specified by the system operator. However, the slice geometry properties, which are required for the build-time estimation, should be calculated via the processing of part geometry information.

2.1 Calculation of slice geometry properties

The best way to evaluate the required geometry properties is to use the slice (data) file of the part. The slice(s) file is the result of slicing operation during which the STL model of the part is intersected with a series of horizontal (parallel to the platform) planes [6]. It includes the exact representation of every slice-layer that will be solidified during fabrication. Depending on the Stereolithography system and the slicing software, the slice file may contain two types of information:

- representation of the borders enclosing the slice area, which consist of oriented polylines/contours - contour file.
- coordinate pairs defining the scan vectors - vector file.

In both cases the length of the contours, that the laser will draw, is calculated as the sum of all vectors defining the slice borders. The hatching length, on the other hand, is calculated either by dividing the hatching area with the hatching space or as the sum of all hatching scan vectors, depending on whether a contour or a vector file respectively is involved.

Sometimes it is desirable to be able to estimate the build-time before the slicing phase. In such cases the STL file (triangles tessellated model) must be used as a source for the required geometric information. In order to evaluate the required total scanning lengths, the (part) volume and the vertical area should be firstly calculated. Assuming that a constant layer thickness Lth is employed during fabrication and noting that the sum of the volume of all layers equals the volume of the part, the total scanning area is the part volume divided by the layer thickness. Then the total hatching length reads:

$$Total\ Hatching\ Length = \frac{Volume}{Lth\ \ hs} \tag{11}$$

Also according to Tata and Flynn [3], who introduced the vertical area concept (the sum of all triangles projected on a vertical, normal to the platform, plane) the total border length of a part is calculated according to the relation:

$$Total\ Border\ Length = \frac{Vertical\ Area}{Lth} \tag{12}$$

Relations (11) and (12) are valid if the part volume is fully cured, i.e. no semi-hollow building styles such as *Skin&Core* [7] or *Quickcast* [5] are adopted. A close observation of these equations reveals that the hatching time which is directly related to the total hatching length, is independent of the particular part orientation during fabrication, while the opposite is true for the border time, since the vertical area is strongly related to the particular placement, orientation, of the part on the platform of the Stereolithography system.

3 Study of Stereos Scanning Mechanism

The equations presented in the previous section, apply, more or less, to most RP systems. However, in order to enhance the accuracy of build-time estimation for a specific system, its operational principles and behaviour should be examined, since even between systems operating on the same principles quite significant differences exist.

In order to test the validity of the above equations for the case of EOS-Stereos systems, simple indicative experiments were conducted. During these experiments a 50 mm square slice was scanned twice (along both X and Y direction), with variable hatching space and constant curing depth, monitoring the scanning time. For each value of hatching space five measurements were taken and an average value for the scanning time was computed. No border scanning was considered, since it involves negligible amount of time. The experiments were carried out with constant hatching velocity, calculated according to equation (10) assuming a laser power value of 30 mW.

The results included in Table 1 show that the actual-measured hatching time is always greater than the predicted one. The observed deviation increases as the value of hatching space decreases. Relatively smaller values for hatching space implicitly mean that the (total) number of hatching vectors and the value for hatching length and velocity is relatively higher. These results indicate that hatching time is strongly related to some (or possibly all) of the above parameters.

Hs (mm)	Hv (mm/s)	No. of Vectors	Hat. Len. (mm)	Meas. Time (s)	Calc. Time (s)	Deviation (s)	Perc. Dev. (%)
0.25	600	400	20000	34.059	33.333	0.726	2.13
0.2	750	500	25000	34.060	33.333	0.727	2.13
0.15	1000	667	33333	34.172	33.333	0.839	2.46
0.1	1500	1000	50000	34.436	33.333	1.102	3.20
0.08	2000	1250	62500	32.397	31.250	1.147	3.54
0.06	2500	1667	83333	34.611	33.333	1.278	3.69
0.05	3000	2000	100000	34.886	33.333	1.552	4.45
0.03	5000	3333	166667	35.545	33.333	2.212	6.22

Table 1. Actual and predicted scanning time for a 50 mm square slice.

A systematic experimental investigation is currently under way to identify the effect of these parameters in EOS-Stereos systems. Preliminary results show that the deviations mostly depend on the number of employed hatching vectors and to a lesser extend, on the values of hatching length and hatching velocity.

4 Software Implementation

Two "build-time" estimation software tools were developed in accordance with the theoretical basis presented in the previous sections. The tools were implemented in C language on an Silicon Graphics Indigo II workstation.

The STL estimation tool (*STL estimator*) is based on the *ADMesh 0.95* STL software tool. This STL software has been developed by A. Martin and is available through the Internet [8] for free use and modification, according to the GNU public software license. Mainly, it deals with STL file manipulation and repair, but it can also calculate basic properties of the model such as part volume or total tessellated triangle area. A subroutine calculating the vertical area was added. This modified version of the software is employed for the evaluation of the STL model properties.

The slice-based estimation tool (*CLI estimator*) uses as a source of information the CLI files, which result from the "slicing" of the STL files with prescribed layer thickness. The CLI format has been developed in the context of the Brite-Euram project as a means for transferring slice-layer information (mainly medical) to RP systems. It is free for use and its syntax description together with sufficient documentation is publicly available through the Internet [9]. The software tool developed does not only calculate scan lengths, but also the estimated build-time according to information prescribed by the user.

5 Accuracy-Comparison of the two Build-Time Estimators

Before the results of the actual testing of both build-time estimators are presented, the problem of laser power prediction should be addressed. The theoretical background presented in the previous section is based on the assumption that the laser power used for the calculation of the scanning velocities is constant during fabrication. However, gas lasers like those installed in EOS-Stereos systems (He-Cd lasers), do not have constant power output during operation. This means that in order to obtain a uniform curing depth for all layers, the scanning velocity should be adjusted to the laser power fluctuations. This is performed automatically by a tuning mechanism, which measures the laser power after the scanning phase of a layer is completed and adjusts appropriately the scanning velocity for the fabrication of the next layer.

Extremely accurate build-time predictions, therefore, imply accurate predictions of the laser power fluctuations. In order to investigate to what extend these power fluctuations can be predicted, the laser power readings obtained by the system for representative jobs, were examined. These readings are included in the report-files, which are automatically generated by the system during fabrication. In agreement with other investigators opinion [3], it seems that is rather difficult, if not impossible, to predict the behaviour of laser power within a job (see Fig. 2). Although the exact prediction of laser power fluctuations during fabrication seems to be impossible, one perhaps could estimate an average value using the power readings of a previously completed job. This will obviously increase the accuracy of build-time estimation.

It is evident that the accuracy of the slice-based build-time estimation tool is impaired only by the uncertainty of the average laser power value. On the other hand, the STL-based estimation will also depend on how accurately the scanning lengths (hatching and border) are estimated from the part volume information. Various STL models and the corresponding CLI files were examined and the value of total hatching area and total contour length were calculated. Results of these calculations employing the two tools are compared in Table 2. It is observed that, with the exception of two parts, *cross* and *turb_09*, in all other cases the estimation of total hatching area and contour length from STL data is very close to that calculated from the CLI data. In fact the average percentage deviation of the to tools (excluding the above two parts) is 0.31 % for the layer hatching area and 0.09 % for contour length, figures which are very satisfactory. These results indicate that, the STL estimator can provide quite accurate predictions in most cases. Indeed, the STL estimator could be very useful for activities such as cost estimation of part fabrication or selection of appropriate orientation and build parameters. The latter should be based especially on STL data, simply because slicing can not be carried out for all possible solutions [10].

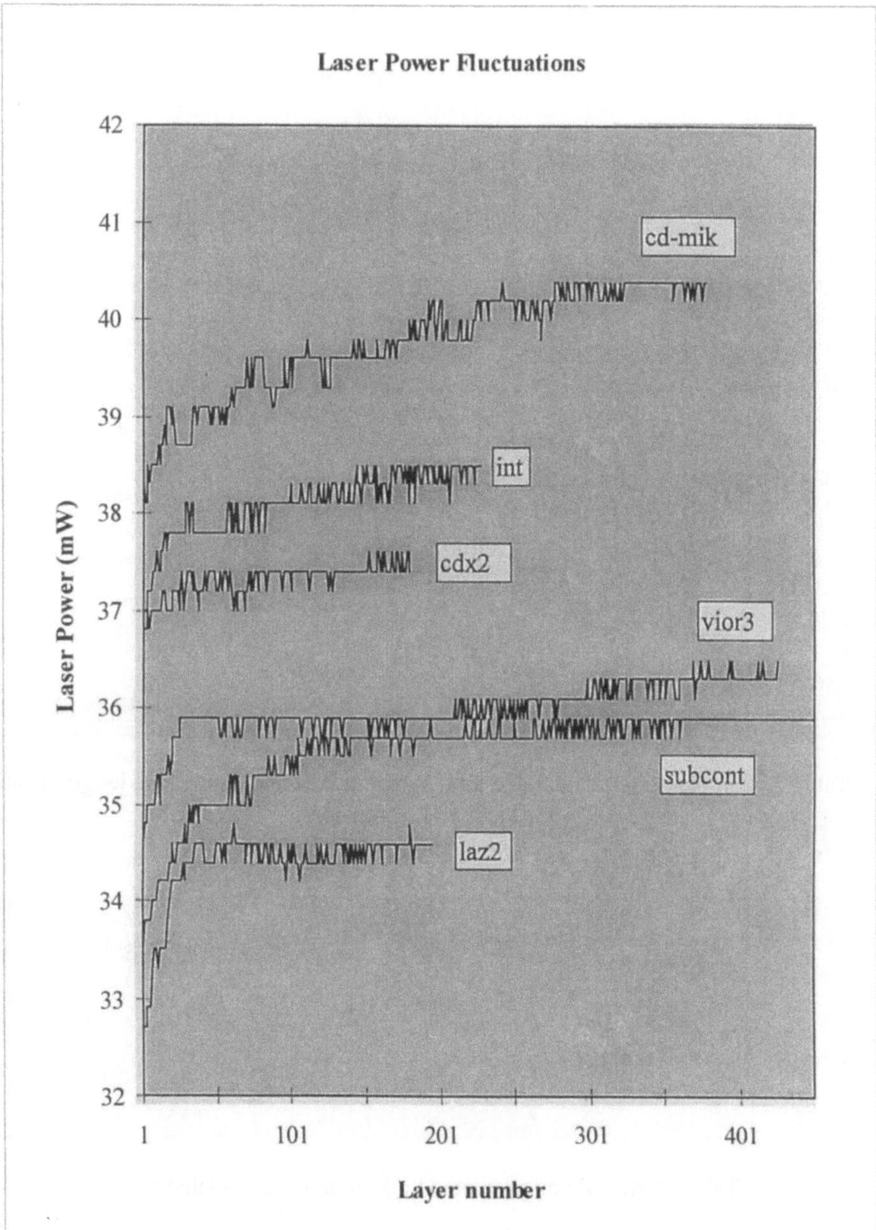

Figure 2. Laser power readings for various jobs.

STL Model	CLI Hatching Area (mm^2)	Diff. (mm^2)	(%)	CLI Contour Length (mm)	Diff. (mm)	(%)
ast18	7370012	9848	0.13	278820	355	0.13
bottle2	1304	6	0.46	9997	33	0.33
cd-bottom	458769	-5222	-1.14	143898	-324	-0.23
cd-left	775296	-8669	-1.12	263190	1270	0.48
cd-top	423142	9678	2.29	146470	183	0.12
cross	6163	6162	99.98	6425	85	1.32
flansch	52668	78	0.15	30137	143	0.47
hand	285435	25	0.01	114677	-4	0.00
housing	743503	97	0.01	124065	40	0.03
lid	3911030	5683	0.15	1464573	123	0.01
mer_down	148226	229	0.15	13860	-269	-1.94
mer__up	147324	241	0.16	13925	-290	-2.08
merc2	5206	7	0.13	7466	8	0.11
mor884	54914	181	0.33	132074	371	0.28
motor	247082	-75	-0.03	154908	198	0.13
pot6	109557	10	0.01	346594	-100	-0.03
ring	322703	1097	0.34	205322	584	0.28
ship2	2283	75	3.29	8269	114	1.38
turb_09	806106	-90	-0.01	263805	295385	111.97
unipi	683878	7242	1.06	61790	450	0.73
wuerfel	13548	-18	-0.13	17125	45	0.26

Table 2. Comparison of calculated total hatching area and contouring length, from CLI and STL estimators.

Job	Actual Build-Time (hr)	CLI Estim. (hr)	Deviation (%)	STL Estim (hr)	Deviation (%)
1	4.088	4.119	-0.76	4.122	-0.83
2	4.136	3.706	10.40	3.705	10.42
3	10.023	9.349	6.72	9.368	6.53
4	5.700	5.666	0.60	6.527	-14.51
5	7.015	6.266	10.68	6.268	10.65

Table 3. Actual and predicted build-time for completed jobs.

Build-time estimations are compared to actual fabrication times in Table 3. The figures included in this table refer to reasonably complicated jobs composed of two or three parts each. It is seen that both build-time estimator tools developed in this chapter perform quite satisfactory. The deviations between predicted and actual build-times are expected to be reduced drastically, if one incorporates in the proposed tools the effect of the number of hatching vectors, hatching length and hatching velocity.

6 Conclusions

In this chapter the problem of build-time estimation for Stereolithography systems and specifically for the EOS-Stereos Desktop S system installed at the University of Piraeus, is examined in detail. Two software build-time estimation tools utilising STL- and CLI-file geometric data were developed and tested. Results presented show that the STL-estimator predictions are quite accurate and reliable besides the fact that the actual part (prototype) fabrication process is not directly related to the STL data. Preliminary investigation indicates that build-time estimations would improve if the effects of the number of hatching vectors, of the hatching length and hatching velocity are included in the mathematical framework of the proposed tools.

Acknowledgements

This work has been carried out as part of the research project "Development of Rapid Prototyping Facilities - NEKTAR" funded by the General Secretariat for Research and Development of the Ministry of Development of Greece in the context of research programme EPET II.

References

1. Chen C, Sullivan P A 1996 Predicting total build-time and the resultant cure depth of the 3D Stereolithography process. *Rapid Prototyping Journal* 2: 27-40
2. McClurkin E J, Rosen D W 1997 Build time estimation for Stereolithography. Technical Report, Georgia Institute of Technology
3. Tata K, Flynn D 1996 A procedure to estimate build time for Stereolithography Machines. Technical Report , Prototype Express, USA
4. Kehagias J, Anagnostopoulos V, Zervos S, Chryssolouris G 1997 Estimation of build times in rapid prototyping processes. In: Dickens P M (ed.) 1997 *Proceedings of the 6th European Conference on Rapid Prototyping & Manufacturing*, Nottingham, pp 137-148
5. Jacobs P 1992 *Rapid Prototyping & Manufacturing: Fundamentals of Stereolithography*. McGraw-Hill, New York

6. Dolenc A 1993 *Software Tools for Rapid Prototyping Technologies in Manufacturing*. Acta Polytechnica Scandinavica, Helsinki (Mathematics and Computer Science No. 62)
7. EOS Gmbh 1995 *Operating Manual: Stereos Desktop S*. Planegg, Germany
8. http://www.engr.csulb.edu/~amartin/
9. http://www.cranfield.ac.uk/coa/rapid/cli/cli_v20.html
10. Cheng W, Fuh J F H, Nee A Y C, Wong Y S, Loh H T, Miyazawa T. 1995 Multi-objective optimization of part-building orientation in Stereolithography. *Rapid Prototyping Journal* 1: 12-23

34

An Efficient PC-Based Environment for the Improvement of Magnetic Cores Industrial Process

P.S. Georgilakis, N.D. Hatziargyriou, N.D. Doulamis, A.D. Doulamis and J.A. Bakopoulos

1 Introduction

Distribution transformer losses constitute a significant amount of the total losses in distribution networks. In Greece, for example, it is estimated that transformer iron losses are about 10% of the total distribution network losses [1]. In an industrial environment, dealing with construction of wound core distribution transformers, prediction of iron losses of individual cores is a crucial task, since the latter significantly affect both the quality and the performance of the finally produced three phase transformers. However, there is no simple analytical relationship for estimating iron losses of individual cores, due to the fact that many parameters, both qualitative and quantitative, are involved in the process.

In the current practice the prediction of specific iron losses of individual cores is based on the loss curve. This curve determines the relation between the rated magnetic induction and the specific iron losses of individual core. This means that all cores which are going to be constructed using the same design, or equivalently using the same magnetic material and working at the same magnetic induction, will have the same theoretical iron losses. This method presents a maximum absolute relative error of approximately 32% in the prediction of iron losses of individual cores.

Intelligent systems conveniently accommodated in modern PC environments that are able to learn and interpret several variations of the same conditions can help in calculating these iron losses with increased accuracy. In particular, artificial neural networks (ANNs) [2-3] with their highly non linear capabilities have been successfully applied in power system load forecasting [4-6], security assessment [7-9] etc. In this chapter the application of multilayer feedforward neural networks trained with a backpropagation variant algorithm is proposed to determine the iron losses of individual cores.

Experimental results indicating the good performance of the neural network scheme as far as the iron losses prediction is concerned are presented. It is shown that the neural network approach produces an average absolute error of 2.3% in the prediction of specific iron losses of individual cores instead of 5.7% obtained by the current practice.

Basic terms of designing wound core distribution transformers are presented in Section 2 while a short description of the adopted ANN architecture is given in Section 3. The applications of multilayer feedforward neural networks to prediction and classification of iron losses of individual cores and the obtained results are described in Section 4. Conclusions are finally presented in Section 5.

2 Design of Wound Core Distribution Transformers

The wound core distribution transformer production includes, at the first stage, the slitting of the raw material into bands of standard width. Then, the slitted sheets are cut to pre-determined lengths and are wound on a circular mandrel. After that a suitable press gives a rectangular shape to the circular core. However, the previously described process significantly deteriorates the core characteristics and especially its physical and electrical properties. To restore these properties, annealing follows at temperatures in a range of 760 to 860^{0}C in a protective environment. The most widely used protective environment is pure, dry nitrogen, containing up to 2% hydrogen. This prevents the steel from oxidation.

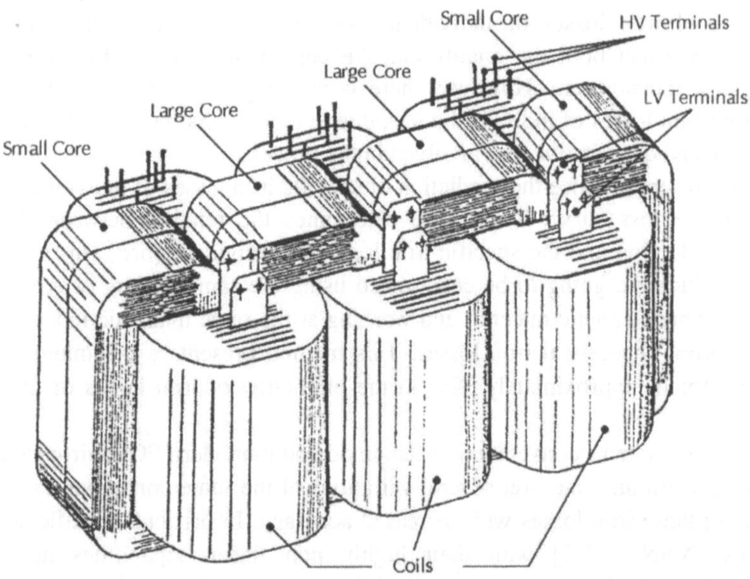

Figure 1: Assembled active part of wound core distribution transformer.

The annealing cycle adopted is divided into four phases: starting and heating up phase (the objective is to avoid oxidation and to normally achieve the temperature of 825^{0}C), soaking phase (to achieve homogeneous temperature distribution for all cores), slow cooling phase (the target is to slowly cool the load in order to avoid the development of internal stresses in the cores) and fast cooling phase (for reduction of the temperature to 380^{0}C, so as to avoid oxidation of cores, when they are exposed to the natural environment).

To construct a three-phase wound core distribution transformer, two small individual cores (width of core window equal to F1) and two large individual cores (width of core window equal to F2) should be assembled (see Figure 1). In general, the width F2 is twice F1.

The theoretical single-phase iron losses, say, W_1 (in Watt), of the small individual core are given by:

$$W_1 = WPK_1 * CTW_1 \quad (1)$$

where WPK_1 are the theoretical individual core specific iron losses at the rated magnetic induction (see Figure 2) and CTW_1 is the theoretical weight of the small core as defined in [10].

Figure 2 presents a typical loss curve where the value of the magnetic induction (B expressed in Gauss) is plotted versus the value of the specific iron losses (WPK_1 expressed in Watt/Kg) of individual cores. In the same figure, the specific iron losses (expressed in Watt/Kg) of transformer are also depicted.

The theoretical single-phase iron losses, say, W_2 (in Watt), of the large individual core are:

$$W_2 = WPK_1 * CTW_2 \quad (2)$$

where CTW_2 is the theoretical weight of the large core [10].

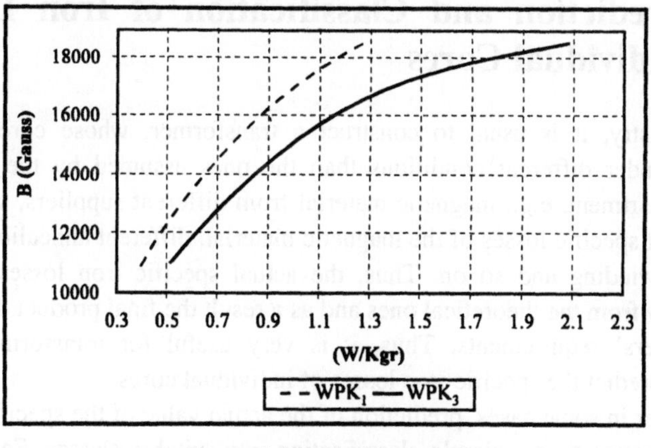

Figure 2: Typical loss curve.

3 The Neural Network Architecture

ANNs are universal approximators, i.e., they can approximate to any desired degree of accuracy any real-value continuous function [3]. In the case of wound core type of transformers no simple linear model (e.g., autoregressive (AR), Moving Average (MA) or ARMA) has been proved capable of estimating the iron losses, even if the involved parameters are known.

In general, the input vectors of an ANN structure are of primary importance. As a result, we use proper features as input vectors so as to achieve good generalisation. These features, also called attributes, comprise characteristic measurements taken during the core production phase. This is due to the fact that the actual values of the production procedure parameters significantly affect the iron losses. Table 1 analytically describes these attributes. Then a learning set (LS) is formed consisting of representative input data and their desired outputs. The performance and the reliability of the ANN architecture is evaluated with independent testing sets (TS) which have the same structure as the LS, i.e., they are generated in exactly the same way, but comprise different samples of transformer designs.

In the rest of the chapter, a multilayer feedforward neural network is proposed for predicting specific iron losses of individual cores. The NNET package [11] has been used for this purpose. After the training procedure the network is able to learn (generalise) the input-output relationship and thus to predict the iron losses to any input vector outside the training set.

In our approach we adopt a backpropogation variant [12] in a constructive framework [13] which begins with a small network size and subsequently adds neurons to improve the network performance. A validation data set has been also used during training to control learning with respect to the generalisation ability of the network.

4 Prediction and Classification of Iron Losses of Individual Cores

In the industry, it is usual to construct a transformer, whose cores have been produced under different conditions than the ones assumed by the Transformer Design Department, e.g., magnetic material from different suppliers, same supplier but different specific losses of the magnetic material, different annealing conditions, quality of winding and so on. Thus, the actual specific iron losses can deviate significantly from the theoretical ones and as a result the final product does not fulfil the customers' requirements. Thus, it is very useful for transformer design to accurately predict the specific iron losses of individual cores.

Moreover in some cases, prediction of the actual value of the specific iron losses is less interesting than a simple classification into suitable classes. For instance, in most cases it is adequate to know if a transformer can pass acceptance tests, according to its iron losses. In addition, it is sometimes desirable to know what is the effect on the final product of changing one or more parameters involved in the process. This is a common problem in the industry, i.e., construction of transformers based on a previous design, where one of the main materials has technical characteristics little different in relation with the previous design. In such cases the Transformers Design Department should check if iron losses are within the limits defined by the international standards and are also in agreement with the guaranteed to the customer iron losses.

4.1 Creation of the Knowledge Base

To improve the quality of individual cores prediction of specific iron losses is required. In particular, the impact of the annealing cycle, the divergence of the actual core weight from the theoretical value and the quality of core magnetic material are taken into consideration as input parameters (attributes). The selection of these attributes was based on extensive research and transformer designers' experience. All tests were done using the same 160 kVA transformer design and also the same supplier of cores magnetic material. The magnetic material was M3, 0.23 mm. For this transformer design the theoretical expected specific iron losses of individual cores are equal to 0.78 Watt/Kg, as provided by the loss curve.

Six attributes have been investigated corresponding to the annealing process, depicted in Table 1.

Symbol	Description	Low	High
ATTR1	Annealing final temperature	825 0C	855 0C
ATTR2	Temperature rising time	3 hours	4 hours
ATTR3	Furnace opening temperature	250 0C	350 0C
ATTR4	Duration of constant temperature	2 hours	3 hours
ATTR5	Position of core in the furnace	Down	Up
ATTR6	Protective atmosphere	100% N2	98% N2 & 2% H2

Table 1: Annealing attributes.

In order to take in account all the combinations of the six attributes with two values (Low and High), 32 experiments are required. Since the use of all combinations is time consuming and expensive, reduction of the implemented experiments can be achieved through the statistical design of experiments method (SDE). According to SDE [14,15] many parameters can by varied at the same time and in a systematic manner, assuring the reliable and independent study of the impacts and interactions of all the main parameters in the production procedure.

Applying this method to our study reduces the essential number of experiments to 8. The parameters characterising each of the 8 tests are shown in Table 2. It can be seen that 4 experiments are carried out with the low value (L) of each attribute and the other 4 with the high value (H) due to the symmetric property.

For every one of the eight tests, 96 (48 small and 48 large) cores were constructed. It should be noticed that all cores were annealed at the same furnace.

Two more attributes considered are the actual over theoretical core weight ratio and the specific losses (Watt/Kg at 15000 Gauss) of core magnetic material. We define these attributes as ATTR7 and ATTR8.

768 samples were collected for the creation of the learning and test sets. The 3/4 (576) of them were used as learning set and the rest (192) as test set. Moreover 1/4 (144) of the samples of the learning set were used as validation set during learning to avoid over-training problems.

	Annealing test number							
	1	**2**	**3**	**4**	**5**	**6**	**7**	**8**
ATTR1	L	H	H	L	L	H	H	L
ATTR2	L	L	H	H	H	H	L	L
ATTR3	L	H	H	L	H	L	L	H
ATTR4	L	L	H	H	L	L	H	H
ATTR5	H	L	H	L	H	L	H	L
ATTR6	H	H	H	H	L	L	L	L

Table 2: Conditions of the various annealing tests.

4.2 Measurements Organization

In order to create the knowledge base of individual cores the industrial measurements are taken using the following systems:

1) *Core Weighing System*

An industrial weighing machine with 4 load cells and serial RS 232 output for transferring data to a computer is used for core weighing.

2) *Core Loss Measuring System*

The core loss measuring system is shown in Figure 3. It comprises a system for automatically applying the appropriate voltage to the core and one single-phase power analyzer with RS 232 output for transferring measurements to a computer.

3) Annealing cycle monitoring and control

In order to monitor and control the annealing cycle PLCs are used.

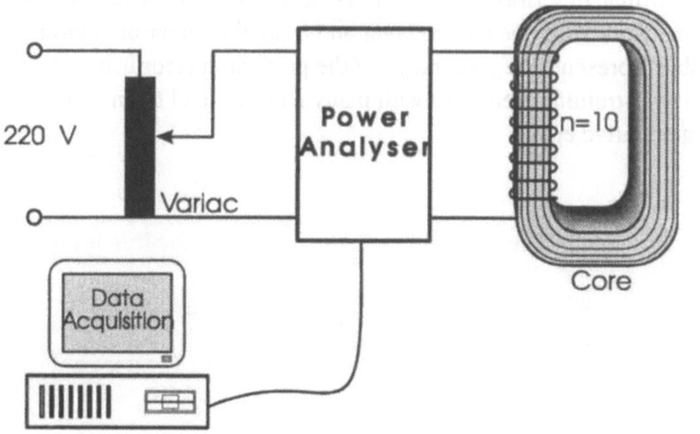

Figure 3: Core Losses Measuring System.

4.3 Prediction Problem

In the prediction problem a multilayer feedforward neural network structure with one output has been used while the input neurons are equal to the number of attributes (8).

After training the ANN, its reliability is evaluated based on the *average absolute relative error* for the test set:

$$Error = \frac{1}{N} * \sum_{\substack{for\ all\ N \\ samples}} \frac{\| ActualSpecificLosses - PredictedSpecificLosses \|}{ActualSpecificLosses} * 100\% \qquad (3)$$

NN Number	Neurons in Hidden Layer 1	Neurons in Hidden Layer 2	Error (%)
1	100	50	2.56
2	100	20	2.39
3	20	0	2.52
4	15	0	2.32

Table 3: Prediction error (%) of specific iron losses of individual cores.

The network output consists of one neuron providing the specific iron losses of individual core. To find the proper network size, a constructive algorithm is applied similar to that described in [13]. It is observed that the proper size comprises a neural network with one hidden layer and a small number of neurons (15).

Table 3 presents the percentage of the prediction error using other network sizes. Due to over-training, networks with many neurons lead to an increase of the average absolute relative error.

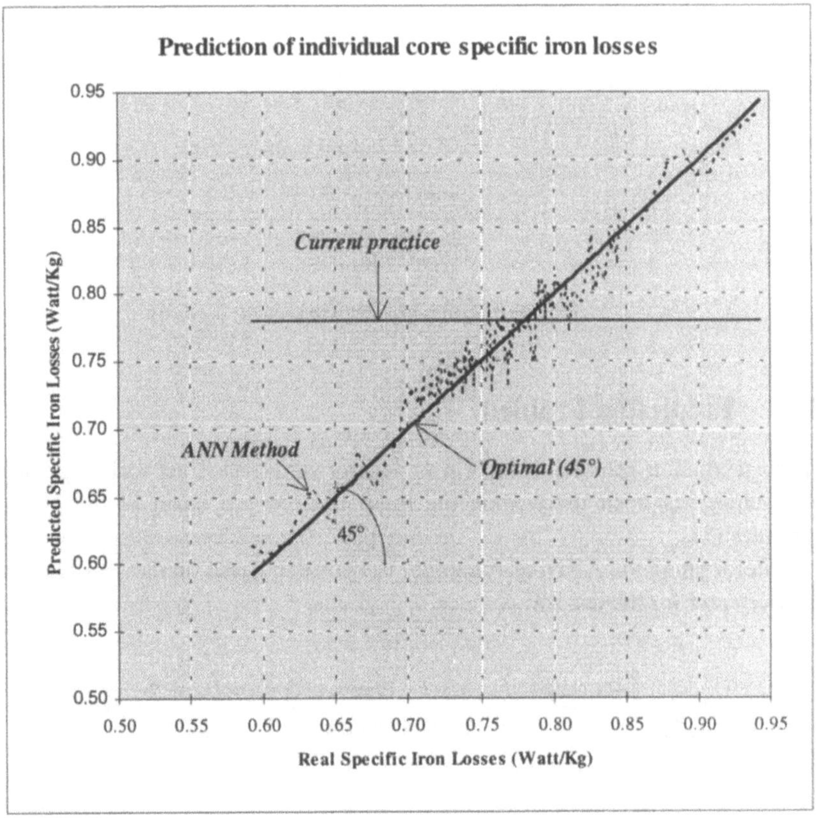

Figure 4: Prediction of specific iron losses of individual core using the typical loss curve (current practice) and the artificial neural network architecture.

Figure 4 presents the fractile diagram or the Q-Q plot (quantile-quantile) of the specific iron losses. According to this method the data of real specific iron losses are plotted versus the predicted ones. Perfect prediction lies on a line of 45^0 slope. It is observed that, the prediction of specific iron losses of individual cores, based only on rated magnetic induction and ignoring all the other parameters, provides a constant (equal to 0.78 Watt/Kg) estimate for all samples belonging to the testing set. Therefore, it significantly diverges from the line of 45^0 providing an erroneous prediction increasing at large or small Watt/Kg values in a range from 0.5925 to 0.9433 Watt/Kg. On the contrary, the proposed ANN method is able to accurately estimate the iron losses of individual cores for all the testing samples, due to the neural network learning capabilities.

Table 4 presents the maximum and average absolute relative error for both the current practice and the ANN method. In particular, the current method shows a maximum absolute relative error of 31.6% and average 5.7% while the proposed ANN method errors are 5.0% and 2.32%, respectively. It is observed that ANN performs better than the conventional method in both average and worst case error.

Parameter	Current practice	ANN method
Maximum absolute relative error (%)	31.6	5.0
Average absolute relative error (%)	5.7	2.32

Table 4: Comparison of predicting results of specific loss values.

4.4 Classification Problem

Classification of specific iron losses of individual cores into two classes is done using the neural network architecture proposed in section 3. The same 768 samples were used split in 576 MS as learning set and 192 MS as test set.

The criterion for classifying core iron losses as non-acceptable is based on the comparison of the actual specific iron losses (measured by the Quality Control Department) to the theoretical expected specific iron losses (used by the Design Department and provided by the loss curve).

If the actual specific iron losses are greater than the product of theoretical specific iron losses and a parameter called "Limit", then the core is characterised as acceptable, otherwise is non acceptable. Parameter "*Limit*" denotes the acceptability limit, which mainly depends on the customer requirements and designer considerations. If parameter "*Limit*" is selected to be 1.10 then:

$$\mathit{if}\ \ ActualSpecificIronLossesofCore \leq 0.78\frac{Watt}{Kg} * 1.10 \Leftrightarrow$$

$$\mathit{if}\ \ ActualSpecificIronLossesofCore \leq 0.86\frac{Watt}{Kg} \qquad (4)$$

then the core is acceptable.

In this framework the first class, say Class 1, corresponds to iron losses less than 0.86 Watt/Kg, while the second class, say Class 2, corresponds to iron losses greater than or equal to this value.

Table 5 illustrates the proportion that both Classes 1 and 2 occupy in the learning and testing sets.

In this experiment the network output consists of two neurons each of them corresponds to one of the two available classes. The input layer neurons are also eight since the same attributes are used as elements of the input vector, while the neurons of the one hidden layer are again 15, as in the previous prediction problem.

Table 6 depicts the classification results obtained by the neural network structure. In particular, 96.4% of the samples belonging to Class 1 have been correctly classified and 91.7% of samples belonging to Class 2. Class 1 presents higher classification success rate than Class 2 due to the fact that Class 1 contains more representatives than Class 2. The total classification success rate is 184/192 or 95.8%.

	Learning Set		**Testing Set**	
Class	Measurement Sets	Percentage (%)	Measurement Sets	Perc. (%)
1	512	88.9	168	87.5
2	64	11.1	24	12.5
Total	**576**	**100.00**	**192**	**100.00**

Table 5: Partition of the learning and testing set in two classes.

Class 1		**Class 2**	
Success	Percentage (%)	Success	Percentage (%)
162/168	96.4	22/24	91.7

Table 6: Performance of the classifier for the individual core specific iron losses.

5 Conclusions

In this chapter, multilayer feedforward neural networks are applied for the prediction and classification of specific iron losses of individual cores. The basic steps in the application of the method, like the generation of the learning and testing sets, the selection of candidate attributes and the derivation of the appropriate ANN structure are presented. The average absolute relative error for the prediction of specific iron losses of individual cores is 2.32%. If two classes are used, the total classification success rate obtained by the ANN method is 95.8%. It is shown that with the LS and TS used and for the selected candidate attribute sets, the ANN method is very suitable for prediction and classification of specific iron losses of individual cores.

Acknowledgements

The authors would like to thank the General Secretariat of Research and Technology of Greece for financing the research programme entitled: "Development of a Quality Control System of Distribution Transformer Magnetic Cores Using Artificial Intelligence Techniques" within the YPER '94 Research Programme. They would also like to thank Dr. D. J. Sobajic for granting permission to use the NNET package. Help from Schneider Electric AE staff in collecting industrial measurements is gratefully acknowledged.

References

1. Papadopoulos M 1994 *Electric Energy Distribution Networks.* NTUA, Athens (in Greek)

2. Beal R, Jackson T 1990 *Neural Computing: An Introduction.* IOP Publishing Ltd., Bristol, UK

3. Haykin S 1994 *Neural Networks: A Comprehensive Foundation.* Macmillan, New York

4. Peng T M, Hubele N F, Karady G G 1992 Advancement in the Application of Neural Networks for Short-Term Load Forecasting. *IEEE Trans. on Power Systems* 7:250-257

5. Lu C N, Wu H T, Vemuri S 1993 Neural Network Based Short Term Load Forecasting. *IEEE Trans. on Power Systems* 8:336-342

6. Park D C, El-Sharkawi M A, Marks R J II, Atlas L E, Dambong M J 1991 Electric Load Forecasting Using an Artificial Neural Network. *IEEE Trans. on Power Systems* 6:442-448

7. Pecas-Lopes J A, Fidalgo J N, Miranda V, Hatziargyriou N 1994 Neural Networks Used for On-Line Dynamic Security Assessment of Isolated Power Systems with a Large Penetration from Wind Production - A Real Case Study. In: *3rd International Workshop on Rough Sets and Soft Computing (RSSC'94).* San Jose, California

8. Sobajic D J, Pao Y H 1989 Artificial Neural-Net Based Dynamic Security Assessment for Electric Power Systems. *IEEE Trans. on Power Systems* 4:220-228

9. Weerasooriya S, El-Sharkawi M A, Dambong M, Marks R J II 1992 Towards Static Security Assessment of a Large Scale Power System Using Neural Networks. *IEE Proc., Part C* 139:64-70

10. Georgilakis P S, Bakopoulos J A, Hatziargyriou N D 1997 A Decision Tree Method for Prediction of Distribution Transformer Iron Losses. In: *32nd Universities Power Engineering Conference (UPEC'97), Vol. 1.* Manchester, pp 257-260

11. AI Ware 1989 *NNET 210 User's Manual.* AI Ware Incorporated, Cleveland, OH

12. Kollias S, Anastassiou D 1989 An Adaptive Least Squares Algorithm for the Efficient Training of Artificial Neural Networks. *IEEE Trans. on Circuits & Systems* 36:1092-1101

13. Kollias S 1996 A Multiresolution Neural Network Approach to Invariant Image Recognition. *Neurocomputing* 12: 35-57

14. Taguchi G, Konishi S 1987 *Taguchi Methods: Orthogonal Arrays and Linear Graphs; Tools for Quality Engineering.* ASI, Dearborn, MI, USA

15. Logothetis N 1992 *Managing for Total Quality.* Prentice Hall International, UK

35

Analysis and Reconfiguration of a Packing Production Process in an Alimentary Pasta Firm

F. Deperini, L. Penso and D. Pozzeto

1 Introduction

The operative cycle in a pasta production plant is divided in four phases:

a) receipt and storing of the raw material (floor),
b) control and mixing of the raw material (floor).
c) hydratation and kneading of dough, extrusion and drying of the pasta,
d) packing, palletization and final storing.

The detailed analysis of the production department has highlighted the criticality of the final packing area. In this zone the pasta coming out from some silos is packed and placed in boxes. The considered system consists in four packing machines each one producing different types of pasta in different boxes. These equipments flow together in specific positions in the same transporter with constant speed that provides the movement of the boxes to the single palletising machine. The insertion in the transporter must be regulated, by photoelectric sensors, in order to avoid placing two boxes, coming out from different machines, in the same position with the subsequent falling of one of them. The transporter flows through the packing area collecting all the produced boxes and arrives to a infrared bar-code detector, which selects the different types of packs and sends them to one of the five buffers, each one associated to a box type. All these five buffers flow together to the final palletising machine which empties the buffers and produces the final pallets.

A preliminary qualitative evaluation on the department layout and configuration has been made without keeping in account the stops of the parts due to failures, since the critical activities are such even in perfect running conditions of the system and are highlighted by hard utilising conditions.

Two critical subprocesses emerged: the box insertion in the transporter and the palletization and associated managing process of the buffers.
The insertion process is of an asynchronous type without any synchronisation; in the case of more than one machine running in parallel, the saturation condition of the transporter is well reached under its theoretic capacity. The simplest way to solve this problem is to increase the speed of the movement and, subsequently its theoretic and real delivering capacity to a value that is enough to absorb the produced boxes flow. This solution does not solve the problem actually, but makes it latent; so a second method could be concerned with a synchronous or "slotted" insertion procedure that keeps in account the positions of the boxes already present in the transporter and inserted by machines that stand before. The use of this procedure induces to the choice of the dimension of the "slot" (timing or physical dimension) and subdivision of the transporter length in slots and, in case the sizes of the boxes are different, may lead to a reduction of its real flow capacity due to wasted space between the packs.
The second critical process is the management of the intermediate buffers and the palletization, with an increasing piling up of boxes in distinct buffers and their palletization by the palletising robot.
In order to analyse this process properly, all the stops of the packing machines and transporter must be taken in account, since all the process is widely affected by the productivity rate.
The buffer selection uses a sequential polling procedure, unsuitable to manage critical situations, in which one of the last buffers is almost full, leading to collapse. Thus it is important to choose a procedure that's more suited to obtain a more stable system.
The subprocess can be generalised as a service that is given by a unique server to different users placed in different waiting queues only if the queue length exceeding a threshold N varying among the queues. Once the service has started, N users (the threshold value) are served, independently from the conditions in the other queues.
In case no queue is over the threshold, the next user arrive is waited to begin a new test on the queue lengths.
The different choosing strategies considered are:

- LQF (longest queue First)

  Is chosen the one with the longest queue in physical dimensions, not number of users, since users of different queues have different sizes, among the ones satisfying the threshold criteria. In case there is still uncertainty in the choice, the First one found is taken.

- SPTF (shortest processing time First)

  Is chosen the one with the lower processing time by the palletizator among the ones satisfying the threshold criteria. In case there is still uncertainty in the choice, the first one found is taken.

- LQF+SPTF

  Same as the LQF procedure regardless the last uncertainty, in which case SPTF criteria is applied to the remaining ones to make the choice.

- SPTF+LQF

  Same as the SPTF procedure regardless the last uncertainty, in which case LQF criteria is applied to the remaining ones to make the choice.
- SEQUENTIAL

  The choice falls on the first one that is found to be satisfying the threshold bound.
- SPTF+LQF Special

  Same as SPTF+LQF but in the case the number of packing machines is less than the number of available buffers, the free ones can be utilised as emergency buffer in case one is full. It might be useful to process first this emergency queue and not the others, even if it has not enough "users" in. The remaining users are taken from the full emergency calling buffer. In this way we hope to keep this emergency queue available for the maximum time, ready to receive surpluses.

Given the need of a quantitative analysis of the effects in the productive system subsequent to the application of the different solutions proposed to rationalise the process, in order to make the strategic choice on the configuration of the department, it has been decided to utilise a discrete event simulation technique, implemented in MODSIM II®.

Regarding the system modelling, each component is described with a well defined procedure that takes in account its behaviour (functional relationships) under conditions given by the values assumed by its parameters.

The choice of these parameters and their influence in the functional relationships is of fundamental importance: more parameters means more real conditions suiting of the model, while the conditions not associated to a parameter will result constant or influent. Obviously the more are the parameters the more data are to be collected and the more complex and slow will be the program, so a compromise must be found.

The definition of the model of each component and its parameters are related to:

a) pasta formats

   the pasta formats are different in various parameters; the most important one have been placed in a record (figure 1)

b) packing machines

   each packing line has been considerated as a box generator which produces the packs corresponding to the format in the field "name". The time between the creation of 2 new boxes is considered to be an aleatory variable distributed as an exponential with "productivity setted" parameters. The packing line produces a determined type of format for a time being also an aleatory variable distributed as a Γ distribution with $\alpha=30$ and means value equal to "lasting in production". After this time a new format is set up, taking in account set-up times. In the format choice the probability associated to each one to go on production is respected. In the case the choice falls on the same format the production and a new run of the choice procedure is done to ensure that there's actually a format change at that time.

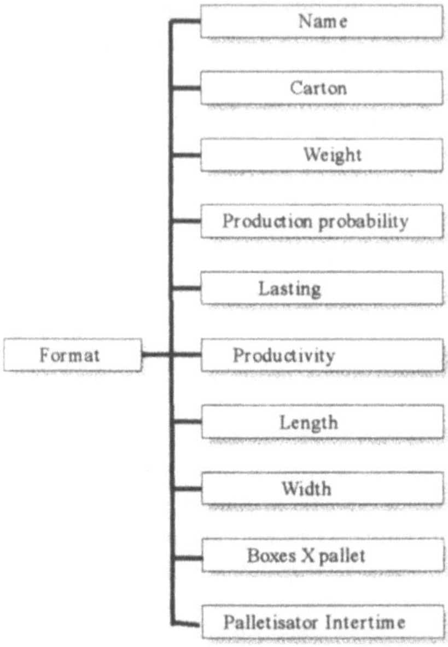

•name:
•carton: code of the type of the box (e.g. C2-250);
•weight;
•probability of being produced;
•mean time of the lasting in production;
•productivity;
•length of the box;
•width of the box;
•number of boxes in each pallet;
•time of interarrivals to the palletizator;

Figure 1

Once the new format has been chosen, the machine is kept paused for an aleatory time associated to the set-up time.
In the case that stops arise during the production time, the inactivity time of the machines is not counted as production time, because the production timing is assumed to be correlated to the number of boxes obtainable in that time. In the case that it is not possible to obtain that amount of production in the given time, the production is kept on going until is reached an effective working time equal to the programmed one. It is possible for some parts of the packing machine to break and a single component failure stops all the involved machine for again an aleatory time depending from the kind of failure. The same failure will reoccur after a aleatory time associated to the MTBF of the part.
An analysis has been made in the collected data to highlight the importance of each failure cause in the total machine stop time, eliminating those of trascurable importance. Subsequently the failure causes have been classified:

- weights failure,
- automatic cellophane roll exchange,
- cartons glued together,
- bundle maker failure,
- elevator failure,
- various.

Each packing unit is placed in different positions and inserts its boxes at a distance from the start of the transporter, recorded in a record field. Each one of them has a short buffer before the transporter insertion section that contains the boxes waiting to be sent to the trasporter.

c) transporter

It has been considerated as a queue that moves the boxes on it at a constant speed, there is also an appropriated number of insertion channels which can accept a single box at the time, each one associated to one of the packing lines which will insert its boxes on it. There are also some critical positions which are the points in which the box affects the insertion procedure of any packing machine and its insertion channel. This critical points correspond to:

1. beginning of the channel occupation by a box that is moving along the trasporter;
2. end of the channel occupation.

The movement of a box on the transporter is made by scheduling the events that change the status of the transporter, so the movement routine can be synthesised as:

1. if the insertion channel is free, insert the new box, else wait until becomes free(asynchronous procedure);

1bis. of the insertion channel is free and synch pulse is on then insert the box, else wait until next synch impulse(synchronous procedure);

2. set the insertion channel busy and wait the littler time needed to transport the box out of it.
3. check the box position (transporter may have stopped in the meantime). If the box is at the end of the insertion channel set it free and wait the time expected for the arrival to the next interesting point, if the box is not already at the end of the insertion channel then wait again the littler time needed to transport the box out of it and repeat point 3.

d) buffers and infrared selector

Buffers have been modelled with FIFO queues, the infrared selector due to its high velocity has been modelled with a simple real time routine that decides where to send the arriving box.

d) palletizator

the palletization process has been modelled taking in account two parameters, depending from pallet box placement, box format and involved buffer. These parameters represent:

- T_p time needed to the first box to reach the palletizator from the buffer. This time depends only from the involved buffer;
- T_s time between the input in the palletizator of two boxes. This time is aleatory because of the unpredictable displacements of the boxes in the trasporter to the palletizator and it has been verified that its is represented by a Gauss distribution with mean and standard deviation depending from the box format only.

The check of the buffers for the discharge condition is made when one of the following conditions arises:

- a new box arrival to the infrared selector in the case that the palletizator is off

•the end of a pallet making procedure in the case that the palletizator is on. After this check, if there are the conditions to start a new pallet then a buffer is emptied choosing it following one of the given procedures.

Even if it is not essential to the simulation model, a graphic output has been developed in order to give a glimpse of the entire process.

2 Simulation results

After the development of the simulation model, collection and analysis of the data, it is necessary to proceed to check the ability of the model to represent reality. So it has been verified that the model will produce a day production in 16 hours close to the real mean one calculated using 6 months of data. The results are shown in figure 2.

The experiments are organised in such a way to give the productivity of the process as a function of the different configurations possible, giving useful data for selecting the best one. Initially it has been considerated the insertion subprocess making some experiments with the objective of:

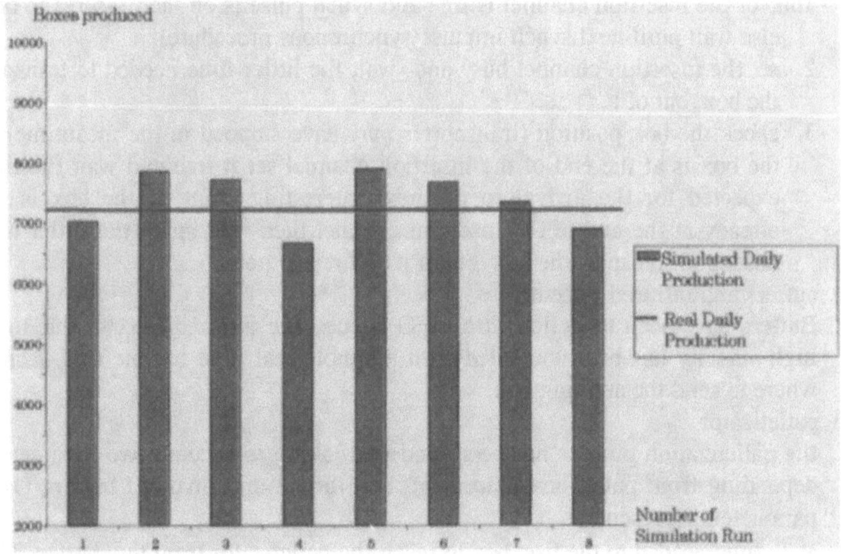

Figure 2

a) make a comparison between the synchronous and asynchronous procedures with 4 packing machines and transporter velocity equal to the real one of 14m/min, each machine producing different size boxes. For each value of single machine top packing productivity 20 runs have been calculated, from whom it has been evaluated the mean value of the transport capacity, measured as the number of boxes brought to destination in a time unit from the transporter. Results are reported in figure 3;

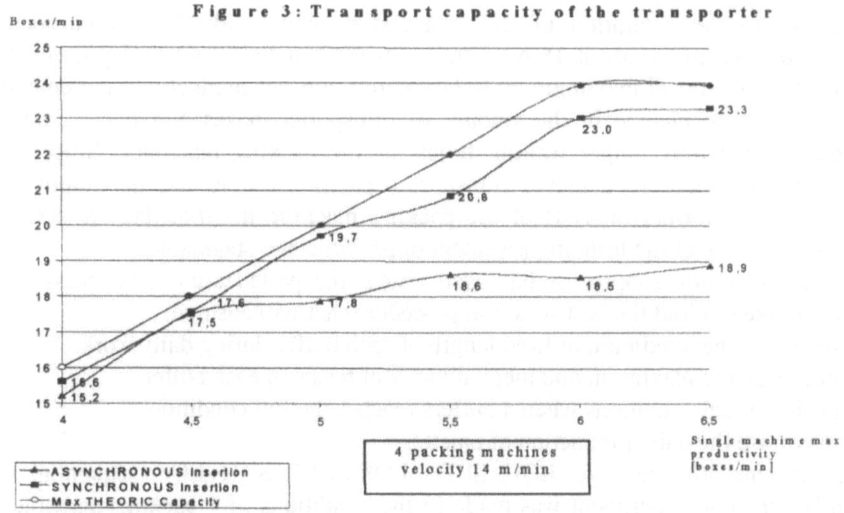

a) determine the transporting capacity with 5 packing machines in the same conditions of the previous experiment. Results are reported in figure 4.

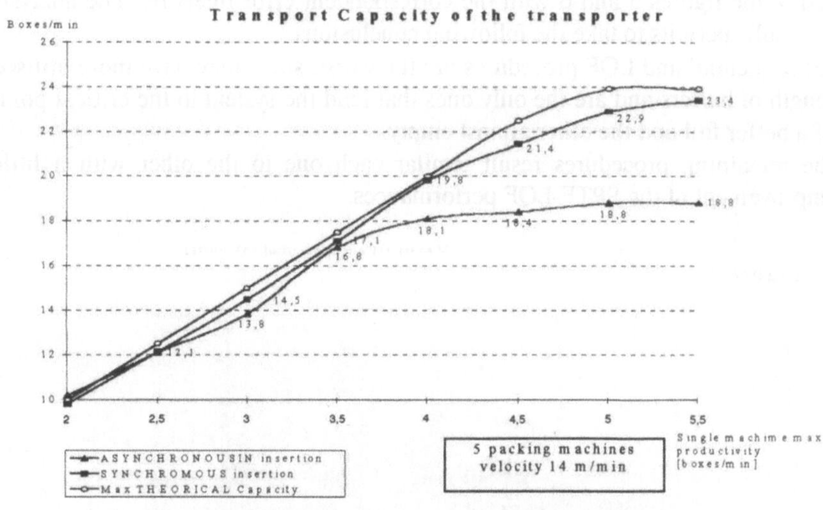

Figure 4: Transporter capacity with 5 packing machines

The two figures 3 & 4 show that the synchronous procedure is always preferable since its performances are better in hard working conditions and equal in soft working condition. The capacity is always under the theoretical maximum value due to the machine stops, which pulled down the number of produced and hence transported boxes during the simulation. Another peculiarity of the curves is the

knee due to the saturation of the transporter; while with the asynchronous procedure it occurs at about 18 boxes/min, for the synchronous one it goes up to almost the theoretical maximum of 24 boxes/min. Further productivity increments will produce no change in the amount of transported boxes and hence in the production but only longer waiting times for the packing machines. Since the synchronous procedure achieves transport capacity close to the maximum 24 boxes/min, no further analysis of the packing machine insertion buffers size is required because of the little improvement of performance obtainable.
Once the insertion subprocess has been tested, the palletization subprocess has been analysed to find the best selection procedure that will ensured:

a) minimise the maximum utilised length of each buffer during daily work,
b) minimise the maximum and mean number of boxes in each buffer,
c) avoid critical conditions when a buffers reaches the full condition,
d) maximise the daily production of palettes.

The choosing strategies are: LQF, SPTF, LQF-SPTF, SPTF-LQF, sequential as seen before. The experiment was made in the conditions of 4 packing machines, real production speed (not maximum and depending from format), 5 buffers before the palletizator, failures enabled and real pallettization speed depending from format. For each strategy 10 runs have been made and the mean results have been plotted in the figures 5 and 6 with the correspondent error intervals. The analysis of the results permits to take the following conclusions:

- the sequential and LQF procedures are the worst, since they have more utilised length of buffers and are the only ones that lend the system to the critical point of a buffer full and the other almost empty.
- the remaining procedures result similar each one to the other with a little improvement of the SPTF-LQF performances.

Maximum occupation of the buffer

Figure 5

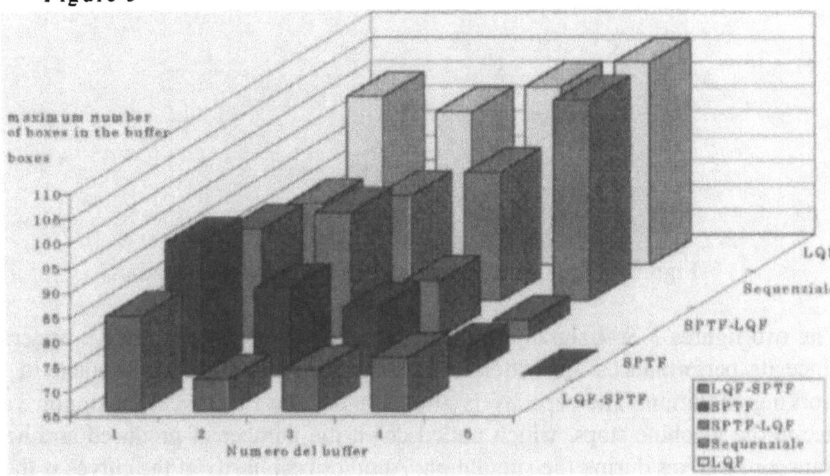

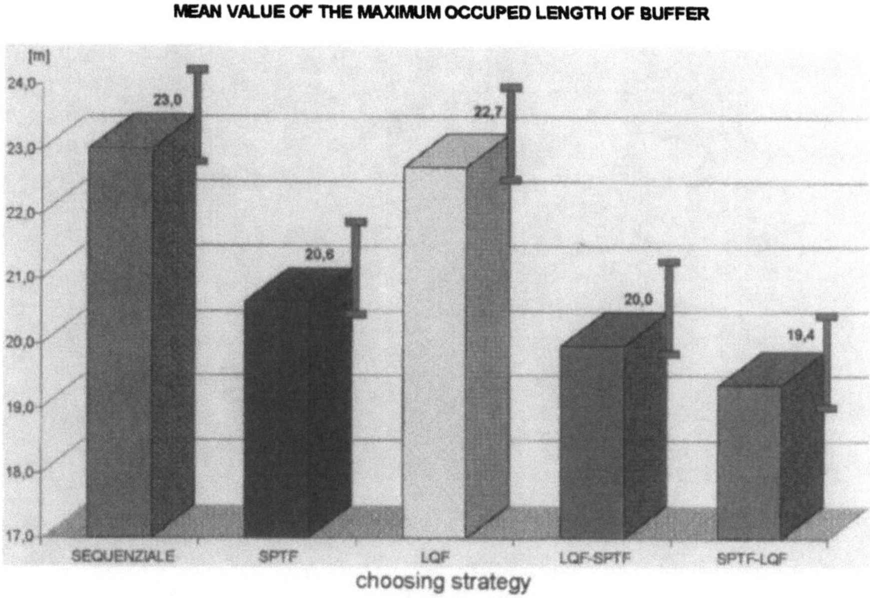

Figure 6

As the last experiment, 5 packing machines have been simulated with the three sequent configurations:

a) improved palletising speed, 5 buffers, SPTF-LQF palletising procedure
b) real palletising speed, 6 buffers, SPTF-LQF palletising procedure
c) real palletising speed, 6 buffers, SPTF-LQF "special" palletising procedure

Again 10 runs for each configuration have been launched and the B configuration has been found to be unfitted to the production needs, stopping the production many times a day due to the critical condition reached of a buffer full, so it is not useful and its results are not reported. In figure 7 the mean results of the other 2 configuration are plotted.

Obviously it results more suitable the A configuration with 5 buffers and improved palletisation speed, since it produces lower occupation of the buffers and a more flexible system.

3. Conclusion

From the simulations made, it has emerged that in the insertion process the transporter capacity reduction due to the asynchronous insertion procedure with respect of the max. theoretical value is about 11% with 4 packing machines running and 22% in the case 5 machines are running. Using the synchronous procedure an improvement of about 10% and 22% respectively in the two cases of 4 and 5 machines running is achieved, bringing the transporter to 97% of its transport efficiency.

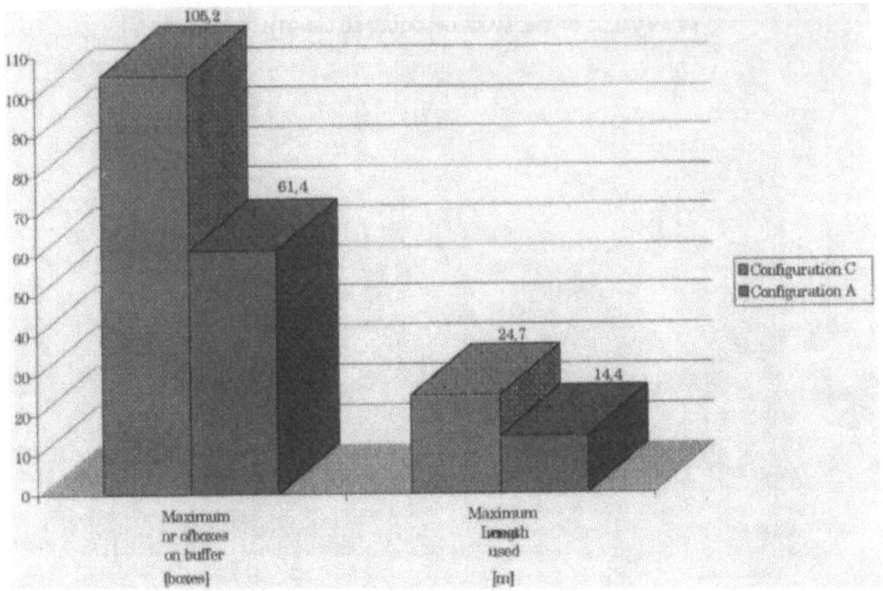

Then it has been verified that the buffer selection procedure in use today is not the best suited. the use of a SPTF-LQF or LQF-SPTF procedures could improve the system reliability and production since they don't lend to a critical point of buffer saturation with the associated dead time to recover. In the case of 5 production machines, a software speed up of the palletizator and the use of a SPTF-LQF special choosing strategy are needed.
The work made inside a alimentary firm for the production of pasta to find the best configuration that can lead to an improvement of the productivity of the department may be a methodological approach for the study, development, design or reconfiguration of any process where logistic flow problems arise.

References

1. Tanner J P 1985 *Manufacturing Engineering - An introduction to the basic functions.* M. Dekker Inc., New York
2. Villa A, Cantamessa M, Rossetto S 1996 *Sistemi logistici: metodi di progetto.* CNR. Roma
3. Law A M, Kelton W D 1991 *Simulation modeling and analysis, McGraw-Hill,* New York
4. Dowlatshahi S 1996 An efficient solution for determining location and size of buffer stocks. Production Planning & Control 18: 282-291
5. Papadopoulos H T, Heavey C, Browne J 1993 *Queueing theory in manufacturing systems analysis and design.* Chapman & Hall, London
6. Hicks C R 1982 *Fundamental Concepts in the design of experiments.* CBS

College Pubblications, New York
7. Villa A, Fassino B, Rossetto S 1986 Buffer size planning versus transfer line efficiency. Transactions of the ASME - Journal of Engineering for Industry 108: 105-108
8. Monti A, Noli F 1993 Modellizzazione: sistemi di produzione. Automazione e Strumentazione 10: 68-71
9. Harrel C, Tumay K 1995 *Simulation made easy.* Institute of Industrial Engineers, USA
10. Guariso G, Rizzoli A 1992 Un ambiente orientato agli oggetti per la modellistica e simulazione su PC. Automazione e Strumentazione 9: 103-107
11. Monte A 1997 *Elementi di impianti industriali.* Libreria Editrice Scientifica Cortina, Torino

Acknowledgements

The paper arises from a research program supported by a contribution from the national research Council (contract no. 96.024.CTY07)

36
Perspectives of Implementing Automated and Robotic-Based Techniques in Road Infrastructure Construction and Maintenance Operations

B. Psarianos, M. Kontaratos, C. Liapakis, M. Lenz and T. Rupp

1 Introduction

In EU the road infrastructure services 90% of the movements of people and goods, whereas the transport services account for ca 5.6 million jobs and contribute to 7-8% of GNP. Huge investments are necessary every year in EU to cope with the three major problems marking the present road network in EU: aging; increased traffic demand; and increased truck traffic. For example in some European countries maintenance costs for motorways amount to about 30,000 ECU/Km and for other major rural roads to about 10,000 ECU/km each year.

To handle these aggravating the fiscal budgets roadway investments new innovative techniques are sought which will reduce the overall cost of Road Infrastructure Construction and Maintenance (RICM) operations without at the same time adversely affecting the quality and performance of the road network. In this context European Commission, Directorate General VII, Transport has initiated and financed a one year research project entitled **Automated and Robotics-based Techniques (ART)-New Solutions for Road Construction and Maintenance**, to address the issue. Ten partners from five EU countries have participated in the project and explored the potential of introducing ART for construction and maintenance purposes for rural and urban roads.

Based on the analysis carried out in this project the implementation features of ART in specific RICM operations selected for European conditions are presented in this chapter.

2 Selected Operations

Operations related to road construction and maintenance activities run into hundreds. Therefore starting initially from a long list of candidate operations for investigating their possibility for consideration in an ART implementation program a shorter list of operations came up finally.

Table 1: Suitability of Implementing ART for Tasks involved in the Paving Operation

Legend: (+) = high priority, (0)= low priority

Tasks for Paving Operation	**Task Characteristics**					
	Hazardous or Dangerous	Repetitive or Tedious	Difficult or High Expertise	Simulta-neous Task	Coordina-ted Task	Suitable for ART
Sweeping	No	Yes	No	Yes	No	Yes (+)
Emulsion Spraying	Yes	Yes	No	Yes	No	Yes (+)
Material Mixing	No	Yes	No	No	No	Yes (0)
Material Loading and Transport	No	Yes	No	No	No	Yes (0)
Paver Feeding	No	Yes	No	Yes	Yes	Yes (+)
Truck Movement to Mixing Plant	No	Yes	No	No	No	Yes (0)
Placing Material	Yes	Yes	Yes	Yes	Yes	Yes (+)
Stopping the Paver	No	Yes	No	No	No	Yes (+)
Compac-tion	No	Yes	No	Yes	No	Yes (+)

These operations include nine operations for urban, thirteen for rural roads and three operations for bridge rehabilitation (urban and rural), altogether 25 operations. The inspection aspect of maintenance operations was excluded from the assessment. Frequency of the operation and corresponding budget allocation have been the primer criteria for entering the list. Paving ranked at the top of the list.

For each operation of the short list a full description was given of the included tasks, their performance and qualitative requirements, needed operator's expertise, the necessary task-oriented machines and working conditions as well as the individual task evolvement within the operation. The description of the operations relied upon a variety of data delivered mainly by two road administrators and a contractor, participating as partners in the project, as well as by other organizations and contractors. For each task of an operation then the suitability of adapting an automation or robotics technology was identified based on the following criteria:

- Task exhibiting a hazard or danger to working personnel;
- Task involving repetitive or tedious work;
- Task being costly;
- Expressed necessity for higher performance rate of task;
- Achievement of higher quality standards of task.

Table 1 comprises for example the suitability assessment data associated with each task involved in the paving operation.

3 Technological Assessment

Using ART in conventionally carried out RICM operations means either a conventional machine is technologically upgraded, for example by building in some sensor systems, or a totally new machine should be developed to cope with the new set ART requirements. To decide upon this a second more detailed assessment was carried out of the tasks included in the selected operations and identified as candidates for deploying ART by taking into consideration the resulting beneficial effects ART would impose to the operation, i.e. diminishing idle times, occupying less space, demanding less personnel etc. This second sorting of tasks pin-pointed the specific tasks or task activities, which should be technologically addressed in order to describe the automation or robotics related technological features needed for a corresponding preprototype development. For the specified task or tasks of each operation a *Technical Scenario* was established, showing the way the task or tasks would be carried out by using ART. Based on this scenario the necessary technological components for implementing ART could be identified and assessed as for their technological status. These components involve the following items:

- Kinematics;
- Gripper System Tools;
- Drive System;

- Undercarriage;
- Sensors;
- Control System;
- Man-Machine Interface;
- Information Flow.

Table 2: Technological Status of System Components needed for the Paving Operation (paver) according to the formulated Technical Scenario

Legend: ± Component exists as product and can be directly used in the machine
MMI = Man-Machine Interface

Component	Description	Research	Development	Product
Kinematics	Kinematics of the conventional screed		●	
Gripper System	None			
Tools	None			
Drive	Electrical drives			±
	Hydraulic drives			±
Power Supply	Diesel-electric (low emission)			●
	Diesel-hydraulic (low emission)			●
Undercarriage	Standard			±
Sensor Technology	Navigation System	●		
Control	Automatic motion: according to the		●	
	navigation data		●	
	Semi-automatic motion: kinematics, sensors,			±
	MMI			●
	Communication: wireless communication			
	Quality Control of the process			
Man-Machine Interface	Communication: display, control panel		●	
	Interaction: steering wheel, joy-stick			±
Information Flow	Wireless Communication			●
	Management System		●	
	Simulation System		●	
	Common Data Base	●		

A basic assumption in addressing these components was that for each machine at least **one human operator** would be involved irrespective of the automation level the investigated task or combined tasks or activities were conceived to achieve. Each of the above technological components could be assigned to one of the three classes "research", "development", and "product". "Product" refers to systems or system components which have attained a prototype status and are or can be readily available i.e. adapted, to the manufacturing industry for building the prototype.

Components were assigned the characterization "development" in case some further work is necessary to existing systems or system components to fulfil the set requirements for a RICM operation.

Finally under the term "research" were all systems or system components included, for which an intensive research work is further needed in order for the component to reach the "product" stage.

Assignment of the components needed to realize the technical scenario to one of the above three technological feasibility classes leads to the definition of the time perspective of deploying ART, i.e. short-, medium- and long-range, for each operation can be directly deducted.

Table 2 illustrates the technological assessment for the paving operation based on the corresponding technical scenario, focusing on the main tasks of laying the asphalt material by the paving machine (road paver).

4 ART Implementation Cost

Development cost of innovative ideas has always been a decisive factor. Therefore in the ART-Project the implementation average cost of the technical scenarios of the selected operations was estimated. This cost estimation reflects solely the man-power needed to apply a product, to develop a system or its components or to conduct the respective research and development summed up for all selected 41 RICM operations.

To estimate the ART implementation cost the following formulae were used

$$aver.(\cos t) = \cos t \bullet \frac{n}{N} \qquad (1)$$

where
aver. (cost)= average cost (weighted) for implementing component [kECU],
cost = actual development cost [kECU],
n = number of times component appears in the technical scenarios,
N= 25 (total number of operations considered),
and

$$total(\cos t) = \sum aver.(\cos t) \qquad (2)$$

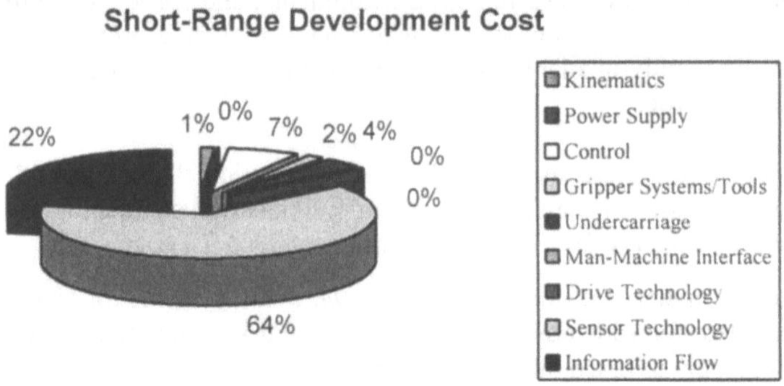

Figure 1: Distribution Cost of Short-Range Development Components of ART

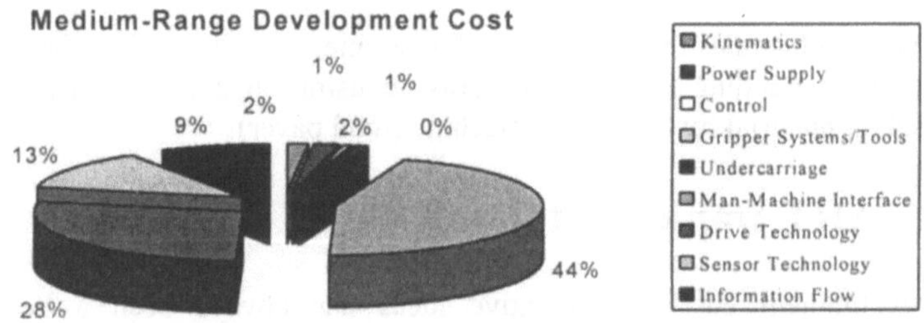

Figure 2: Distribution Cost of Medium-Range Development Components of ART

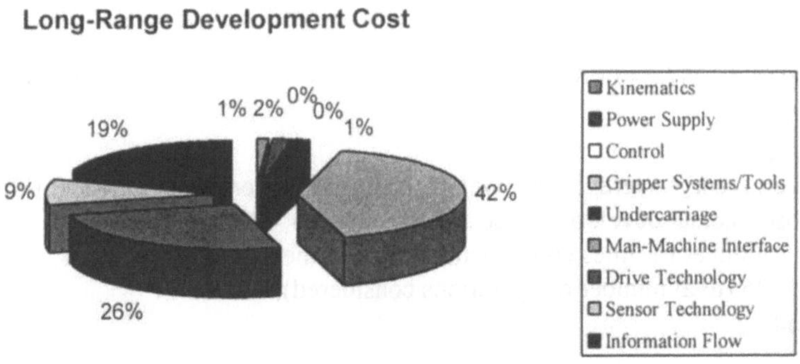

Figure 3: Distribution Cost of Long-Range Development Components of ART

Both formulae were applied for the extremes cost values, *minimum* and *maximum* estimated cost, and separately for the development classes "product", "development", and "research". Table 3 shows the cost estimation procedure for the case of **kinematics**.

Table 3: Cost Estimation of the Kinematics Component

Cost Estimation Steps		Product		Development		Research	
		Min	Max	Min	Max	Min	Max
Case 1: 6-axes Kinematics		4 x		2 x		0 x	
Case 2: Operation depending Kinematics		2 x		9 x		0 x	
No of Operations or Machines		6 x		11 x		0 x	
Single Cost	Case 1	0	50	0	200	0	0
Average Cost		0	6.9	0	82.8	0	0
Single Cost	Case 2	0	100	50	300	0	0
Average cost		0	6.9	15.5	113.8	0	0
Total Cost		0	13.8	15.5	196.6	0	0

Table 4: Investment Funds needed to Implement ART in high ranked RICM operation in the EU

Implementation Period	**Implementation Cost [kECU]**
Short-range	279.0 .. 680.0
Medium-range	1584.1 .. 9095.6
Long-range	2290.9 .. 12699.0

Having determined the development cost of every system component in order to implement the technical scenario an overall investment funds needed to apply ART in RICM operations in the short-, medium-, and long-range periods can be predicted. The results are shown in Table 4.

Figures 1, 2 and 3 illustrate the distribution cost of each technological component needed to be invested in order to implement ART in RICM operations in the EU on a short-, medium-, and long-range basis respectively.

5 The Paving Operation

As stressed above the paving operation ranked as the operation with the highest priority for adopting ART.

ART machines could contribute to alleviate the shortcomings that this operation suffers. First, the required high accuracy of height determination is satisfied by placing a stringline along the roadbed. This construction hinder the free movements of the machines in the work area, producing delays and difficulties in managing the vehicle movements and the material flow. Second, the structural characteristics of the pavement are deteriorated from stops of feeding the paver with material, as well as by insufficient compacting due to poor control of the compaction levels. Finally, the operation is slow due to idle times, to extensive surveying work, and to performance controlling work.

Thus, the main objectives of introducing ART techniques in alsphalt paving operation aim at:

- Releasing the operation for surveying infrastructure that blockade the vehicles movements.
- Improving the performance quality.
- Reducing the operation time.

For this specific operation a design scheme was developed, for which a description of the various technological aspects is given, as determined in the technical scenario for the automated paver.

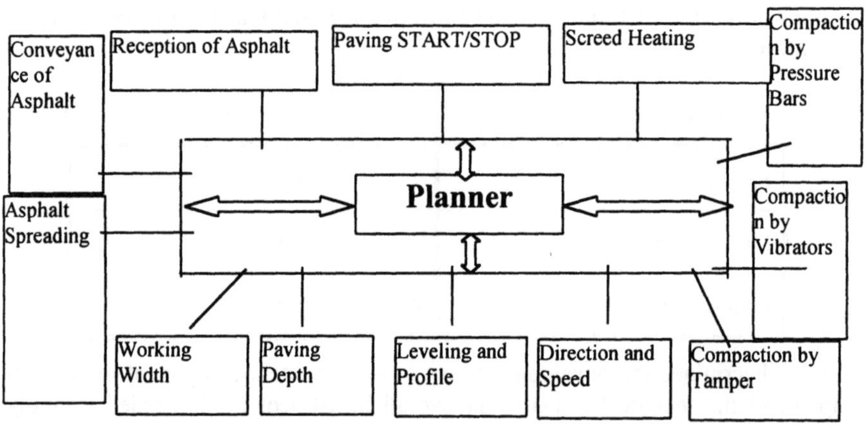

Figure 4: Block Diagram of Functional Modules for the Automated Road Paver

The functional modules needed for an automated the paver sum up to twelve and are illustrated in Figure 4. These twelve modules operate in an autonomous mode. Initial values are fed to the planner, which is controlling the paving individual tasks by linking the input information to the correct module. The positioning and navigation of the paver is performed in the *rural area* through a GPS (Global Positioning System) and especially a technique called Real Time Kinematics (RTK) with On the Fly (OTF) capabilities, a system that utilizes the position of satellites to calculate the position in space of an antenna receiving signals from a constellation of satellites. Since the road pavement surface is a three-dimensional one, known in differential geometry as *ruled surface*, a three-GPS-antenna concept is proposed to be built on the paver, which will determine not only the position in space of the paver, and especially its screed, but also the rotating angles of the machine in reference to its longitudinal, radial and vertical (normal) axes (roll, pitch and lead angles). Expected construction accuracy of the pavement surface is better than **0.5 cm horizontally** and **1 cm vertically**, which fully comply with required construction standards of the road pavement surface.

Table 5: Expected Beneficial Features of the Automated Road Paver Prototype

Legend: + : lightly beneficial, ++ : enough beneficial, (-)/= : same as before

Criteria	Description	Benefit
Working time duration	Time operation is expected to fall down by 20%. The simultaneous construction of two layers contributes considerably to this outcome.	++
Idle time	Idle time is reduced, since the use of feeder ensures continuous material flow. No "stop & go" process takes place anymore.	+
Quality	The elimination of "stops' during the material placing improves the quality in terms of structural characteristics. They also are affected positively by controlling in place the compaction levels, which optimizes the rolling task. The automatic navigation enhances the result in terms of final geometric shaping.	+
Total operation cost	Total cost is expected to decrease due to time and labor reductions.	+
Machines	(Due to better performance more trucks are in the worksite simultaneously).	(-)/=
Workers	Aiding personnel on the paver and on the roadway for machine guiding are not needed anymore.	++
Space	The occupied roadway space is reduced, since no stringline is required for height determination. Also, there are no working personnel moving on the roadway, so the relevant safety buffer space is not occupied.	+
Safety	Safety is significantly improved, since no workers are moving on the roadway.	++

Figure 5 shows for example the concept of the first functional module, the reception of asphalt. According to this concept two sensors S1 and S2 are provided at the front of the paver, which aim at sweeping the rear of the feeding truck and at detecting the position of the truck in relation to the paver. A force sensor S3 is further provided on the push roller to determine the force required for pushing the truck and its position in relation to the road paver. The sonic sensor S4 serves to pick up the asphalt level in the material hopper. Two further sensors S5 and S6 are provided on the hopper sides to pick up the positions of the hopper sides when folded out. Two actuators A1 and A2 for the electrohydraulic adjustment serve to fold the hopper sides in and out. Two displays D1 and D2 on the left and right side of the road paver inform the operator of the machine on:

(1) forwards motion,
(2) reverse motion,
(3) stop,
(4) releasing brake,
(5) dumping,
(6) dumping stop.

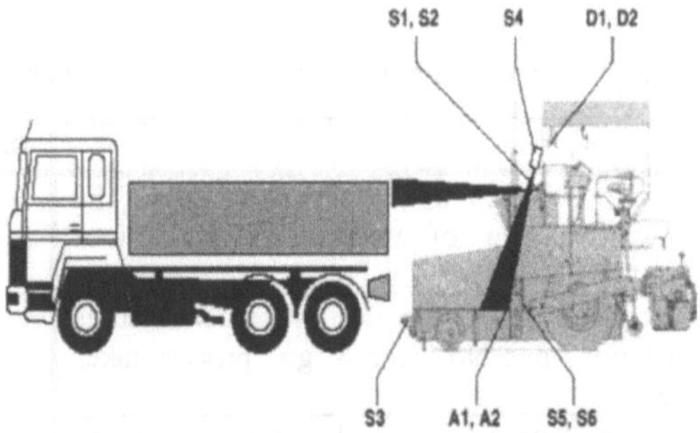

Figure 5: Paving Task "Reception of Asphalt" Implementing ART

A computer module controls the reception of the asphalt, whereas the planner regulates the asphalt reception at a high level according to a block diagram of sensors, actuators, displays and a selector switch corresponding to the above given task discreption and of Figure 5.

Similar concept descriptions are existing for the remaining eleven functional modules.

In Table 5 the beneficial features of the new automated road paver are illustrated, which are expected to result when prototypes will be built and introduced in road construction and maintenance.

6 Conclusion

The European Project ART – Automated and Robotics-based Techniques investigated the feasibility of implementing automation and robotics technology for 25 selected Road Infrastructure Construction and Maintenance operations. There are two main recipients of this information. The first one is the manufacturing industry, which can directly assess the results of the ART project and take the appropriate decisions on its future development plans. The second recipient is the transport policy makers in Europe, which can identify the benefits of implementing Automation and Robotics Technology for Road Construction and Maintenance activities in the European Continent and make the necessary investments to get the best of the technology on the road the as far as possible.

Acknowledgement

This work has been supported by the European Commission, Directorate General VII, Transport, under the Transport RTD Program of the 4th Framework Program.

REFERENCES

1. ART consortium 1997 *Selection of RICM Operations.* 1st Internal Report, EC/DG VII, Brussels

2. ART consortium 1998 *New Approach to RICM: Introduction of Automated and Robotics-based Technologies* 2nd Internal Report, EC/DG VII, Brussels

Index

MIX
Papier aus verantwortungsvollen Quellen
Paper from responsible sources
FSC® C105338

If you have any concerns about our products,
you can contact us on
ProductSafety@springernature.com

In case Publisher is established outside the EU,
the EU authorized representative is:
Springer Nature Customer Service Center GmbH
Europaplatz 3, 69115 Heidelberg, Germany

Printed by Libri Plureos GmbH
in Hamburg, Germany